Iron, Siderophores, and Plant Diseases

NATO ASI Series

Advanced Science Institutes Series

A series presenting the results of activities sponsored by the NATO Science Committee, which aims at the dissemination of advanced scientific and technological knowledge, with a view to strengthening links between scientific communities.

The series is published by an international board of publishers in conjunction with the NATO Scientific Affairs Division

A Life Sciences **B Physics**	Plenum Publishing Corporation New York and London
C Mathematical and Physical Sciences	D. Reidel Publishing Company Dordrecht, Boston, and Lancaster
D Behavioral and Social Sciences **E Engineering and Materials Sciences**	Martinus Nijhoff Publishers The Hague, Boston, and Lancaster
F Computer and Systems Sciences **G Ecological Sciences**	Springer-Verlag Berlin, Heidelberg, New York, and Tokyo

Recent Volumes in this Series

Volume 109—Central and Peripheral Mechanisms of Cardiovascular Regulation
edited by A. Magro, W. Osswald, D. Reiss, and P. Vanhoutte

Volume 110—Structure and Dynamics of RNA
edited by P. H. van Knippenberg and C. W. Hilbers

Volume 111—Basic and Applied Aspects of Noise-Induced Hearing Loss
edited by Richard J. Salvi, D. Henderson, R. P. Hamernik, and V. Colletti

Volume 112—Human Apolipoprotein Mutants: Impact on Atherosclerosis and Longevity
edited by C. R. Sirtori, A. V. Nichols, and G. Franceschini

Volume 113—Targeting of Drugs with Synthetic Systems
edited by Gregory Gregoriadis, Judith Senior, and George Poste

Volume 114—Cardiorespiratory and Cardiosomatic Psychophysiology
edited by P. Grossman, K. H. Janssen, and D. Vaitl

Volume 115—Mechanisms of Secondary Brain Damage
edited by A. Baethmann, K. G. Go, and A. Unterberg

Volume 116—Enzymes of Lipid Metabolism II
edited by Louis Freysz, Henri Dreyfus, Raphaël Massarelli, and Shimon Gatt

Volume 117—Iron, Siderophores, and Plant Diseases
edited by T. R. Swinburne

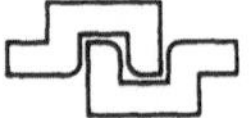

Series A: Life Sciences

Iron, Siderophores, and Plant Diseases

Edited by

T. R. Swinburne

East Malling Research Station
Maidstone, Kent, England

Plenum Press
New York and London
Published in cooperation with NATO Scientific Affairs Division

Proceedings of a NATO Advanced Research Workshop on
Iron, Siderophores, and Plant Diseases,
held July 1-5, 1985,
in Wye, Kent, United Kingdom

ISBN 978-1-4615-9482-6 ISBN 978-1-4615-9480-2 (eBook)
DOI 10.1007/978-1-4615-9480-2

Softcover reprint of the hardcover 1st edition 1986
A Division of Plenum Publishing Corporation
233 Spring Street, New York, N.Y. 10013

PREFACE

The importance of competition for iron in the interactions between saprophytic microorganisms, pathogens and plants has been recognised for almost a decade. This has been reflected in an upsurge of publications on the topic over the last five years. Paradoxically, the subject was only touched upon during the International Congress of Plant Pathology held in 1983. In response to this apparent omission, a few of those most closely associated with the topic met one evening during which they resolved to organise a symposium devoted solely to the various aspects of iron uptake and its relation to plant disease. It was my privilege to be asked to undertake the task of convenor.

Early correspondence brought a wealth of positive replies to the proposal, particularly from Bob Schippers in Baarn. With the increasing costs of international symposia the need for a sponsor soon became apparent and an application to NATO was favourably received, following helpful advice from Dr. di Lullo, Advanced Research Workshop Programme Director, to whom all the participants in this Workshop owe a debt of gratitude.

Plant pathologists are not the only scientists expressing an interest in iron uptake mechanisms. Chemists, biochemists, microbiologists and geneticists have made great progress in the elucidation of these mechanisms at the molecular level. Happily, the leading exponents accepted my invitation to participate in our proceedings and the plant pathologists gained much from their experience. Moreover, as iron also features in the interactions between microorganisms and plants, we were fortunate to attract scientists from these fields. Thus an assembly of scientists from many different fields of experience was drawn together at the Centre for European Agricultural Studies, Wye College, Ashford, Kent from 1st-5th July, 1985, skilfully looked after by Mr. Austen, the Centre's Director.

The quality of the contributions was excellent and I wish to express my gratitude to all those who made formal presentations, poster demonstrations and took part in the discussion, which was lively, constructive and stimulating.

The papers gathered here represent the formal contributions of the speakers. In order to reduce the time between conference and publication it was necessary to carry out editorial work without proper consultation with authors. If there are errors these are entirely mine. Happily the text has been typed by experts who have also been able to correct some of my faults. To Mrs. Jean Allen-King goes our profound thanks for an excellent job, and also to Mrs. Angela Pain who helped in the final stages. Our thanks are due also to Mr. John Carder for the magnificient job he did in producing the index.

The organisation of the conference was done from East Malling Research Station and many of my colleagues there gave invaluable assistance. In particular, Mr. Brian Self (Liaison Officer) and Mrs. Kay Webster who helped with audiovisual aids and some of the illustrations in this text; Dr. Simon Slade who did sterling service in the projection room during the conference. Finally, but by no means least, we all thank Mrs. Joan Mutti, who took on the responsibility for most of the organisation and without whose help the conference would never have taken place.

T.R. Swinburne

CONTENTS

SIDEROPHORES AND BIOLOGICAL SYSTEMS; AN OVERVIEW

B.R. Byers

Department of Microbiology
University of Mississippi Medical Center
Jackson, Mississippi, 39216-4505, U.S.A.

INTRODUCTION

The two stable valences of iron impart considerable range to the potential of the metal for participation in oxidation-reduction reactions. Most biological systems make use of this potential in the catalysis of many different critical steps. By placing iron within domains on enzyme molecules, the catalytic properties of the atom are disciplined in order to permit only a single reaction with strictly limited types of reactants, thereby achieving the specificity necessary for metabolism. Biological dependence on iron means that high affinity iron chelating agents should have an impact on most life forms. The microbial siderophores are such chelators. The present paper sketches some of the roles (or effects) of siderophores in several biological systems. It is intended as a survey or "overview", not as an exhaustive review and references to original literature (which could total several hundred) are kept to a minimum. For preparation of the overview, use was made of extensive revies by (Braun, 1985; Byers and Arceneaux, 1977; Crosa, 1984; Emery, 1982; Hider, 1984; Lankford, 1973; Neilands, 1981a: 1981b: 1982; Raymond et al., 1984; Weinberg, 1984; Williams and Carbonetti, 1984) and these should be consulted for citations of research papers supporting most of the information presented in the overview. Recently research describing a role for certain siderophores in plant iron nutrition and plant diseases has blossomed; extensive discussion of these topics is correctly reserved for the remaining authors.

THE IRON ACQUISITION PROBLEM

Before the cyanobacteria began photosynthetic oxygen evolution, iron gathering was much simpler for living cells (thought to be all prokaryotic at that time). Ferrous iron (present in the Earth's early anaerobic conditions) is much more soluble than ferric iron, which forms virtually insoluble hydroxyaquo complexes. The early iron uptake systems probably were adjusted to ferrous uptake from the prevailing iron concentration. The switch to ferric iron may have (in the words of J.B. Neilands) produced a "crisis on a grand scale". To survive, life forms would have to do at least one of the following: 1) retreat to the remaining anaerobic niches, 2) replace iron with other cofactors (certain members of the genus *Lactobaccilus* may have done this), 3) develop a new system to solubilize ferric iron and transport it, 4) devise a means to externally reduce iron and immediately transport it by the old ferrous uptake system. All of these alternatives appear to have been followed.

Infectious microorganisms that invade vertebrate hosts are also faced with serious iron acquisition problems. Availability of iron is one of the fulcrums on which the final outcome of the invasion is balanced. Vertebrates have developed an elaborate strategy for retention of iron. The two iron-binding glycoproteins lactoferrin and transferrin have broad spectrum anti-microbial activity, withholding iron from many microorganisms. An important property of lactoferrin is its capacity to bind iron at low pH, making the protein a good iron scavenger in areas of infection. Transferrin has dual function; it delivers iron to host cells and acts as a defense protein. In response to a microbial invasion, the vertebrate host also lowers its serum iron level (transferrin saturation), temporarily storing the iron in ferritin. As some bacteria can use heme and hemoglobin in wounds and abscesses, vertebrates also produce haptoglobulin; this agent will combine with hemoglobin and render it unavailable to many microorganisms. These defense measures are nonspecific with respect to invading pathogen. It is possible that immune systems may produce antibodies directed specifically against certain components of the microbial iron uptake system, thereby blocking its function.

SIDEROPHORE MEDIATED IRON UPTAKE

Many bacteria and fungi excrete low molecular weight, high affinity, ferric chelating siderophores (microbial iron transport cofactors). Siderophores will capture oxidized iron, delivering it to the micro-organisms. In the case of pathogenic microbes, certain siderophores will extract iron from host defense proteins, making siderophore a critical virulence factor for some pathogens. More than forty structurally different siderophores have been described; however, almost all are phenolates (catecholates; derivates of 2,3-dihydroxybenzoic acid) or secondary hydroxamates, containing these groups in their iron binding centers, and most siderophores chelate ferric iron in the hexadentate configuration. All siderophores have high affinity for ferric iron, but weakly chelate ferrous iron, an important chemical property. Detailed discussions of structural chemistry and metal coordination properties can be found in the reviews cited above. Although most siderophores are phenolates or hydroxa-mates, the definition of a siderophore includes any chelating agent that functions as a transport cofactor and in some organisms citrate may be included in this category. Recently a nonphenolate, nonhydroxamate siderophore has been characterized in the nitrogen fixing plant symbiont *Rhizobium meliloti* (Smith et al., 1985).

Iron starvation usually increases deferrisiderophore (iron-free) excretion and concomitantly increases synthesis of several proteins that are placed at the microbial cell surface in its membrane. One of the pro-teins is the receptor involved in uptake of iron from the siderophore. An excreted deferrisiderophore will bind iron atoms (usually one per sidero-phore molecule) to form the ferrisiderophore chelate. The ferrisiderophore then associates with its receptor in the cell membrane (outer membrane in gram negative bacteria). Specific receptors that recognize a given ferri-siderophore (or a structurally similar group of ferrisiderophores) probably are obligatory for uptake of iron from the ferrisiderophore. It is important to remember that a microorganism often can utilize not only its own ferri-siderophore but structurally different ferrisiderophores produced by other microorganisms as well. However, to use such exogenous ferrisiderophores, specific membrane receptors for these ferrisiderophores must be present. If a microorganism lacks a receptor for a siderophore, it may suffer iron starvation if all the iron in the environment is chelated by that siderophore.

Membrane

1) Chelate transported

Ferrisiderophore (receptor bound) → 2) Fe transferred to cell-associated chelator

3) Fe reduced and transported

Iron uptake from Ferrisiderophores

After binding of a ferrisiderophore to its cognate receptor, several alternative routes for delivery of iron to metabolism may be followed. Some microbial types may make exclusive use of only one of these pathways, while other types might simultaneously use more than one. The intact chelate may be transported into the cytoplasm, or iron may be incorporated from the ferrisiderophore without uptake of the chelate. If the chelate is not transported, some microorganisms may employ a second, entire cell associated siderophore-like molecule that accepts iron from the external ferrisiderophore and delivers the metal internally. Iron also may be reductively removed as the ferrous atom from an external ferrisiderophore.

Ferrisiderophores that are transported (internalized) are subsequently processed to remove iron for assimilation of the metal into metabolism. Evidence suggests that iron is removed from the chelate by ferrisiderophore reductase enzymes that employ physiological reductants. During the process of iron release, some siderophores may be degraded (by esterases, the synthesis of which also is increased by low-iron cultivation). It was concluded earlier that the low redox potential of one of the siderophores (enterobactin, also called enterochelin), which should prevent its reduction by physiological reductants, made depolymerization of this siderophore essential for iron removal. However, more recent evidence (obtained with living cultures and with isolated ferrisiderophore reductases) using analogs of enterobactin which are not susceptible to hydrolytic destruction by the esterase has re-opened this question. Iron is reductively removed, without destruction of the siderophore, from low redox ferrienterobactin analogs by ferrisiderophore reductase employing physiological reductants at pH levels near 7. How this difficult step is biochemically accomplished is uncertain. There is also an example of covalent modification (possibly lowering iron binding capacity) of one of the siderophores during its function in iron transport. Other siderophores appear to be released in unmodified form after removal of iron from their chelates and the resulting deferrisiderophores may function again to deliver an iron atom. In iron assimilation, a required step (but one which occurs immediately after iron removal from the chelate) may be either degradation of the siderophore, its modification, or its rapid removal from the cell. Enzyme kinetic studies with ferrisiderophore reductases show that high concentrations of deferrisiderophores will slow removal of iron from ferrisiderophores (Arceneaux and Byers, 1980).

Genetic studies of siderophore production and utilization (particularly in gram negative bacteria) have advanced quickly in recent years as the powerful tools of molecular genetics have been applied to the problem. In *Escherichia coli*, the genes for enterobactin synthesis, the enterobactin esterase, and the enterobactin receptor are linked in an operon carried on the chromosome. Each of these genes is being cloned, and the precise functions of the gene products are being studied (Pierce et al., 1985). A significant advance was made by the discovery that invasive, highly virulent *E. coli* synthesize, in addition to enterobactin an hydroxamate siderophore aerobactin, and that the genes for aerobactin synthesis and its receptor are located in an operon that is often encoded on the ColV plasmid. The presence of this plasmid is crucial for proliferation of *E. coli* in the host's blood and tissue. Recent evidence indicates that aerobactin but not enterobactin, is the siderophore functional in the host for invasive *E. coli* (Konopka and Neilands, 1984). Aerobactin also is produced by certain other gram negative pathogenic bacteria. Restriction enzyme digestion and hybridization studies with the cloned aerobactin operon show that the operon is flanked by insertion sequence IS1 (Lawlor and Payne, 1984; McDougall and Neilands, 1984; Perry and Brubaker, 1979). The presence of IS1 might allow spread of the aerobactin operon to other bacterial strains, a finding of medical significance.

Iron starvation in *E. coli* initiates high level transcription of genes for enterobactin, aerobactin (if present), receptor proteins for exogenous siderophores used by *E. coli*, and several other membrane proteins required for transport of ferrisiderophores. All of this activity is controlled through a single chromosomal locus designated fur. Presently, the mechanism by which fur controls expression of the iron regulated operons is unknown. The fur gene has been mapped and cloned(Bagg and Neilands, 1985) and its nucleotide sequence determined (Schaffer et al., 1985). The fur gene product may be a repressor protein, with iron acting as a corepressor.

The gram negative bacterium *Vibrio aguillarum* causes a fatal hemorrhagic septicemia in marine fish. High virulence in the microorganism is associated with a plasmid; strains lacking the plasmid are of low virulence. Molecular genetic studies showed that encoded on this virulence plasmid is an operon for a siderophore and its membrane receptor and that this siderophore was essential for the high virulence phenotype.

MEDICAL USE AND POTENTIAL USES OF SIDEROPHORES

The siderophore ferrioxamine B (Desferal) currently is the only approved drug for treatment of human iron overload. Iron overload (often resulting from repeated blood transfusions that are required by some persons, such as those suffering from Cooley's anemia) is usually fatal unless reversed. There are some potential medical uses of siderophores. Derivatives of certain siderophores might become useful antimicrobial agents. Although siderophores have highest affinity for ferric iron, some other metals can form stable complexes with siderophores. The chromium and gallium complexes have been useful in studying uptake of siderophores in the absence of metal assimilation, as these two metals cannot be reductively removed from the chelates. The scandium and indium complexes have antibacterial activity for some bacteria during an infection (Rogers et al., 1982). Cupric siderophores have bactericidal activity and evidence suggested that reduction of this complex by the ferrisiderophore reductase system rapidly generated toxic levels of the cuprous ion in microbial cells (Arceneaux et al., 1984). Siderophores also might be used as vaccines. Although the molecular weights of the known siderophores are too low for the siderophores to be good immunogens, some siderophores might be coupled to larger molecules for vaccine preparation. Immunoglobulin specific for at least one siderophore has been found in the serum of normal persons, and these antibodies inhibit siderophore function (Moore and Earhart, 1981). It also might be possible to use siderophore receptor proteins as vaccines and thereby strengthen the iron withholding defenses of animals.

NON-SIDEROPHORE MEDIATED IRON UPTAKE.

Production of a siderophore is not the only means of microbial iron uptake. In some pathogenic microorganisms, direct acquisition of iron from transferrin and possibly lactoferrin, without aid of a siderophore, may be accomplished. The obligatory parasitic *Neisseria* species produce no detectable siderophores and can derive iron from transferrin, lactoferrin, heme and hemoglobulin (West and Sparling, 1985, and papers cited therein). *Haemophilus influenzae* utilizes transferrin (but not lactoferrin) without intervention of a siderophore (Herrington and Sparling, 1985). Other exanples of pathogenic, iron-requiring forms that produce no siderophore are *Yersinia* species, *Legionella pneumophila,* and *Listeria monocytogenes* (Cowart and Foster, 1985; Perry and Brubaker, 1979; Reeves et al., 1983). Little is known about the iron transport mechanisms in these microorganisms; some of them may have receptors for transferrin (Dyer et al., 1985).

Some siderophore producing microorganisms, as well as some of the types lacking siderophore production, may have the capacity to transport ferrous iron (possibly maintained since the period before the atmosphere became oxidizing). For a ferrous iron uptake system to function in an aerobic environment, the cells must be able to externally reduce ferric iron. Some fungi may do this (Winkelmann, 1974) and *L. monocytogenes* excretes a reductant that effectively reduces iron and mobilizes it from transferrin (Cowart and Foster, 1985). *Streptococcus mutans* does not produce a siderophore, transports only ferrous iron, and probably is able to externally reduce iron (D. Evans, J. Arceneaux, H. Aranha, M. Martin and B.R. Byers, unpublished data.

SIDEROPHORE AND PLANTS

Many of the participants in this conference will discuss in detail the observation that the presence of certain siderophores (or the siderophore producer) in the environment of plant roots may protect the plant from several plant pathogens, possibly by chelating all available iron and withholding it from the pathogen. This raises many fundamental questions on the roles of siderophores in plant iron nutrition; discussion of these questions is reserved to the other authors.

REFERENCES

Arceneaux, J.E.L., Boutwell, M.E., and Byers, B.R., 1984. Enhancement of copper toxicity by siderophores in *Bacillus megaterium.* Antimicrob. Agents Chemother. 25:650-652.

Arceneaux, J.E.L., and Byers, B.R., 1980. Ferrisiderophore reductase activity in *Bacillus megaterium.* J. Bacteriol. 141:715-721.

Bagg, A. and Neilands, J.B., 1985. Mapping of a mutation affecting regulation of iron uptake systems in *Escherichia coli* K-12. J. Bacteriol. 161:450-453.

Braun, V., 1985. The iron-transport systems of *Escherichia coli.* In: "The enzymes of biological membranes", Vol. III. 617-651. (A.N. Martonosi, ed.), Plenum Publishing Corp., New York.

Byers, B.R., and Arceneaux, J.E.L., 1977. Microbial transport and utilization of iron. In: "Microorganisms and minerals", p. 215-249. (E.D. Weinberg, ed.), Marcel Dekker Inc., New York.

Cowart, R.E., and Foster, B.G., 1985. Differential effects of iron on the growth of *Listeria monocytogenes*: minimum requirements and mechanism of acquisition. J. Infec. Dis., 151:721-730.

Crosa, J.H., 1984. The relationship of plasmid-mediated iron transport and bacterial virulence. Ann. Rev. Microbiol., 38:69-89.

Dyer, D.W., Blackman, E.Y., and Sparling, P.F., 1985. Isolation of a transferrin-specific iron-uptake mutant of *Neisseria meningitidis.* Abstracts Annual Meeting American Society for Microbiology, p.31.

Emery, T., 1982. Iron metabolism in humans and plants. Am. Scient. 70: 626-632.

Herrington, D.A., and Sparling, P.F., 1985. *Haemophilus influenzae* can use human transferrin as a sole source for required iron. Infec. Immun., 48:248-251.

Hider, R.C., 1984. Siderophore mediated absorption of iron. Struct. Bond. 58:25-87.

Konopka, K. and Neilands, J.B., 1984. Effect of serum albumin on siderophore-mediated utilization of transferrin. Biochemistry, 23: 2122-2130.

Lankford, C.E., 1973. Bacterial assimilation of iron. Crit. Rev. Microbiol. 2:273-331.

Lawlor, K.M., and Payne, S.M., 1984. Aerobactin genes in *Shigella* spp. J. Bacteriol., 160:266-272.

McDougall, S., and Neilands, J.B., 1984. Plasmid- and chromosome-coded aerobactin synthesis in enteric bacteria: insertion sequences flank operon in plasmid-mediated systems. J. Bacteriol. 159:300-305.

Moore, D.G., and Earhart, C.F., 1981. Specific inhibition of *Escherichia coli* ferrienterochelin uptake by normal human serum immunoglobulin. Ifec. Immun. 31:631-635.

Neilands, J.B., 1981a. Iron absorption and transport in microorganisms. Ann. Rev. Nutr., 1:27-46.

Neilands, J.B., 1981b. Microbial iron compounds. Ann. Rev. Biochem., 50:715-731.

Neilands, J.B., 1982. Microbial envelope proteins related to iron. Ann. Rev. Microbiol., 36:285-309.

Perez-Casal, J.F., and Crosa, J.H., 1984. Aerobactin iron uptake sequences in plasmid ColV-K30 are flanked by inverted IS1-like elements and replication regions. J. Bacteriol. 160:256-265.

Perry, R.D., and Brubaker, R.R., 1979. Accumulation of iron by *Yersiniae*. J. Bacteriol. 137:1290-1298.

Pierce, J.R., Earhart, C.F., and Pickett, C.L., 1985. Identification of proteins required for ferric enterochelin transport of *Escherichia coli* K-12. Abstracts Annual Meeting American Society for Microbiology, p. 173.

Raymond, K.M., Müller, G., and Matzanke, B.F., 1984. Complexation of iron by siderophores. A review of their solution and structural chemistry. and biological function. Topics Curr. Chem., 123:49-102.

Reeves, M.W., Pine, L., Neilands, J.B. and Balows, A., 1983. Absence of siderophore activity in *Legionella* species grown in iron-deficient media. J. Bacteriol., 154:324-329.

Rogers, H.J., Woods, V.E., and Synge, C., 1982. Antibacterial effect of scandium and indium complexes of enterochelin on *Escherichia coli*. J. Gen. Microbiol., 128:2389-2394.

Schaffer, S., Hantke, K., and Braun, V., 1985. Nucleotide sequence of the iron regulatory gene *fur*. Molec. Gen. Genet. (in press).

Smith, M.J., Schoolery, J.N., Schwyn, B., Holden, I., and Neilands, J.B., 1985. Rhizobactin, a structurally novel siderophore from *Rhizobium meliloti*. J. Am. Chem. Soc., 107:1739-1743.

Weinberg, E.D., 1984. Iron withholding: a defense against infection and neoplasia. Physiol. Reviews. 64:65-102.

West, S.E.H., and Sparling, P.F., 1985. Response of *Neisseria gonorrhoeae* to iron limitation: alterations in expression of membrane proteins without apparent siderophore production. Infec. Immun., 47:338-394.

Williams, P.H., and Carbonetti, N.H., 1984. The plasmid-specified aerobactin iron uptake system of *Escherichia coli*, In: "Plasmids in bacteria," p. 741-757. (D.R. Helsinki, S.N. Cohen, D.B. Clewell, D.A. Jackson, and A. Hollaender, eds.), Plenum Publishing Corp., New York.

Winkelmann, G., 1974. Uptake of iron by *Neurospora crassa*. Arch. Microbiol., 98:39-50.

IRON UPTAKE SYSTEMS IN FUNGI

G. Winkelmann

Institut für Biologie I, Mikrobiologie I
Universität Tübingen, Auf der Morgenstell 1
D-7400 Tübingen, Federal Republic of Germany

INTRODUCTION

The highly selective barrier action of cellular membranes applies to all hydrophilic molecules, to which almost all essential ions belong. The strength of this barrier is underscored by its very high resistance and capacitance. However, the barrier action applies much less to hydrophobic, lipophilic molecules. As a consequence, living organisms have developed special mechanisms in order to translocate ions across the lipid bilayer. The majority of fungi synthesize and excrete iron chelating agents to facilitate the permeation of ferric ions across the cytoplasmic membrane. The ferric specific chelating agents of microbial origin are now collectively called siderophores, although there is only limited analogy to the classical ionophores, such as valinomycin, enniatins, nigericins etc.

Obviously a first requirement for iron uptake is the substitution of the hydrate shell of ferric ions by complexing to a siderophore, yielding molecules with altered solubility properties. For example, ferric ions can be extracted after complexing to siderophores from an aqueous environment with benzyl alcohol, chloroform-phenol mixtures or several other solvents. Today the most convenient method for isolating siderophores is by adsorption to a styropor matrix, which generally binds lipophilic molecules dissolved in water mixtures.

FUNCTIONS OF SIDEROPHORES

Another important aspect of iron uptake is the necessity of solubilizing iron hydroxides. It is well known that the amount of soluble ferric ions in water is determined by the solubility product of iron hydroxide, which is approximately 10^{-38}M at pH7. The solubility product, however, as it is cited in so many papers, does not describe the actual amount of available iron in chemically defined or complex media usually used for growth of fungi. Generally, we find iron concentrations of approximately 10^{-6} - 10^{-7} M, which are far above the solubility product of iron hydroxide. Therefore, we may assume that iron is bound to a variety of other constituents prevailing in culture media and presumably also in the natural environment. Surface bound iron hydroxides, as well as iron bound to complexes such as amino acids, organic acids or phosphates, are equally well suited for iron acquisition by excreted siderophores. The ability to be excreted and to be regulated by the environmental and cellular iron content seems to be the main advantage of siderophores in fungi. We should perhaps make the obvious point that fungi often colonize surfaces of compact materials, on which the more water soluble siderophores can mobilize more iron because of the greater diffusion area. More lipophilic siderophores may be of greater value when fungi live in an aqueous environment, where greater diffusion may be a disadvantage. The variety of siderophores produced by fungi may be an adaption to the frequently varying environmental conditions. In fact, most fungi produce more than one siderophore, some produce more than a dozen structurally different siderophores. *Ustilago sphaerogena* produces ferrichrome and ferrichrome A (Emery, 1971; Winkelmann, 1983), *Neurospora crassa* produces coprogen and ferricrocin (Keller-Schierlein and Diekmann, 1970; Horowitz et al., 1976), *Aspergillus ochraceus* produces ferrichrysin and ferrirubin as its main siderophores (Jalal et. al., 1984). Unquestionably, the changing environmental conditions are the main reason for the maintenance of a great variety of different iron complexing siderophores in fungi. The prime focus of this contribution is the analysis of the various iron transport systems in fungi.

TRANSPORT MODELS

From the statements of the preceding paragraph, we can readily progress to a general transport mechanism, which is called simple diffusion. After excretion of desferrisiderophores under iron deprivation, high local siderophore concentrations may occur, which give rise to a concentration gradient. If we use the expression for the chemical potential of siderophores (μ_s):

$$\mu_s = \mu_s^0 + RT \ln [S]$$

the driving force for siderophore diffusion results from the difference of the chemical potential of two compartments. If we analyze the distribution of siderophores between the outer and inner aqueous compartment of the cell, assuming identical standard chemical potentials, we can predict a solute flow according to the concentration gradient. However, if we consider the distribution between water and the membrane lipid phase, assuming unequal standard chemical potentials, an accumulation of siderophores in the membrane compartment may occur, even if the concentration of the outer compartment is lower. The foregoing thermo-dynamic description merely illustrates that siderophores may be enriched to a certain degree in the cytoplasmic membrane after enhancing the lipophilicity of ferric ions by complexing with desferrisiderophores. It is an open question, whether this model of simple siderophore diffusion would allow for the observed rapid growth of fungal mycelia and whether this mechanism indeed functions *in vivo*. In every case in which siderophores are added in high amounts (50 μM) to mycelial suspensions *in vitro*, however, the diffusion effect has to be considered. Fig.1 schematically illustrates four possible mechanisms: 1) diffusion of siderophores into the membrane (M), 2) diffusion into the interior of the cells, 3) transport of siderophores to the membrane with subsequent reduction (R), 4) uptake of siderophores into the cell by a transport system requiring energy (E).

RECOGNITION OF SIDEROPHORES

Simple diffusion is not specific as to structure. We have earlier shown that uptake of ferrichrome in *Neurospora crassa* and *Penicillium parvum* is highly stereospecific (Winkelman, 1979; Winkelmann and Braun, 1981). Using Fe-55 labeled synthetic enantio-ferrichrome and enantio-ferricrocin, synthesized from D-orithine instead of from L-orinthine, we

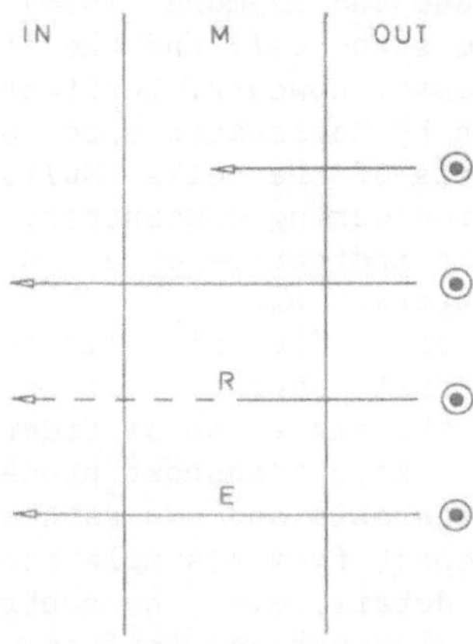

Fig. 1. Possible mechanisms of siderophore transport. (1) Diffusion into the cell membrane (M). (2) Diffusion into the interior of the cell. (3) Transport to the membrane with subsequent reduction (R). (4) Transport into the interior with energy consumption(E).

were able to demonstrate that siderophores which are not recognized by membrane located recognition sites (receptors) show non-saturable concentration dependent uptake kinetics (Fig. 2). The enantiomeric ferrichromes behaved exactly according to the laws of diffusion, e.g. the amount of siderophore associated with the cells was always proportional to the concentration of the incubation medium. The natural siderophore, however, possessing the correct configuration of the peptide backbone and a Λ - cis absolute configuration about the metal center, revealed typical saturation kinetics with a Km-value for half maximal transport rates of approximately 5 μM. The superiority of the natural compound is obvious, showing at least a ten-fold higher uptake rate in the range of the Km, whereas at higher concentrations there was no difference in the magnitude of uptake. We may summarize these observations by recognizing the substantially better biological activity of the natural compound at low concentrations, confirming the view that simple diffusion is by no means the actual uptake mechanism of ferrichrome-type siderophores in fungi. There is evidence that the uptake of ferrichrome A in *Ustilago sphaerogena* (Ecker et al., 1982 and the uptake of ferric rhodotorulic acid in

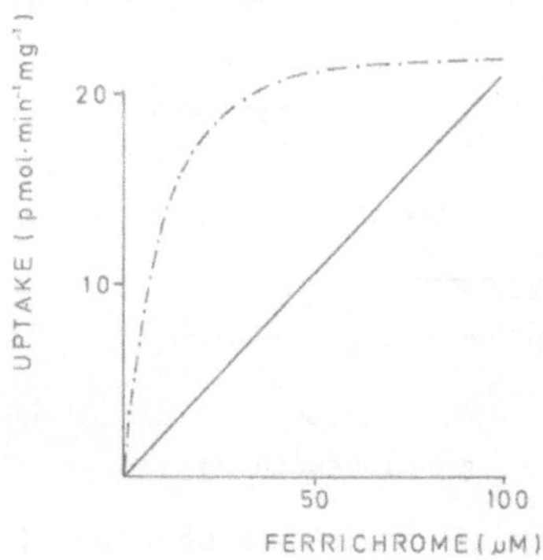

Fig. 2. Rate-concentration curves of recognized (—·—) and non-recognized (——) siderophores in fungi.

Rhototorula pilimanae (Carrano and Raymond, 1978) function as an "iron taxi", in that the iron enters the cell and the ligand remains extracellular. In *Neurospora crassa,* however, ferrichrome and coprogen are accumulated inside, as shown by Moessbauer spectroscopy (Matzanke and Winkelmann, 1981) and by lysis of the cells (Muller and Winkelmann, 1981). The observed saturation during concentration dependent uptake of siderophores in *Neurospora* is indicative of a mediated transport process, in which the siderophores interact with internal components of the membrane, enabling a facilitated diffusion. The initially incomplete association of siderophores with these internal components is supposed to be finally complete and may then limit the migration of siderophores at high concentrations, resulting in a saturable transport process. Transport kinetics are sometimes mixtures of saturable and non-saturable events. In this case, rate-concentration curves depart from a simple rectangular hyperbola and have to be analyzed in more detail, e.g., by subtraction. It is our experience that the transport of siderophores in fungi is fast enough, so that a correction for the non-saturable component is negligible in most cases and can be omitted. The uptake of a variety of ferrichrome- and coprogen-type siderophores has been studied in a siderophore-free mutant of *Neurospora crassa* (Winkelmann, 1974). The mutant *Neurospora crassa* (arg-5 ota aga) allows the investigation of uptake of Fe-55 lebeled siderophores without any transfer of Fe-55 to endogenous siderophores. Fig. 3 shows the biosynthesis pathway of ornithine in *Neurospora crassa* and the three blocked enzymes of the mutant.

As iron (III) siderophores are kinetically labile complexes, competition studies with Fe-55 labeled siderophores are not conclusive. Therefore, we have prepared C-14 lebeled coprogen by N-acetylation of coprogen B (desacetyl coprogen B) with 1-C14 labeled acetic anhydride. In *Neurospora crassa* uptake of C-14 coprogen was competitively inhibited in the presence of ferrichromes, such as ferrichrome, ferricrocin, ferrichrysin and tetraglycylferrichrome, indicating the presence of a shared transport system in the cytoplasmic membrane (Muschka et. al., 1985). The pronounced structural specificity of siderophore uptake ruled out a competition for the same recognition site. Thus, we suggested the presence of a shared transport system but distinct recognition sites (Fig. 4). A much simpler system seems to operate in *Penicillium parvum,* being unable to transport coprogen. In

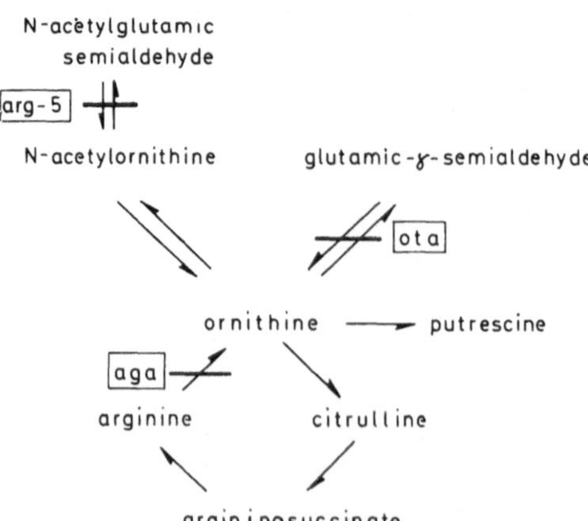

Fig. 3. Biosynthesis pathway of orinthine showing the three blocked enzymes of the triple mutant *Neurospora crassa* (arg-5 ota aga): arg-5: acetylglutamic transaminase, ota: ornithine-δ-transaminase, aga:arginase.

this fungus the transport of ferrichrome could not be inhibited by coprogen, confirming that *Penicillium parvum* possesses only ferrichrome-type specific receptors.

THE METAL CENTER AND ITS N-ACYL RESIDUES

Although the process of recognition of siderophores in fungi is still under investigation, several lines of evidence point to the importance of the iron coordination center and its surrounding N-acyl residues for recognition. As the transport of ferrichromes possessing structural alterations of the peptide portion, is nearly identical, and as semisynthetic ferrichromes, possessing extensive side chains at the peptide backbone show only slightly reduced uptake rates, the peptide backbone seems to be of minor importance for recognition of siderophores in fungi. It has recently been shown that a retrohydrohydroximate ferrichrome lacking the N-acetyl groups was not recognized as a siderophore in *Utilago sphaerogena* (Emery et al., 1984). Tetraglycyl ferrichrome (Deml et al., 1984), a heptapeptide siderophore, containing four instead of three glycine residues in the peptide backbone, was taken up at similar rates to ferrichrome, confirming the structural independence of that part of the molecule (Huschka et al., 1985). The proposed model for siderophore transport in *Neurospora crassa* (Fig. 4) implies that there is only one receptor for those ferrichromes possessing three identical ornithyl-N-acetyl residues. An additional receptor is postulated for coprogen and related compounds. The two receptors would be sufficient to take up all the different siderophores produced by this fungus. Generally, the restriction to few receptors would explain why the iron surrounding area of siderophores had to be the most structurally conserved part.

THE DRIVING FORCE FOR SIDEROPHORE TRANSPORT

It is a characteristic feature of siderophore transport that metabolic energy is required. In the presence of respiratory inhibitors, such as cyanide, or in the presence of uncouplers, such as 2,4-dinitrophenol, siderophore transport is severely inhibited. Furthermore, there is no outflow of siderophores from poisoned cells, indicating that movement of siderophores is intimately connected to an energy consuming process. Early attempts to identify the source of energy have pointed to ATP. However, as is the case with various other transport systems, a direct reaction of ATP

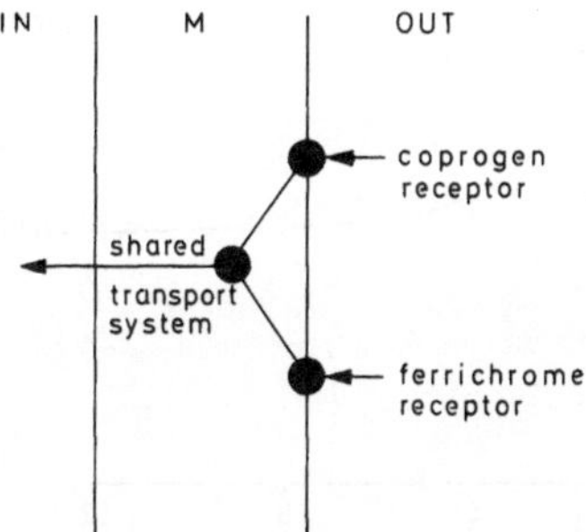

Fig. 4. Proposed model of siderophore transport in *Neurospora crassa* showing two siderophore recognition sites (coprogen receptor and ferrichrome receptor) and a shared transport system in the membrane (M).

with the transported molecules or with proteins involved in siderophore transport has never been shown. We have obtained evidence that ATP is only indirectly involved in siderophore transport by maintaining a high membrane potential. The membrane potential of *Neurospora crassa* is approximately 200 mV inside negative, as determined by microelectrodes. The membrane potential of glucose starved cells is rapidly depolarized by adding high amounts of glucose (Slayman and Slayman, 1974). This depolarization is explained by a proton symport mechanism of glucose taken up by the high affinity glucose (II) transport system. We have taken advantage of this depolaization method to demonstrate the dependence of siderophore transport on the membrane potential (Huschka et al., 1983). In fact, not the decreased ATP level but the transient depolarization of the membrane is the actual reason for the inhibition of siderophore transport. According to the chemiosmotic theory, the membrane potential is rapidly restored by the action of a proton translocating cytoplasmic membrane ATPase. For that reason the siderophore transport is only transiently inhibited. A permanent inhibition of membrane depolarization is achieved if glucose and a membrane ATPase inhibitor, such as dicyclohexyl carbodiimide or diethylstilbestrol, are added simultaneously. Under these conditions the depolarization remains irreversible and the transport of siderophores is not resumed. Parallel determinations of the ATP level confirmed that in depolarized cells ATP is still present in detectable amounts (30-50%), indicating that the depolarization is the primary event and that the decrease in ATP is secondary. The membrane ATPase inhibitors alone have no effect on siderophore transport. A further proof that the membrane potential is the driving force for siderophore transport in *Neurospora crassa* was obtained by cytoplasmic membrane vesicles prepared from protoplasts according to the method of Stroobant and Scarborough (1979). The transport of Fe-55 siderophores from loaded vesicles to the outside was studied. Based on the fact that significant numbers of membrane vesicles are functionally inverted (inside-out) vesicles, the addition of ATP to the incubation medium will energize the proton pumping ATPase, which in inverted vesicles is accessible from the outside. The added ATP generates a transmembrane electrochemical proton gradient giving rise to a transmembrane electrical potential (inside positive). The addition of ATP to loaded membrane vesicles resulted in an immediate efflux of labeled ferricrocin (Fig.5). As a control, the presence of thiocynate autons (SCN) abolished the membrane potential by compensating the positive charges at

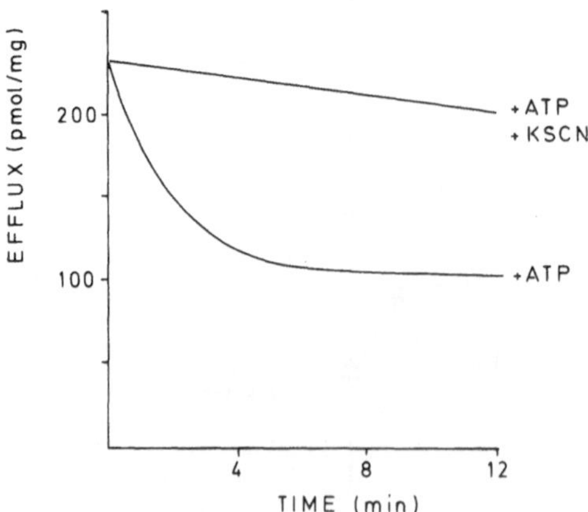

Fig. 5. Efflux of Fe-55 ferricrocin from inverted membrane vesicles of *Neurospora crassa*. Loaded vesicles were incubated in the presence of ATP or in the presence of ATP plus KSCN.

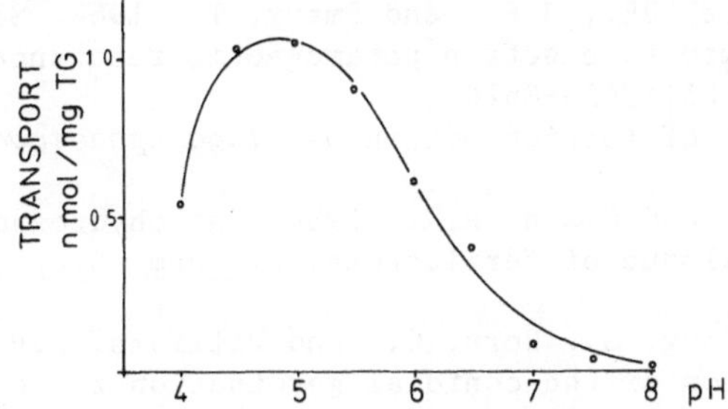

Fig. 6. Dependence of siderophore uptake on the pH of the incubation medium, determined with *Neurospora crassa* and C-14 coprogen.

the inside and consequently prevented the translocation of siderophores from the inside to the outside. It is interesting to note that siderophore uptake in fungi requires an acid pH of the incubation medium. The optimum is pH 5 in *Neurospora crassa* (Fig. 6). Therefore, it may be assumed that a protonation of amino acid residues in proteins with pKa at about 5 (aspartic or glutamic acid) in proteins is required, which enables binding or translocation of siderophores.

OTHER MECHANISMS

Not all fungi studied so far synthesize hydroxamate siderophores. *Saccharomyces cerevisiae, Mucor, Phycomyces, Cryptococcus* and a variety of other opportunistic and systemic fungal pathogens have been analyzed. Although colour formation has been reported in some attempts, no definite siderophore structure has been reported until now. The identification of coprogen B in *Histoplasma capsulatum* has been reported (Burt, 1982). It may be that in some cases the conditions of iron deprivation were not achieved or that in some fungi other iron acquisition mechanisms may occur which do not depend on siderophores. Using a siderophore-free mutant of *Neurospora crassa* we have shown earlier that hydroxy acids such as citrate and malate support the growth in low iron cultures (Winkelmann, 1979). On the other hand, the production of internal iron storage proteins may possibly render some fungi independent of the siderophore uptake system.

ACKNOWLEDGEMENTS

Profound thanks are due to my collaborators, in particular Hans-Georg Huschka, Gertraud Müller and Berthold Matzanke. The research summarized here was supported by the Deutsche Forschungsgemeinschaft (SFB 76).

REFERENCES

Burt, W.R., 1982. Identification of coprogen B and its breakdown products from *Histoplasma capsulatum*. Infect. Immun. 35:990-996.

Carrano, C.J., and Raymond, K.N., 1978. Coordination chemistry of microbial iron transport compounds: Rhodotorulic acid and iron uptake in *Rhodotorula pilimanae*. J. Bacteriol., 136:69-74.

Davis, R.H., 1968. Utilization of exogenous and endogenous ornithine by *Neurospora crassa*. J. Bacteriol., 96:389-395.

Deml, G., Voges, K., Jung, G., and Winkelmann, G., 1984. Tetraglycylferrichrome - The first heptapeptide ferrichrome. FEBS Lett., 173:53-57.

Ecker, D.J., Lancaster, JR., J.R., and Emery, T., 1982. Siderophore iron transport followed by electron paramegnetic resonance spectroscopy. J. Biol. Chem., 257:8623-8626.

Emery, T., 1971. Role of ferrichrome in *Ustilago sphaerogena*. Biochemistry, 10:1483-1488.

Emery, T., Emery, L., and Olson, R.K., 1984. Retrohydroxamate ferrichrome, a biomimetic analogue of ferrichrome. Biochem. Biophys. Res. Commun., 119:1191-1197.

Horowitz, N.H., Charlang, G., Horn, G., and Williams, N.P., 1976. Isolation and Identification of the conidial germination factor of *Neurospora crassa*. J. Bacteriol., 127:135-140.

Huschka, H., Naegeli, H.U., Leuenberge-Ryf, H., Keller-Schierlein, W., and Winkelmann, G., 1985. Evidence for a common siderophore transport system but different siderophore receptors in *Neurospora crassa*. J. Bacteriol., 162:715-721.

Huschka, H., Müller, G., and Winkelmann, G., 1983. The membrane potential is the driving force for siderophore iron transport in fungi. FEMS Microbiol. Lett., 20:125-129.

Jalal, M.A.F., Mocharla, R., Barnes, C.L., Hossain, M.B., Powell, D.R., Eng-Wilmot, D.L., Grayson, S.L., Benson, B.A., and Van Der Helm, D., 1984. Extracellular siderophores from *Aspergillus ochraceous*. J. Bacteriol., 158:683-688.

Keller-Schierlein, W., and Diekmann, H., 1970. Stoffwechselprodukte von Microorganismen. Zur Konstitution des Coprogen. Helv. Chim. Acta. 53:2035-2044.

Matzanke, B., and Winkelmann, G., 1981. Siderophore iron transport followed by Mössbauer spectroscopy. FEBS Lett., 130:50-53.

Müller, G., and Winkelmann, G., 1981. Binding of siderophores to isolated membranes of *Neurospora crassa*. FEMS Microbiol. Lett., 10:327-331.

Slayman, C.L., and Slayman, C.W., 1974. Depolarization of the plasma membrane of *Neurospora crassa* during active transport of glucose: Evidence for a proton-dependent cotransport system. Proc. Nat. Acad. Sci., USA, 71:1935-1939.

Stroobant, P., and Scarborough, G.A., 1979. Active transport of calcium in *Neurospora* plasma membrane vesicles. Proc. Nat. Acad. Sci., USA. 76:3102-3106.

Winkelmann, G., and Zähner, H., 1973. Stoffwechselprodukte von Microorganismen Eisenaufnahme bei *Neurospora crassa*. I. Zur spezifität des eisentransportes. Arch. Mikrobiol. 88:49-60.

Winkelmann, G., 1974. Metabolic products of microorganisms. Uptake of iron by *Neurospora crassa* III. Iron transport studies with ferri-chrome type compounds. Arch. Microbiol., 98:39-50.

Winkelmann, G., 1979. Surface iron polymers and hydroxy acids. A model of iron supply in sideramine-free fungi. Arch. Microbiol., 121:43-51.

Winkelmann, G., 1979. Evidence for stereospecific uptake of iron chelates in fungi. FEBS Lett., 97:43-46.

Winkelmann, G., and Braun, V., 1981. Stereoselective recognition of ferrichrome by fungi and bacteria. FEMS Microbiol. Lett., 11:237-241.

Winkelmann, G., 1983. Specificity of siderophore iron uptake by fungi. In: "The biological chemistry of iron," 107-116. (H.B. Dunford et al., Eds.) D. Reidel Publishing Co., Dordrecht.

ABSORPTION, TRANSPORT AND METABOLIC SIGNIFICANCE OF IRON IN PLANTS

P.C. DeKock

Macaulay Institute for Soil Research
Aberdeen, U.K.

The soil contains large quantities of iron, often in sesqui-oxide form. Very little iron is in solution in agricultural soils under normal circumstances but in the lower layers some iron occurs in solution and under anaerobic conditions this iron, in ferrous form may leach from the soil, and reoxidize in other locations. In more acid soils iron deficiency is very rarely observed in crops but in calcareous soils of which there is a preponderance in the drier parts of the world, crops are often pale yellow or even severely chlorotic to cause withering (DeKock, 1955). This is a result of the high pH which allows any ferrous iron to be oxidised very readily. Hence in such soils practices are followed which increase the solubility of iron such as mulching with organic matter or peat or spraying crops with ferrous sulphate or iron chelates, or even adding native sulphur to the soil (Wallace et al., 1982).

Certain plant species have become adapted to growing on calcareous soils and are known as 'calcicoles' whereas others are adapted to very acid soils or conditions where iron is readily available such as plants growing on peaty soils - the heathers or rhododendrons to name a few garden favourites, are known as 'calcifuges'.

It is thus apparent that plants under some conditions such as a high soil pH cannot obtain adequate iron from the soil whereas others flourish under these conditions. Conversely where iron is adequate or plentiful, the 'calcicoles' wither whereas the 'calcifuges' flourish.

The reactions of ferrous iron are well known and its affinity for oxygen under alkaline conditions is made use of in stoichiometric reactions. Iron also forms very insoluble compounds with phosphates. Under alkaline conditions the bicarbonate ion must be taken into account. So it can be seen that the plant root is exposed to a number of factors which influence the availability of iron not the least of which is water, calcareous soils being rare in wetter areas, and soil oxygen and carbon dioxide playing a major role in iron availability. Under very wet conditions, ferrous iron may become plentiful and plants exposed to such conditions may necessarily have aeration mechanisms in their roots which can influence the ambient solution. Hence it is evident that iron toxicity is important to plants as well as iron deficiency.

Mechanisms whereby plants obtain iron from the soil have been closely researched. Compounds which will bind iron are ubiquitous in plants and are well known. Citric acid forms a soluble complex with ferric iron which could readily pass through cell walls and the complex move up the plant to the leaves, there to be metabolized to ferrous iron. Phenolic compounds such as tannins and simpler phenols such as salicylic and caffeic acid are universal constituents of plants and will similarly complex with iron in both ferric and ferrous states (Olsen et al., 1981). Reductants such as ascorbic acid are present in every living cell, and will influence the redox potential as must also glutathione and the other sulphydryl compounds. Very complex molecules will pass through cell membranes and iron chelates of the EDTA type are no exception, but this is ongoing research and no doubt any statement of categorical nature will be hotly debated (Bienfait, 1985).

Has the root an exclusion mechanism for iron, as iron is very toxic and readily kills plants? The root hairs of plants, are the main absorption sites of nutrients and are very ephemeral, probably only existing for a few days before withering. Toxic conditions result in death of all the root hairs and if persistent, death of the particular root. However, other exclusion mechanisms must exist which will be discussed later.

Translocation in plants is remarkably rapid. The water columns from root surface to leaf surface are continuous and under tension. Water loss occurs from the leaf surfaces and so root metabolites are transported to the leaves by way of the xylem vessels. At the same time translocation downwards proceeds through the phloem. Compounds sprayed on the leaves will influence root growth. The sugars and sugar acids are good chelators of iron and may be involved in iron translocation.

The growing areas of plants, the meristems, are rich in iron and there is good evidence to show that shortage of iron affects the meristems, slowing down cell division. Expansion of the leaf bud depends on many factors but can best be described as a "package deal". If several conditions are fulfilled the leaf bud rapidly proliferates and the iron in the leaf is fixed into compounds of a structural nature, such as cytochromes and other haem compounds, phyto-ferritin, iron-protein complexes and ribonucleic acids. Rapid cell expansion causes the leaf blade to form with a dilution of the main cations, phosphorus and potassium, but iron and calcium continue to increase until senescence.

The young leaf, at first pale and fragile, depends for the formation of chlorophyll on that portion of the iron in the leaf which is metabolically active and so has been termed 'active' iron. Where this fraction is inadequate, the leaf stays yellow or yellow-green. Oserkowsky (1933) demonstrated that in pear leaves the iron extracted from a dried leaf powder by normal hydrochloric acid bore a linear relationship to the amount of chlorophyll which could be extracted. Subsequently Bolle-Jones (1955) used etherized tenth-normal hydrochloric acid on potato leaves with similar results but recently Katyal and Sharma (1984) have demonstrated that 1:10 _o_-phenanthroline will effectively extract this fraction. In all cases of iron deficiency chlorosis, this fraction has been found to be lower than in normal green leaves.

The fact that there is a linear relationship between the active iron fraction and chlorophyll shows that some carefully maintained equilibrium is operative; hence a complex molecule, itself containing no iron but magnesium, is dependent upon the ferrous iron fraction of the leaf. No relationship or at best a very poor relationship could be found between total iron and chlorophyll, most of the iron in the leaf appearing inactive in chlorophyll formation. The amount of iron involved in haem formation is also very small, about one-hundredth or less of the total iron, but a linear relationship also exists between total haem pigments and chlorophyll (DeKock et al., 1960).

Some years ago Biddulph and Woodbridge (1952) showed that the phosphate ion would interfere with the translocation of iron, retarding upward movement. A characteristic of chlorotic tissue is the enhanced content of phosphorus, and it appeared likely that the phosphate in the leaf was affecting the iron in some way. It could be shown that the phosphorus-iron ration bore a linear relationship to chlorophyll, and also to the active iron fraction. Also results of workers who had failed to relate total iron to chlorosis could be used to demonstrate significant phosophorus-iron ratios (DeKock, 1979; Little, 1971).

Failure to find significant relationships using total iron led Bennett (1945) to propose the use of another parameter, the calcium-potassium ratio or more conveniently, the potassium-calcium ratio as this is usually greater than one. It is indeed surprising to observe that chlorotic leaves contain less calcium and more potassium than green leaves and this potassium-calcium ratio can be used as an indicator of iron chlorosis. A century or so ago, this ratio was used to guard against liming of potassium-deficient soils so that it has some very fundamental significance, being used by animal physiologists also.

So if the phosphorus-iron ratio and the potassium-calcium ratio are both parameters of chlorosis, they should show some inter-relationship. It could also be demonstrated that chlorotic leaves contained more citric acid and relatively less malic acid than green leaves.

Shortly after Krebs (1946) proposed the Tricarboxylic acid cycle, it was demonstrated that the enzyme aconitase which converts citric acid into iso-citric acid through cis-aconitic acid was ferrous-iron dependent so that the cycle was blocked at the citric acid stage and malic acid tended to decrease. This was found to be the case in plants also (Bacon et al., 1959). The roots of iron deficient plants are known to contain more citric acid and so this acid may be involved in solubilization of iron from the soil.

Some plants contain oxalic acid as a major constituent acid and linear relationships have been found between calcium content and oxalic acid content. This acid was also known to be adversely affected by iron deficiency but the metabolic connexion was not apparent until Kornberg and Krebs (1957) demonstrated the iso-citric acid shunt whereby iso-citric acid was split to succinic and glyoxylic acid. The enzyme iso-citritase was similarly ferrous iron dependent. Glyoxylic acid will condense with Acetyl CoA to form malic acid, so that metabolic links are apparent between malic and oxalic acids as glyoxylic acid itself is in equilibrium with oxalic acid. Linear relationships between calcium and malic acid have been demonstrated in cruciferous and other plants (Kirkby and DeKock, 1965).

It may well be worth considering the implications of these ratios at this stage. If in fact iron metabolism is inter-related with that of phosphorus it means that the phosphate-iron ratio controls the amount of ferrous iron in the cell. Support for this view is accorded by the work of Taborsky (1979) who showed that ferrous ions will autoxydize in the presence of cytochrome C and phosphate and that the reaction is of a 'Mass Action' nature. The phosphorus-iron ratio also means that phosphate and iron control the uptake of each other and that phosphate deficiency is in fact iron toxicity. Symptoms of phosphate deficiency are dark green colours with rusty-purple patches or edges to the leaves. Potassium deficiency similarly is characterized by dark green colours and rusty-reds in the older plant leaves. Calcium deficiency on the other hand is very similar to iron deficiency, pale or even albino patches occurring in the young tissues.

It must now be apparent that it is hardly possible to consider iron metabolism without considering every other element and in fact every other physiological factor because where deficiency is mentioned, an excess of another element is implied. Hence confusion can arise very easily. For instance it is claimed that potassium fertilization can ameliorate iron deficiency chlorosis, but analyses show that iron-deficient tissues already contain enhanced amounts of potassium. It would appear that the added potassium in these circumstances is buffering an overdose of phosphate, thereby releasing iron. However, the field is enormous and in no way can every facet of plant growth and metabolism be considered here.

One more aspect however deserves notice. Iron deficient tissues contain greater amounts of free amino acids and addition of iron to the leaf causes these amino acids to disappear, presumably due to protein synthesis. A search for a particular metabolic block in synthesis of amino acids showed that each individual species usually had its array of amino acids and any one of these may accumulate in greater amounts. In most plants glutamic acid showed enhanced accumulation (DeKock and Morrison, 1958).

Nitrogen is especially problematical in plant nutrition because, although all forms must finally appear as ammonium and react with keto groups, the ammonium ion and the nitrate ion are both involved in plant nutrition and have different physiological effects. Thus application of ammonia-fertilizers result in plants with increased potassium and phosphorus and decreased calcium, so that they may induce calcium deficiency. Nitrate fertilizers cause an accumulation of organic acids, higher calcium concentrations in the tissues and a depression of potassium and phosphorus uptake (DeKock, 1981).

Rapidly growing plants are characterized by pale colours but as tissues mature the green colours darken and remain so for a period of time until senescence sets in. Such young tissues contain high phosphorus and potassium concentrations and less iron and calcium. As maturity is attained phosphorus and potassium concentrations decrease and iron and calcium contents increase. Thus the stage of maturity as well as other physiological conditions are necessary for full assessment of mineral status (Kirkby and DeKock, 1965).

Growth rate is largely dependent on hormonal balance. For convenience let us only consider two hormones, benzyladenine (BA) a growth promoter and abscissic acid (ABA) an inhibitor of growth. Thus by using the duckweed *Lemna gibba* in axenic conditions, mixtures of BA and ABA can be administered. BA being a growth promoter, will induce higher ratios of potassium to calcium and high ratios of phosphorus to iron. Conversely ABA, the inhibitor of growth will cause potassium and phosphorus to be lost from tissues. In the duckweed, changes in morphogenesis are induced resembling the autumn to winter changes (DeKock et al., 1978).

Any constraint of growth will thus induce changes lowering the ratios of phosphorus to iron and potassium to calcium. Water supply is a very variable factor in plant growth in natural conditions and shortage of water seemingly causes abscissic acid to be generated and growth to cease. The opening and closing of stomata are dependent on such hormonal balances.

More recently work has been done on the ATP content of duckweed grown with BA or ABA, showing that when plants are growing rapidly, there is little detectable ATP in the tissues but when retardation of growth occurs, ATP can be demonstrated.

The chlorophyll molecule closely resembles that of haem and so their synthesis must be closely related. Estimations of total haem in plant tissues showed linear relationships to the chlorophyll content. Hence again equilibrium reactions are involved. Steps in the synthesis of chlorophyll are also dependent on ferrous iron such as the condensation of γ amino laevulinic acid (ALAD) to form the pyrole structure. It is interesting to speculate whether increased complexation is essential in chlorophyll synthesis. Thus the activity of peroxidase is much enhanced where chlorophyll is deficient whereas catalase activity is always enhanced in green tissues. The ratio of peroxidase to catalase can similarly be used as a parameter of chlorosis (DeKock et al., 1960).

Superoxide dismutase (SOD) has recently sprung into prominence as an important enzyme in cell metabolism. In the plant the enzyme molecule contains ferrous iron and zinc, protecting the cell against free radicals. This may explain why zinc deficiency closely resembles iron deficiency.

Although magnesium is the constituent metal of chlorophyll, the amount involved is very small whereas the content of magnesium in plant cells is quite large and characteristically varies within only narrow limits, so being rather different from potassium and calcium which may vary quite widely. It would appear that magnesium content is very closely regulated in some manner, showing as it does metabolic similarities with calcium.

In general plant growth is affected to some extent by almost every factor and by almost every compound to which it may be exposed. Thus although galactose is a constituent of plant cell walls, feeding duckweed plants galactose in axenic culture soon induces growth stasis even though galactose is easily transformed into glucose (DeKock et al., 1978). It is even more amazing to discover that the sugar amine, galactosamine is several hundred fold more toxic than galactose.

It is thus evident that consideration of the factors which control infectivity of plants may be very complex indeed. Thus vesicular-arbuscular fungi attack and enter plant roots mainly at the region of cell extension behind the region of mitosis. Generally during extension cell walls are plastic and there is little cross-linkage but later cross-linkage occurs between phenolic constituents and calcium. Thus calcium deficiency may ensure that those phenolic constituents which are normally involved in cross-linkage remain unaltered and may serve as either fungistatic or bactericidal agents. Calcium deficient tissues are high in phenols and fluoresce under ultra-violet light. Similar effects of increased phenols are found under conditions of iron deficiency and these phenolic componds may be excreted from iron-deficient plant roots, so protecting the plant from bacterial attack.

REFERENCES

Bacon, J.S.D., DeKock, P.C., and Palmer, M.J., 1959. Aconitase levels in the leaves of iron deficient mustard plants. Biochem. J., 80:64-70.

Bennett, J.P., 1945. Iron in leaves. Soil Sci., 60:91-105.

Biddulph, O., and Woodbridge, C.G., 1952. The uptake of phosphorus by plants with particular reference to the effects of Fe. Plant Physiol., 27:431-444.

Bienfait, H.F., 1985. Regulated redox processes at the plaslemma of plant root cells and their function in iron uptake. J. Bionerget. Biomembr., 17:73-83.

Bolle-Jones, E.W., 1955. The interrelationships of iron and potassium in the potato plant. Plant Soil, 6:129-173.

DeKock, P.C., 1955. The iron nutrition of plants at high pH. Soil Sci., 79:167-175.

DeKock, P.C., 1979. Active iron in plant leaves. Ann. Bot., 43:737-740.

DeKock, P.C., 1981. Iron nutrition under conditions of stress. J. Plant. Nutr., 3:513-521.

DeKock, P.C., Cheshire, M.V., Mundie, C.M. and Inkson, R.H.E., 1978. The effect of galactose on the growth of *Lemna*. New Phytol., 82:679-685.

DeKock, P.C., Commisiong, K., Farmer, V.C., and Inkson, R.H.E., 1960. Interrelationships of catalase, peroxidase, hematin and chlorophyll. Plant Physiol., 35:599-604.

DeKock, P.C., and Morrison, R.I., 1958. The metabolism of chlorotic leaves 1. Amino Acids. 2. Organic Acids. Biochem. J., 70:266-278.

DeKock, P.C., Vaughan, D., and Hall, A., 1978. The effect of abscissic acid and benzyl adenine on the inorganic and organic composition of the duckweed, *Lemna gibba* L. New Phytol., 81:505-511.

Katyal, J.C., and Sharma, B.D., 1984. Association of soil properties and soil and plant iron to iron deficiency response. Commun. Soil Sci. Plant Anal., 15:1065-1081.

Kirkby, E.A., and DeKock, P.C., 1965. The influence of age on the cation-anion balance in the leaves of Brussels sprouts. Z. Pflanzenernahr., Düngung Bodenkd., 111:197-203.

Kornberg, H.A., and Krebs, H.A., 1957. Synthesis of cell constitutents from C_2-units by a modified tricarboxyllic acid cycle. Nature, 179:988-990.

Krebs, H.A., 1946. Cyclic processes in living matter. Enzymologia, 12:88.

Little, R.C., 1971. The treatment of iron deficiency. Tech. Bull. GB Minist. Agric. Fish Food, 45-61.

Olsen, R.A., Bennett, J.H., Blume, D., and Brown, J.C., 1981. Chemical aspects of the Fe stress response mechanism in tomatoes. J. Plant. Nutr., 3:905-921.

Oserkowsky, J., 1933. Quantitative relation between chlorophyll and iron in green and chlorotic pear leaves. Plant Physiol., 8:449-468.

Taborsky, G., 1979. Interaction of cytochrome C, ferrous iron and phosphate. J. Biol. Chem., 254:5246-5251.

Wallace, A., Samman, Y.S., and Wallace, G.A., 1982. Correction of lime-induced chlorosis in soybeans in a glasshouse with sulphur and an acidifying iron compound. J. Plant Nutr., 5:949-953.

IRON-EFFICIENCY REACTIONS OF MONOCOTYLEDONOUS AND DICOTYLEDONOUS PLANTS

H.F. Bienfait

Laboratory for Plant Physiology
University of Amsterdam
Kruislaan 318
1098 SM Amsterdam
The Netherlands

INTRODUCTION

Iron deficiency in plants is reflected in a decreased capability to make chlorophyll, thus resulting in yellowing of the leaves, chlorosis. A characteristic feature of iron deficiency chlorosis is that the leaf tissues bordering the veins remain green. Cells in this region utilise traces of iron arriving from the roots. Chlorosis induced by the action of toxins usually occurs uniformly in all tissues.

Even before chlorosis becomes apparent plants respond to iron deficient conditions by bringing into play one of two strategies. In Strategy I, found in dicotyledons and non-grass monocotyledons, several events can be detected:

(a) Acidification of the rhizosphere
(b) An increase in reductive capacity for ferric chelates
(c) A probable increase in the ability to take up ferrous ions.

Strategy II <u>found only in grasses</u> (so far), involves the excretion of siderophores, a mechanism comparable with that found in microorganisms.

This review explores these strategies and indicates, with the help of models, how they may relate to the activities of microorganisms at the root surface.

MECHANISMS

<u>Strategy I</u>

Rhizosphere acidification concomitant with iron-chlorosis was first observed in 1956 (Fuss, 1956). It is not caused by excretion of organic acids (Fuss, 1956; Venkat Raju et al., 1972), although leaves and roots of chlorotic plants do accumulate substantial amounts of acids, most characteristically citrate (Landsberg, 1979), most probably, protons are excreted by an ATP-dependent pump which can locally attain very high rates (Römheld, 1984). According to Römheld and Kramer (1983), the acidification

Table 1. Properties of two possible transplasmamembrane electron-transfer systems in plant cells (from Bienfait, 1985)

	Constitutive 'Standard'	Inducible 'Turbo'
Presence	All plants	Dicots and nongrass monocots
Localization	Ubiquitous	Epidermis of young lateral roots
Active with	Ferricyanide	Ferricyanide and Fe-EDTA
Natural electron acceptor	Oxygen	Ferric chelate
Electron donor	NADPH (NADH)	NADPH
K_m for electron donor	Low	High
Function	Membrane polarization	Iron uptake

activity is located in transfer cells in the root epidermis, the formation of which is induced by iron deficiency (Kramer et al., 1980). Transfer cells are characterized by labyrinth-like involutions of the plasma membrane, with concomitant enhanced deposition of cell wall material. The resulting increased plasma membrane surface is thought to be the site of the proton pumping ATPases.

The advantage to the plant of a more acid environment for the roots is evident; the ferric ion and its hydroxylated complexes become more soluble (Lindsay and Schwab, 1982). Another effect of acidification is that the roots tend to become leaky, so that compounds of low molecular weight are excreted at increased rates (Römheld and Marschner, 1983). This leakage will lead to a higher concentration of ferric chelating compounds around the roots, which is of course helpful in bringing soil iron into solution. High concentrations of organic matter in the rhizosphere will also stimulate the growth of microorganisms.

Reduction activity towards ferric chelates is present at a low level in root tips of dicotyledonous plants. The ferrous chelates resulting from reduction will mostly have a lower stability than the original ferric chelate and free ferrous ions are readily taken up (Chaney et al., 1972). Upon iron shortage the zones with ferric reducing capacity are lengthened, and their reducing activity is strongly increased. This activity resides exclusively in the epidermis (Sijmons and Bienfait, 1983), and when root hairs are formed, the activity is concentrated there. Moreover, in some species, the formation of root hairs is strongly stimulated by iron deficiency (Kramer et al., 1980).

In epidermal cells, the reducing activity towards ferric chelates resides in the plasma membrane (Bienfait et al., 1983), where probably a shortened form of the respiratory chain shuttles electrons from cytosolic NADPH to extracellular acceptors (Sijmons et al., 1984). The substrate specificity of this system is low (Bienfait et al., 1983), and this is, of course, in accordance with the function of the system: to reduce any ferric chelate which may present itself at the cell surface. The ferric complexes of desferrioxamine (Bienfait et al., 1983; Römheld and Marschner, 1983), aerobactin (Bienfait et al., 1983), and two siderophores excreted by pseudomonads (Schippers and Bienfait, unpublished) however, are reduced only at scarcely measureable rates.

The pH optimum for the reducing system is low; depending on the substrate, between pH 3-5 (Bienfait et al., 1983; Römheld and Marschner, 1983). The induced acidification of the rhizosphere therefore assists by enhancing ferric reducing capacity.

Grasses do not reduce ferric chelates at appreciable rates, whether iron-deficient or not. However, they do reduce ferricyanide, but this activity is not influenced by the iron status of the plant. Iron-sufficient dicotyledons also reduce ferricyanide, and this activity is increased upon iron deficiency. I recently proposed that plants may contain two electron transferring systems in the plasma membrane: a 'Standard' system present in all cells of all plants, active with ferricyanide as an artificial acceptor but presumably with oxygen as the natural acceptor, and another the 'Turbo' system, present in the root epidermis of all plants except the grasses, its activity being regulated by the iron status of the plant, and active with a large range of ferric chelates (Bienfait, 1985). The properties of these two systems are summarized in Table 1.

The high level of NADPH found in roots of iron-deficient beans (Sijmons et al., 1984) can, in a metabolism which is actively turning over, only be realized by an enzyme which is still working at strongly reduced NADP levels. Studies in our laboratory with enzyme preparations from bean roots

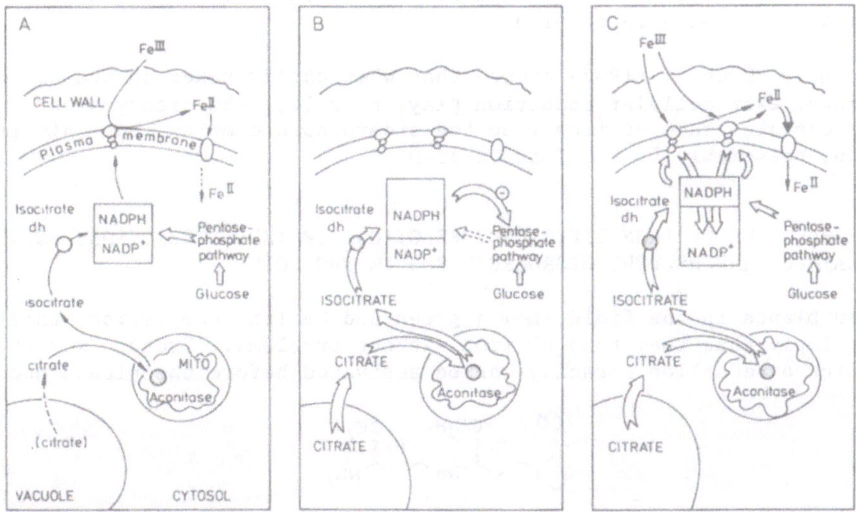

Fig. 1. Flow of reducing equivalents in root epidermis cells of a dicotyledonous plant. A. Control; B. Iron-deficient plant; C. Iron-deficient plant after addition of a reducible ferric chelate (Lubberding et al., 1985).

showed that isocitrate dehydrogenase could fulfil this condition (Lubberding et al., 1985). Experiments with glucose labelled at C-1 or C-6 showed that, once the system started donating electrons to an added extracellular acceptor, and the NADP redox level was lowered to a new steady-state (Sijmons et al., 1984), the pentose phosphate pathway could also contribute to the electron flow (Lubberding et al., 1985).

Iron-deficient plants have a pronounced tendency to accumulate citrate, especially in the roots (Landsberg, 1979). The cause of this accumulation is still unknown; it has been suggested that lowered levels of the iron-containing enzyme aconitase could be the clue (Bacon et al., 1961; Venkat Raju et al., 1972). However, in bean roots, at the degree of iron deficiency where the Turbo activity was optimally induced, levels of aconitase and other iron-containing enzymes are not yet affected (De Vos and Bienfait, 1985). Thus, the route of reducing equivalents from accumulated citrate, via mitochondrial aconitase and cytosolic isocitrate dehydrogenase, to NADP is free to feed the potential high flux of electrons to the Turbo system. Figure 1 shows our view on fluxes of reducing equivalents in iron-deficient and control dicotyledons.

Strategy II

Grasses do not acidify the rhizosphere, nor do they reduce ferric chelates (Brown, 1978). Both statements are not 100% correct: Landsberg showed that iron-deficient maize, under conditions where nitrate reduction is low, could acidify the nutrient medium (Landsberg, 1979). Also, grasses show some reduction of ferric chelates but the rates are 10 to 100 fold lower than those of the dicotyledons. They have apparently elected to use the microbial-type of reaction to iron deficiency in excreting siderophores such as mugineic and avenic acid (Fushiya et al., 1980; Sugiura et al., 1981). These compounds, neither catechols nor hydroxamates, are related to the non-protein amino acid nicotianamine (Ripperger and Schreiber, 1982, see Figure 2), a siderophore which is probably used by all higher plants as a ferrous carrier over short (cell to cell) distances (Scholz et al., 1985). The affinity of mugineic acid for ferric ions (10 , Sugiura et al., 1981) is not as high as that of the microbial siderophores like ferrioxamine or enterochelin.

Recently, Römheld (1985) showed that when barley takes up its own Fe-siderophore, extracellular reduction plays no role; the dicotyledon cucumber can also take up iron from the siderophore complex, but only after reduction, presumably by the Turbo system.

THE POSSIBLE ROLE OF IRON-EFFICIENCY REACTIONS IN THE INTERACTIONS BETWEEN ROOTS, HARMFUL AND HELPFUL MICROORGANISMS IN THE SOIL

When plants in the field show a green and healthy complexion, this does not imply that they take up iron without problems. Proton extrusion and the Turbo reduction capacity can be activated before the plants become

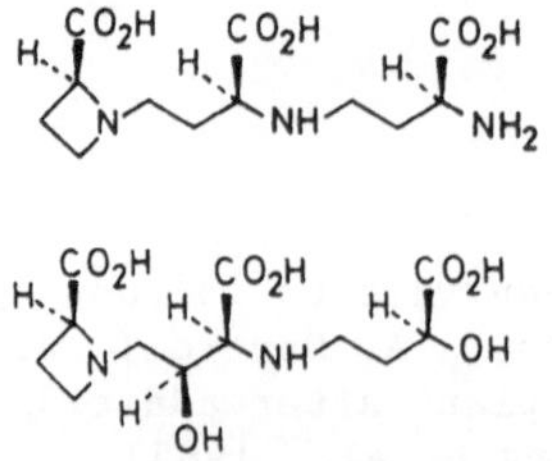

Fig. 2. Structures of nicotianamine (top) and mugineic acid (bottom)

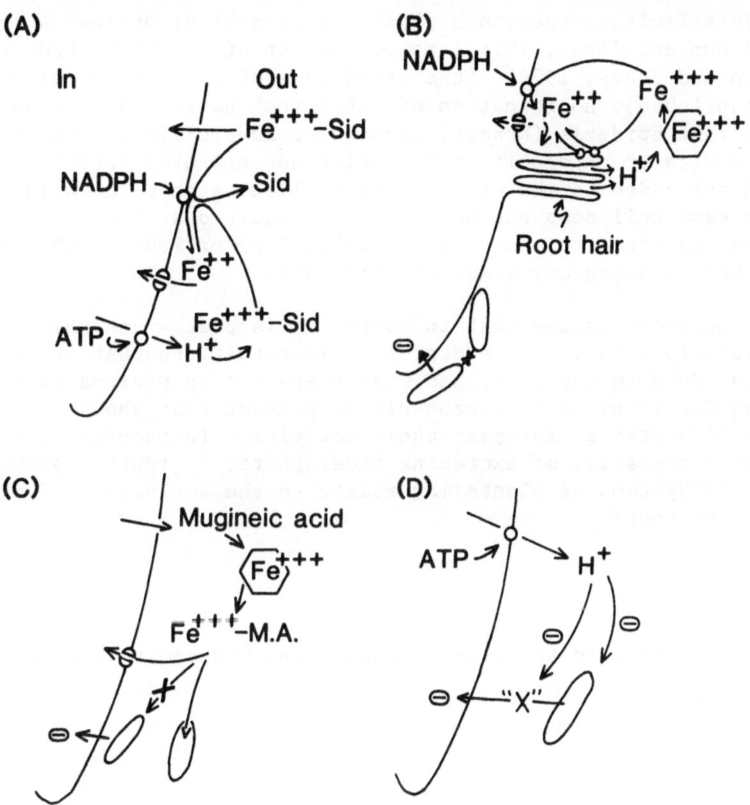

Fig. 3. Fe-efficiency mechanisms of plant roots in relation to the growth of microorganisms. A. Utilization of microbial siderophores by the plant. Microbial siderophores might be taken up as such, or the ferrous ion only, after reduction (however, see the text). Protons, excreted by the roots, may lower the rhizosphere pH to 3 or lower and, in doing so, both increase ferric solubility and decrease the strength of the ferric-siderophore bond. B. Spatial non-competition for iron. Iron uptake is located in a zone which begins behind the elongation zone (0.3 - 1 cm). This uptake zone may be lengthened upon iron deficiency. There, proton extrusion and root hair formation may increase iron availability for the plant without an effect on competition for iron between microorganisms at the root tip. The same may hold for phytosiderophore excretion and uptake. C. Exclusive use of a phytosiderophore. A plant (grass) may excrete a phytosiderophore which cannot be taken up by one or more of the microorganisms living at the root surface. In doing so, it may influence any microbial competition for iron. Taking into consideration the relatively low affinity of, e.g., mugineic acid for ferric, this would only be relevant when the microorganism involved would produce no or inferior siderophores. D. Elicited interference. Protons excreted by roots upon a beginning iron deficiency lower the rhizosphere pH to 3 or lower and thereby influence both the growth of microorganisms and their production of toxic substances, as well as the stability of these substances. $\ominus$: Negative effect.

chlorotic. Probably, iron-efficiency reactions are constantly being switched on and off during growth, and this may, in many soils, result in a well-regulated flux of iron to the growing parts (see, e.g., Römheld and Marschner, 1981). For example the tomato mutant fer lacks the capacity to induce the Fe-efficiency reactions characteristic of dicotyledons: no proton extrusion (Brown and Jones, 1974), no activation of the Turbo reduction system (Brown and Jones, 1974), (the Standard system being normally active, Bienfait unpublished), no formation of extra root hairs and of transfer cells in the root epidemis (Kramer, personal communication). The mutant can only grow without chlorosis when massive amounts of a ferric chelate-like FeEDDHA are added to the soil. This implicates that the wildtype FER, which on the same soil does not need any iron additions, manipulates its Fe-efficiency reactions to maintain a regular flow of iron to the leaves, and that without showing any signs of chlorosis.

We can therefore assume that in normal soils plants use their iron-efficiency reactions to variable degrees. In situations that extra micro-organisms are added to the soil, and that there are suspicions that these are competing for iron, it is reasonable to presume that the roots of the plants will, if anything, increase their activities in pumping protons, reducing ferric chelates, or excreting siderophores. Figure 3 shows how the iron uptake systems of plants may relate to the activities of micro-organisms at the roots.

ACKNOWLEDGEMENT

The author wishes to thank Drs. Sijmons and Lubberding for stimulating discussions.

REFERENCES

Bacon, J.S.D., DeKock, P.C., and Palmer, M.J., 1961. Aconitase levels in the leaves of iron-deficient mustard plants (*Sinapis alba*). Biochem. J., 80:64-70.

Bienfait, H F., 1985. Regulated redox processes at the plasmalemma of plant root cells and their function in iron uptake. J. Bioenerget. Biomembr., 17:73-83.

Bienfait, H.F., Bino, R.J., Van Der Bliek, A.M., Duivenvoorden, J.F., and Fontaine, J.M., 1983. Characterization of ferric reducing activity in roots of Fe-deficient *Phaseolus vulgaris*. Physiol. Plant., 59: 196-202.

Brown, J.C., 1978. Mechanism of iron uptake by plants. Plant, Cell Environ., 1:249-257.

Brown, J.C. and Jones, W.E., 1974. pH changes associated with iron-stress responses. Physiol. Plant., 30:148-152.

Chaney, R.L., Brown, J.C. and Tiffin, L.O., 1972. Obligatory reduction of ferric chelates in iron uptake by soybeans. Plant Physiol. 50:208-213.

De Vos, C.H. and Bienfait, H.F., 1985. Role of organic acids in Fe-efficiency reactions of bean plants. Plant Physiol. (in press).

Fushiya, S., Sato, Y., Nozoe, S., Nomoto, K., Takemoto, T. and Takagi, S., 1980. Avenic acid, a new amino acid possessing an iron chelating activity. Tetrahedron Lett., 21:3071-3072.

Fuss, K., 1956. Die Ansäuerung der Nährlösung durch *Lupinus luteus* und ihre papierchromatografische Untersuchung auf saüre Wurzelausscheidungen. Flora, 144:1-46.

Kramer, D., Römheld, V., Landsberg, E. and Marschner, H., 1980. Induction of transfer-cell formation by iron deficiency in the root epidermis of *Helianthus annuus* L. Planta, 147:335-339.

Landsberg, E.C., 1979. Einfluss des Saurestoffwechsels und der Nitratreduktion auf Eisenmangel-bedingte Veranderungen des Substrat-pH-Wertes bei mono- und dikotyle Pflanzenarten. Ph.D. Thesis, Technische Universitat Berlin.

Lindsay, W.L., and Schwab, A.P., 1982. The chemistry of iron in soils and its availability to plants. J. Plant Nutr., 5:821-840.

Lubberding, H.J., De Graaf, F.H.J.M., and Bienfait, H.F., 1985. Regeneration of reducing equivalents for extracellular Fe(III) reduction in roots of iron-deficient bean plants. Plant Physiol. (in press).

Ripperger, H., and Schreiber, K., 1982. Nicotianamine and analogous amino acids, endogenous iron carriers in higher plants. Heterocycles, 17: 447-461.

Römheld, V., 1985. Specific uptake of Fe phytosiderophores by grasses. Plant Physiol. (in press).

Römheld, V., and Kramer, D., 1983. Relationship between proton efflux and rhizodermal transfer cells induced by iron deficiency. Z. Pflanzenphysiol., 113:73-83.

Römheld, V., and Marschner, H., 1981. Rhythmic iron stress reaction in sunflower at suboptimal iron supply. Physiol. Plant., 53:347-353.

Römheld, V., and Marschner, H., 1983. Mechanism of iron uptake by peanut plants. I. Fe(III) reduction, chelate splitting, and release of phenolics. Plant Physiol., 71:949-954.

Römheld, V., Muller, C., and Marschner, H., 1984. Localization and capacity of proton pumps in roots of intact sunflower plants. Plant Physiol., 71:603-606.

Scholz, G., Schlesier, G., and Seifert, K., 1985. Effect of nicotianamine on iron uptake by the tomato mutant 'chloronerva'. Physiol. Plant., 63:99-104.

Sijmons, P.C., and Bienfait, H.F., 1983. Source of electrons for extracellular Fe(III) reduction in iron-deficient bean roots. Physiol. Plant., 59:409-415.

Sijmons, P.C., Van Den Briel, M.L., and Bienfait, H.F., 1984. Cytosolic NADPH is the electron donor for extracellular Fe(III) reduction in iron-deficient beans. Plant Physiol., 75:219-221.

Sugiura, Y., Tanaka, H., Mino, Y., Ishida, T., Ota, N., Inoue, M., Nomoto, K., Yoshioka, H., and Takemoto, T., 1981. Structure, properties and transport mechanism of iron(III) complex of mugineic acid, a possible phytosiderophore. J. Am. Chem. Soc., 103:6979-6982.

Venkat Raju, K., Marschner, H., and Römheld, V., 1972. Effect of iron nutritional status on iron uptake, substrate pH and production and release of organic acids and riboflavin by sunflower plants. Z. Pflanzenernahr Bodenkd., 132:178-191.

SIDEROPHORE INVOLVEMENT IN PLANT IRON NUTRITION

C.P. Patrick Reid[1], Paul J. Szaniszlo[2], and
David E. Crowley[1]

Department of Forest and Wood Sciences[1]
Colorado State University
Fort Collins, Colorado 80525

Department of Microbiology[2]
University of Texas at Austin,
Austin, Texas 78712

INTRODUCTION

Iron is one of the most important and abundant, yet least available of all trace elements required by living organisms (Emery, 1982), a paradox partially explained by iron's chemistry and availability during evolution of life (Neilands, 1982; Lewin, 1984). Although iron is abundant, comprising 4-5% of the average soil (Lindsay, 1979), iron deficiency of crop plants is common in calcareous soils, which represent over one-third of the world's land surface area (Chen and Barak, 1982; Vose, 1982; Wallace and Lunt, 1960) and some 44 million acres of cropland in the United States (Clark, 1982).

Problems in iron availability are directly related to the solubility of iron as governed by chemical equilibria for dissociation of solid phase iron minerals (Lindsay, 1979). Iron availability is most limiting in aerated, well-oxidized soils where soluble iron exists in two valence states, as ferrous and ferric hydrolysis species, and solubility is controlled by the ferric mineral, ferric hydroxide (Lindsay, 1979; Lindsay and Schwab, 1982).

To acquire and transport iron in aerobic environments, living organisms often complex iron with organic chelates that combine with inorganic iron and greatly increase its solubility (Emery, 1977). Perhaps the most important of the naturally-occurring chelates are the microbial siderophores, produced by nearly all microorganisms under iron stress. Although the importance of siderophores in the iron nutrition of microorganisms has been realized for some time (see reviews: Emery, 1977, Lankford, 1973; Neilands, 1980), little attention has been given to their potential importance in the mineral nutrition of plants in soil. However, Estep et al., (1975) and Murphy et al., (1976) have presented evidence that hydroxamate siderophores are an important factor in determining iron acquisition by plants in aquatic ecosystems.

In this paper we present evidence suggesting that microbial siderophores are likely to be extremely important in the utilization of iron by higher plants. In documenting the potential importance of siderophores to plants, we have relied heavily on our own research results, and for the purpose of this review have placed them into the context of what appears to be the current concepts of iron use by plants. Our treatment of siderophores is almost entirely concerned with the hydroxamate siderophores (HS), produced by both fungi and bacteria, and only briefly mentions the catecholate siderophores primarily produced by bacteria. To recognize the way in which siderophores are likely to function in soil systems, we necessarily examine several aspects of the control of iron solubility in soil and then consider how plants acquire iron. Where appropriate, we indicate how we view siderophores functioning in relation to these factors.

CHEMISTRY AND AVAILABILITY OF IRON TO PLANTS

For iron to be taken up by plants, soluble iron concentrations must be in the range of 10^{-8}M 'Lindsay and Schwab, 1982). Of the ferric hydrolysis species, only $Fe(OH)_2^+$ and $FeOH^{2+}$ ever meet or exceed this minimum concentration, and then only at pH 5 or less. At neutral or alkaline pH, only Fe^{2+} and its hydrolysis species ever reach the critical concentration required by plants. In equilibrium with ferric hydroxide, concentrations of ferric hydrolysis species are pH dependent, whereas, concentrations of ferrous iron species are controlled by both redox and pH (Schwab and Lindsay, 1983). Therefore, redox relationships are extremely important in controlling availability of iron to plants.

In solution, inorganic iron can be complexed by organic molecules to form organic complexes with much greater solubility than inorganic iron. The affinity of a chelating agent for iron (the stability or formation constant) is expressed by the log K of the equilibrium reaction for metal binding. Stability of iron chelates with iron is dependent not only on the formation constant, but also the affinity of the chelating agent for other ions that can compete for and displace iron from the chelate. By using equilibrium relationships for metal chelates, one can determine chelate stability under given redox and pH conditions in soil and in nutrient solution (Lindsay, 1979; Lindsay et al., 1967; Lindsay and Norvel, 1969; Norvell, 1972; Sommers and Lindsay, 1979).

Iron chelates also increase solubility and availability of iron to plants in iron-limited soils and hydroponic culture. For example, synthetic chelating agents have been used in agriculture for over thirty years (Brown, 1969, Wallace, 1962, 1982), and the mechanisms by which synthetic chelates increase iron availability to plants have been examined. In soils, chelates increase the pool of soluble iron and diffusion of iron from the bulk soil to depleted microsites in the rhizosphere (O'Connor et al., 1970). At the root surface, chelates buffer concentrations of soluble inorganic iron in accordance with equilibrium relationships for dissociation of the metal ligand. In plant iron uptake from synthetic chelates it has been shown that some chelates function by transporting iron to the root surface (shuttle mechanism) where iron dissociates from the chelate to replenish soluble inorganic iron depleted by plant uptake (Lindsay and Schwab, 1982). However, depending on chemical equilibria in the rhizosphere, chelates with high affinity for iron can actually compete with plant roots for inorganic iron in solution and inhibit plant iron uptake (Becker et al., 1985; Brown, 1961, Cline et al., 1984). In addition to the shuttle mechanism, there is evidence that iron may be removed from chelates by enzymic reduction of chelated ferric iron at sites on the plasmalemma of root cortical cells (Bienfait et al., 1983; Romheld and Marschner, 1983). Such a mechanism may be particularly important for obtaining iron from chelates which have very high affinity for iron in relation to other ions competing for chelation, and thereby, do not support critical levals of soluble inorganic iron by extracellular dissociation. Without direct mechanisms for acquiring iron from chelates, the usefulness of any particular chelate depends upon its stability with iron and the ability of plant roots to alter chemical equilibria to favor chelate dissociation.

OCCURRENCE OF SIDEROPHORES IN SOIL

In soils, many naturally occurring organic compounds could be important in iron chelation, including the organic matter decomposition products, humic and fulvic acids, other organic acids such as citrate and malate, and microbially-produced siderophores. However, not all naturally occurring chelates are useful under conditions where iron is most limiting. Stability diagrams compiled in our laboratories by Cline et al. (1982, 1983) reveal that while hydroxamate siderophores are stable with iron over the entire pH range expected to be found in soils, naturally-occurring organic acids such as citrate and malate are unstable with iron above pH 6 (Fig. 1). Similarly, humic and fulvic acids have a low stability constant for iron in comparison to other ions such as copper and calcium (Spositio et al., 1981). Therefore, the potentially most important naturally occurring chelates for increasing iron availability are microbially-produced siderophores. Sensitive and specific bioassay techniques, based on growth of different siderophore auxotrophs, have demonstrated that siderophores exist

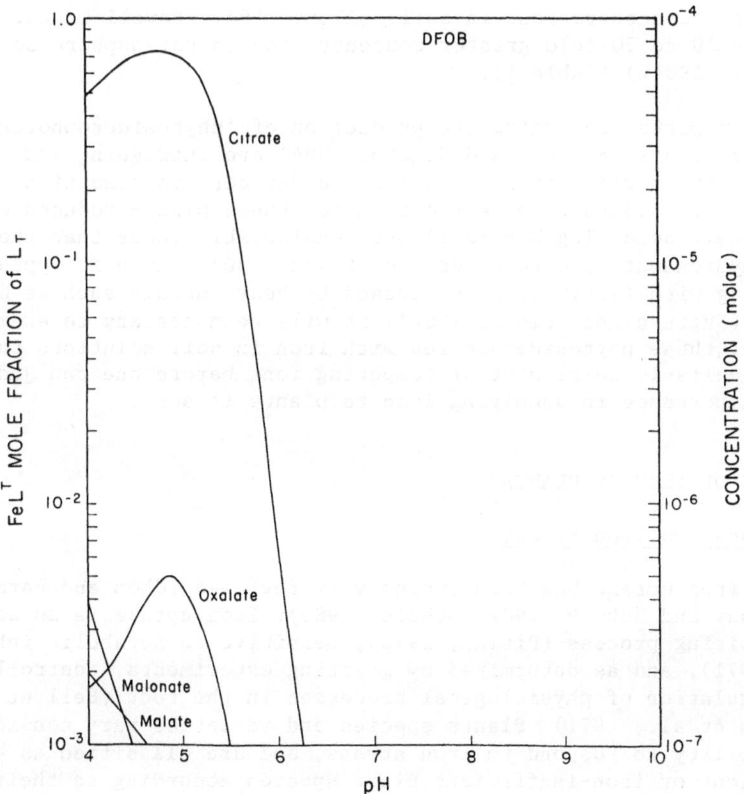

Fig. 1. Summary diagram predicting the mole fraction of organic ligands chelated with Fe in soils of pe + pH = 18 and CO_2 = 0.003 atm. FeL^T = sum of all Fe-chelated species of ligand. L_T = total ligand = 10^{-4}M (from Cline et al., 1983).

Table 1. Hydroxamate siderophore concentrations (nM DFOB equivalents) and R:S values for (a) soil samples collected from five locations in ponderosa pine nursery beds; and (b) soil samples collected from five locations in a lodgepole pine forest stand in Colorado (modified from Reid et al., 1984b)

	Sample	Rhizosphere Soil (R)	Bulk Soil (S)	R:S
(a)	PP1	ND	0.59	ND
	PP2	11.9	0.70	17.0
	PP3	15.7	1.42	11.1
	PP4	20.1	0.36	55.8
	PP5	45.4	1.07	42.4
(b)	LP1	834	305	2.73
	LP2	203	54.3	3.74
	LP3	43.6	0.80	54.5
	LP4	285	300	0.95
	LP5	378	51.7	7.31

ND = Not determined
R:S = Ratio of siderophore concentration in rhizosphere to bulk soil.

in significant concentrations in soils (Akers, 1981; Powell et al., 1980, 1983) and in 10 to 50-fold greater concentration in rhizosphere soil (Reid et al., 1984b) (Table 1).

Recent reports concerning the production of 'phytosiderophores' by graminaceous plants (Sugiura and Nomoto, 1984) are intriguing and may prove to be important in plant iron acquisition under certain conditions. However, the stability constants with ferric iron for these plant-produced compounds (e.g., mugineic acid, log K = 18.1) are considerably lower than those of microbial hydroxamate siderophores (log K = ca. 30). Since it appears that their binding with ferric iron is blocked by heavy metals such as Cu(II) and Zn(II) (Sugiura and Nomoto, 1984), it will be necessary to examine the stability of these phytosiderophores with iron in soil solutions that contain a realistic complement of competing ions before one can judge their potential importance in supplying iron to plants in soil.

UTILIZATION OF IRON BY PLANTS

Plant Response to Iron Stress

Plant iron uptake has been extensively reviewed (Chen and Barak, 1982: Lindsay and Schwab, 1982; Scholz, 1983). Iron uptake is an active, energy requiring process (Pitman, 1976), sensitive to metabolic inhibitors (Kannan, 1971), and as determined by grafting experiments, controlled by genetic regulation of physiological processes in the root (Bell et al., 1962; Brown et al., 1971). Plants species and varieties vary considerably in their ability to respond to iron stress, and are classified as either iron-efficient or iron-inefficient plant species according to their ability to acquire iron (Marschner et al., 1978).

Plants, like microorganisms, have specific mechanisms for responding to iron stress, such as production of specialized iron transfer cells in rhizodermal tissue and root hairs (Kramer et al., 1980; Landsberg, 1984; Romheld and Marschner, 1981) and release of protons and reductants into the rhizosphere to increase iron solubility in the solution next to the root (Olsen and Brown, 1980). However, there are many unanswered questions in regard to plant iron nutrition, particularly in understanding how chelates are utilized in iron uptake.

Iron efficiency in plants is based on the ability of plant roots to reduce pH and redox in the root environment, and thereby increase iron solubility (Ambler and Brown, 1972; Bell et al., 1958; Brown, 1961, 1978; Brown et al., 1971). Reduction of Rhizosphere pH is dependent upon relative rates of release of protons and hydroxyl ions from plant roots (Van Egmond and Atkas, 1977) and is significantly influenced by nitrogen supply as ammonium or nitrate nitrogen (Marschner and Romheld, 1983). The ability of plants to alter rhizosphere pH is also restricted by the presence of bicarbonate ions (Venkate Raju and Marshner, 1981), the principle ion buffering pH in calcareous soils.

Another important component of the iron stress response is the production and release of reductant compounds such as caffeic and chlorogenic acids (Heather et al., 1984; Olsen and Brown, 1980; Olsen et al., 1981, 1982), which may provide electrons for reduction of ferric to ferrous iron in solution (Venkat Raju and Marschner, 1972). Both pH and redox reduction occur primarily along young lateral roots and behind the root tip, corresponding to areas where iron is taken up in greatest quantities (Ambler et al., 1971; Brown and Ambler, 1974, Clarkson and Sanderson, 1978). In general, monocotyledonous plant species have a less efficient iron stress response than dicotyledonous plants (Christ, 1974; Olsen and Brown, 1980), although the ability of monocot plant species to predominate on calcareous, iron-limiting soils (Odum, 1971) suggests other, perhaps unknown, mechanisms are also important for iron acquisition.

Uptake of Iron from Synthetic Chelates

Although there is general consensus that the use of chelated iron by plants requires external separation of iron from the chelate and that iron is taken up as Fe(II) (Scholz, 1983), the mechanisms by which plants take up chelated iron are only partially understood. In work with synthetic chelates, Chaney et al. (1972) used a synthetic ferrous chelating agent to trap ferrous iron and demonstrate the requirement for extracellular dissociation and reduction of chelated iron. However, this work did not consider competition between ferric and ferrous chelating agents when the ferrous chelating agent was added in large quantities. More recently, Schwab and Lindsay (1983) demonstrated the relationship between plant iron uptake and inorganic ferrous iron in solutions maintained at fixed concentrations by addition of different synthetic chelating agents. Results of these experiments allowed determination of critical quantities of soluble inorganic iron required for plant uptake, and provided quantitative data for modeling the shuttle mechanism of chelate utilization by plants.

In studies of reductive removal of iron from chelates, there has been much emphasis on the role of plant-produced reductants. However, there is much evidence that at least some plants also have a direct mechanism for iron reduction, involving an enzyme system on the plasmalemma of root cortical cells. Such a process for direct transfer of electrons by a membrane bound protein was first suggested by Chaney et al. (1972), but was later over-shadowed by work examining plant reductants (Ambler et al., 1971; Brown, 1978; Brown and Ambler, 1973; Brown et al., 1971, Olsen et al., 1981, 1982). While no doubt important, there are, however, problems in regulating an exclusive role for iron reduction by plant-produced reductants. Reductants are released into solution in greatest quantities at pH 4.5 or less (Brown, 1977), and therefore are probably not readily released in significant quantities at neutral and alkaline pH in calcareous soils where iron is most limiting. Also, the addition of reductants did not increase iron uptake by iron-inefficient soybean in nutrient solution (Brown and Ambler, 1973), and reductants have been found in roots of both iron-efficient and inefficient plants (Heather et al., 1984), suggesting that if reductants are important, they are only important when accompanied by pH reduction. Other experiments also exclude a role for plant-produced reductants in iron uptake by peanut (*Arochis hypogeae L.*) (Romheld and Marshner, 1983) and bean (*Phaseolus vulgaris*) (Bienfait et al., 1983).

While the idea of membrane binding sites for reductive removal of iron from chelates has only recently resurfaced (Romheld and Marschner, 1983), good experimental data are available to support the existence of such a mechanism. In sunflower (*Helix annuus L.*), an iron-efficient plant species, there is substantial reductive activity on the root surface, but no release of reductants at any pH (Romheld and Marschner, 1981). Recent histochemical studies with bean (Sijoms and Bienfait, 1984) indicate a requirement for cytosol NADH or NADPH in iron reduction and transport. Furthermore, in unrelated experiments (Lin, 1982), an NADH oxidation system has been isolated from the plasmalemma of corn root cortical cells, although use of this same system in iron reduction and transport remains to be demonstrated.

Uptake of Iron from Siderophores

It has been shown that hydroxamate siderophores can supply iron to a number of different plants including, tomato (Stutz, 1964), duckweed (Orlando and Neilands, 1982), oat (Powell et al., 1982; Reid et al., 1984a), and sunflower and sorghum (Cline etal., 1984). However, there are apparent differences in the ability of plants to acquire iron from hydroxamates depending upon plant species and the relative concentrations of iron and siderophore in solution (Cline et al., 1984). In previous research (Cline et al., 1984) we demonstrated that iron-inefficient sorghum could acquire iron from the siderophore, ferrioxamine B, only when iron was supplied in large excess of the chelating agent (Table 2), whereas sunflower, an iron-efficient plant species, could acquire iron from ferrioxamine B even when the desferri-ferrioxamine B was supplied in large excess of iron (Table 3). Such different plant responses to the same siderophore may indicate that plant release of reductants, as well as a membrane-bound system for reductive removal of iron, are important in plant iron acquisition from chelates. In one of the few experiments examining iron uptake from both synthetic chelates and siderophores, Bienfait et al. (1984) found that iron supplied by synthetic chelates precipitated as an extracellular iron pool in the root apoplast of bean. However, iron supplied by the siderophore, ferrioxamine B, was used more directly, without formation of an extracellular iron pool. These data suggest that the use of siderophores involves chelate binding on the membrane surface and subsequent direct removal of iron, whereas the use of iron from synthetic chelates involves extracellular reduction. With synthetic chelates, plant-produced reductants may aid in mobilization of the extracellular iron pool, but have little effect on iron uptake from siderophores by a direct use mechanism. This view is supported by work with peanut (Romheld and Marschner, 1983). Although peanut did not produce significant quantities of reductants, addition of reductants to the nutrient solution enhanced iron uptake from synthetic chelates, but not from the siderophore, ferrioxamine B.

In our laboratories, we have calculated the experimental conditions required for the maintenance of critical concentrations of the inorganic iron required for active plant uptake, based on the equilibrium relationships for iron and siderophores as controlled by redox and pH. When

Table 2. Effects of various molar ratios of DFOB/Fe on the growth of sorghum in one-third strength modified Hoagland nutrient solution buffered at Ph 7 to 7.5 (modified from Cline et al., 1984)

Fe	DFOB	Total Plant Dry Weight	Severity of Chlorosia[a]
μM		g $\pm$ SE	
100	0	0.24 $\pm$ 0.02	5.1
100	10	0.29 $\pm$ 0.04*	3.6
100	100	0.18 $\pm$ 0.02*	6.1
200	100	0.43 $\pm$ 0.04**	2.2

[a]Mean of all plants per treatment which were individually rated as: 0 (dark green) to 10 (yellow with brown areas).
*Significantly (α = 0.05) different from 100 μM Fe and 0 M DFOB treatment as determined by Student's t test
** Significantly (α = 0.05) different from both the 100 μM Fe, 100 μM DFOB and the 100 μMFe, 0 M DBOB treatments

Table 3. Effects of various molar ratios of DFOB/Fe on the growth of sunflower in a full strength modified Hoagland nutrient solution buffered at pH 7 to 7.5 (modified from Cline et al., 1984).

Fe	DFOB	Dry Weight	Severity of Chlorosis[a]
μM		g $\pm$ SE	
10	0	0.34 $\pm$ 0.05	7.1
10	5	0.50 $\pm$ 0.06*	3.7
10	100	0.59 $\pm$ 0.10*	3.5

[a]Mean of all plants per treatment which were individually rated as described in Table 2
*Significantly (α = 0.05) different from the 10 μMFe and 0 M DFOB treatment as determined by Student's t test

deferrated chelate is at 1% of the concentration of ferrated chelate i.e., in excess, soluble inorganic ferric iron is unavailable above pH 3 (Fig. 2). In contrast, ferrous iron may be available at the critical concentration at pH 7 if redox is sufficiently lowered in the rhizosphere (Fig. 3). However, under aerated conditions, at neutral or alkaline pH, inorganic iron is essentially unavailable for plant uptake. Yet, in prior experiments with oat (Reid et al., 1984a), in which such conditions were imposed with chelating agent approximately 500-fold in excess of total iron, we have shown that oat can acquire iron from the HS, ferrichrome, but less

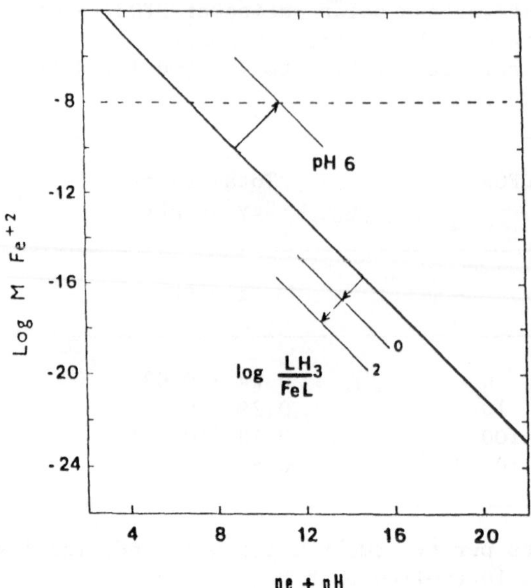

Fig. 2. Soluble Fe^{3+} in equilibrium with ferrichrome as controlled by pH. Desferri-ferrichrome is at 1% of the concentration of ferrated ferrichrome with shifts indicated for increasing concentration of non-ferrated ligand (from Crowley et al., 1985).

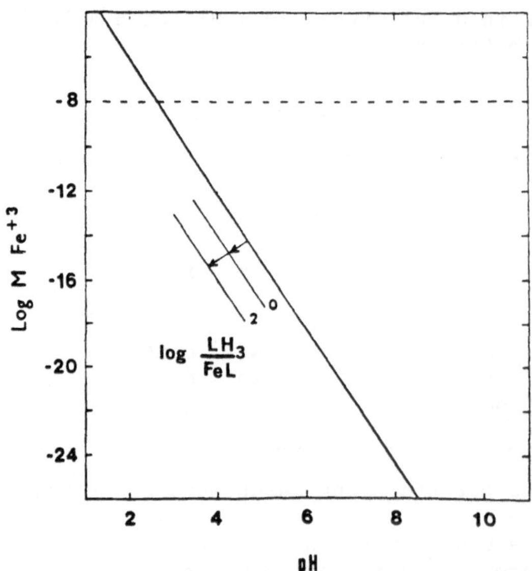

Fig. 3. Soluble Fe^{2+} in equilibrium with ferrichrome as controlled by redox (pe + pH) at pH 7. Desferri-ferrichrome is at 1% of the concentration of ferrated ferrichrome with shifts indicated for decrease in pH and increase in concentration of non-ferrated ligand(from Crowley et al., 1985).

effectively from the synthetic chelate, EDDHA, both of which maintain similar levels of inorganic iron in solution. These data suggest that oat has a mechanism for direct utilization of HS-chelated iron which is not operable with synthetic chelates such as EDDHA.

Table 4. Effect of addition of desferri-ferrichrome or Cr-ferrichrome on shoot content of ^{55}Fe in hydroponically-grown oat provided with $10^{-7.4}M$ ^{55}Fe-ferrichrome in the presence of excess chelating agent (from Crowley et al., 1985).

	Treatment		Shoot Content
	Desferri-ferrichrome (log M)	Cr-ferrichrome (log M)	Fe (dpm/mg)
I.	10^{-6}	-----	4,100 a
II.	2×10^{-5}	-----	980 b
III.	10^{-6}	2×10^{-5}	740 b

Different letters indicate a significant difference at 0.1 level of significance by Tukey's HSD.

Although there is substantial indirect evidence for chelate binding sites on plant roots, there are no experiments directly examining the existence of these hypothesized binding sites. In recent work in our laboratories (Crowley et al., 1985), we have taken the same direct approach used to demonstrate HS receptors in bacteria and fungi (Neilands, 1982) to determine whether chelate-binding sites exist in plant roots. These experiments are based on methodology previously employed for study of HS receptors in fungi (Ecker and Emery, 1983). Inert analogs of the siderophores, ferrichrome and ferrioxamine B, were constructed by irreversible chelation with chromium (Raymond et al., 1984). Ten-day-old, iron-stressed oat seedlings, hydroponically-grown in aerated, $CaCO_3$ pH-buffered nutrient solution (Reid et al., 1984a) were provided with $10^{-7.4}$ M Fe-55 (10 µCi) as ferrated siderophore with excess deferrated siderophore at $10^{-6}M$, in the presence or absence of Cr-siderophore. When $2 \times 10^{-5}M$ Cr-siderophore was provided in the nutrient solution, shoot content of Fe-55 was 4-fold less than in control plants (Table 4). Similar results were obtained for ferrioxamine B.

Table 5. Effect of addition of Cr-ferrichrome on shoot content of ^{55}Fe in in hydroponically-grown oat provided with $^{55}FeCl_3$ or ^{55}Fe-ferrichrome in the presence of excess chelating agent (from Crowley et al., 1985).

	Treatment		Shoot Content
	Log M $FeCl_3$ or Fe-ferrichrome	Log M Cr-ferrichrome	Fe (dpm/mg)
I.	$10^{-7.4}M$ Fe-ferrichrome	-----	9,646 a
II.	$10^{-7.4}M$ Fe-ferrichrome	$10^{-5}M$	1,136 c
III.	$10^{-7.4}M$ $FeCl_3$	-----	6,399 b
IV.	$10^{-7.4}M$ $FeCl_3$	$10^{-5}M$	304 c

Different letters indicate a significant difference at 0.05 level of significance by Tukey's HSD.

In other experiments, Cr-siderophore was found to block uptake of inorganic iron in solutions lacking siderophore other than the inert Cr-analog (Table 5). Inhibition of iron uptake by possible chromium toxicity was considered unlikely after determination of plant Cr content by IPC analysis of acid digest of hydroponically-grown plants provided with Cr-siderophore, inorganic Cr, and non-radioactive iron. Results of these studies strongly support the existence of siderophore binding sites on oat roots and further suggest that these sites also serve as the ion channels used in uptake of inorganic iron.

Plant mechanisms for reductive removal of iron from chelates may have similarities with microbial mechanisms for reductive release of siderophore iron at the cell membrane. Both catechol and hydroxamate siderophores have very low affinity for ferrous iron, and reduced iron within the chelate complex is readily displaced by protonation of the ligand (Raymond et al., 1984). Redox potentials for hydroxamate siderophores, obtained by cyclic voltametry, are within the range of reductants such as NADH and ferredoxin (Brockway et al., 1980). Mechanisms for reductive removal of iron from catechols, which have much higher affinity for iron, are less certain (Raymond et al., 1984).

There is little information on how ferric iron contained within the siderophore molecule may be reduced to ferrous by enzymic electron transfer. Perhaps the best analogy may be the reduction of iron within cytochromes which use NADH and NADPH. Electron transfer properties of siderophores may be quite different from those of synthetic chelates and therefore affect their potential use by an enzymic reduction system. It is also possible that pH may be important in reductive removal of iron, with low pH providing increased availability of protons for displacement of iron, as occurs in ligand exchange between siderophores (Raymond et al., 1984). In peanut, reductive removal of iron has a pH optimum around 5 (Romheld and Marschner, 1983). Iron uptake from the synthetic chelate, EDDHA, is pH dependent (Jeffreys et al., 1961; Jeffreys and Wallace, 1968), and in sunflower, occurs at a 50% greater rate at pH 3.8 than at pH 6.5 (Romheld and Marschner, 1983).

In microbial use of siderophore iron, chelate recognition at membrane binding sites has been shown (Winklemann and Braun, 1981), and microorganisms have differing affinities for various chelates (Wiebe and Winklemann, 1975). Raymond et al., (1984) hypothesize that recognition of siderophores by membrane receptors is based on recognition of the iron center co-ordination site with octahedral bonding and adjacent functional groups. However, Ecker and Emery (1983) suggest that it is probably iron and not the chelate that is recognized by the receptor. In plant use of siderophores, low specificity would be advantageous for using diverse siderophores produced by rhizosphere microorganisms.

In conclusion, the occurrence of hydroxamate siderophores in a wide range of soils, their stability with ferric iron over a broad range of pH in the presence of competing cations, and the demonstrated ability of plants to utilize hydroxamate siderophores, all suggest that these microbial compounds are undoubtedly involved in the iron nutrition of plants under certain natural soil conditions and may be of particular importance in the availability of iron to plants under calcareous conditions.

ACKNOWLEDGEMENT

We wish to acknowledge funding support from National Science Foundation grants DEB-7700883, DEB-7911276, BSR-8307349 and BSR-8415991.

REFERENCES

Akers, H.A., 1981. Multiple hydroxamic microbial iron chelators (siderophores) in soils. Soil Sci., 135:156-159

Ambler, J.E. and Brown, J.C., 1972. Iron-stress response in mixed and monoculture of soybean cultivars. Plant Physiol., 50:675-678.

Ambler, J.E., Brown, J.C., and Gauch, H.G., 1971. Sites of iron reduction in soybean plants. Agron. J., 63:95-97.

Becker, J.O., Hedges, R.W., and Messens, E., 1985. Inhibitory effect of pseudobactin on the uptake of iron by higher plants. Appl. Environ. Microbiol., 49:1090-1093.

Bell, W.D., Bogorad, L., and McIlrath, W.J., 1958. Response of the yellow stripe maize mutant ys, to ferrous and ferric iron. Bot. Gaz., 120:36-39.

Bell, W.D., Bogorad, L., and McIlrath, W.J., 1962. Yellow stripe phenotype in maize. I. Effects of ys locus on uptake and utilixation of iron. Bot. Gaz., 124:1-8.

Bienfait, H.F., Bino, R.J., van der Bliek, A.M., Duivenvoorden, J.F., and Fontaine, J.M., 1983. Characterization of ferric reducing activity in roots of Fe-deficient *Phaseolus vulgaris*. Physiol. Plant., 59: 196-202.

Bienfait, H.F., van den Briel, M.L., Mesland-Mul, N.T., 1984. Measurement of the extracellular mobilizable iron pool in roots. J. Plant Nutr., 7:659-665.

Brockway, D.J., Murray, K.S. and Newman, Peter J., 1980. An electrochemical study of thiohydroxamate and hydroxamate complexes of iron (III). J. Chem. S. Da., 1112-1117.

Brown, J.C., 1961. Iron chlorosis in plants. In: Adv. in Agron. 13:329-369. (A.G. Norman ed.).

Brown, J.C., 1969. Agricultural use of synthetic chelates. Soil Sci. Soc. Am. Proc., 33:59-61.

Brown, J.C., 1977. Genetically controlled chemical factors involved in absorption of iron by plants. In: Bioinorganic Chem. II., pp. 93-103. (K.N. Raymond, ed.). Am. Chem. Soc. Washington.

Brown, J.C., 1978. The mechanism of iron uptake by plants. Plant, Cell and Environ., 1:249-257.

Brown, J.C. and Ambler, J.E., 1973. "Reductants" released by roots of Fe-deficient soybeans. Agron. J., 65:311-324.

Brown, J.C. and Ambler, J.E., 1974. Iron stress response in tomato. 1. Sites of Fe reduction, absorption, and transport. Physiol. Plant., 31:221-224.

Brown, J.C., Chaney, R.L., and Ambler, J.E., 1971. A new tomato mutant inefficient in the transport of iron. Physiol. Plant. 25:48-53.

Chaney, R.L., Brown, J.C. and Tiffin, L.O., 1972. Obligatory reduction of ferric chelates in iron uptake by soybean. Plant Physiol., 50: 208-213.

Chen, Y., and Barak, P., 1982. Iron nutrition of plants in calcareous soils. Adv. in Agron., 35:217-240.

Christ, R.A., 1974. Iron requirement and iron uptake from various iron compounds by different plant species. Plant Physiol., 54:582-585.

Clark, R.B., 1982. Iron deficiency in plants grown in the Great Plains of the U.S., J. Plant Nutr., 5:251-268.

Clarkson, D.T. and Sanderson, J., 1978. Sites of absorption and translocation of iron in barley roots, tracer and microautoradiographic studies. Plant Physiol., 61:731-736.

Cline, G.R., Powell, P.E., Szaniszlo, P.J., and Reid, C.P.P., 1982. Comparison of abilities of hydroxamic, synthetic, and other natural organic acids to chelate iron and other irons in nutrient solution. Soil Sci. Soc. Am. J. 46:1158-1164.

Cline, G.R., Powell, P.E., Szaniszlo, P.J., and Reid, C.P.P., 1983. Comparison of the abilities of hydroxamic, synthetic, and other natural organic acids to chelate iron and other ions in soil. Soil Sci., 136:145-157.

Cline, G.R., Reid, C.P.P., and Szaniszlo, P.J., 1984. Effects of a hydroxamate siderophore on iron absorption by sunflower and sorghum. Plant Physiol., 76:36-39.

Crowley, D.E., Reid, C.P.P. and Szaniszlo, P.J., 1985. Plant root binding sites for microbially-produced hydroxamate siderophores. 85th Ann. Mtg. Am. Soc. Microbiol. Las Vegas, Nevada. (Abstr.).

Ecker, D.J., and Emery, T., 1983. Iron uptake from ferrichrome A and iron citrate in *Ustilago sphaerogena*. J. Bacteriol., 155:616-622.

Emery, T., 1977. The storage and transport of iron. In: "Metal Ions in Biological Systems. Vol. 7. Iron in Model and Natural Compounds" pp. 77-125. (H. Sigel, Ed.). Marcel Dekker, Inc., New York.

Emery, T., 1982. Iron metabolism in humans and plants. Am. Scient., 70:626-632.

Ernst, J., and Winklemann, G., 1977. An attempt to localize iron-chelate binding sites on cytoplasmic membranes of fungi. FEBS Lett., 76:71-76.

Estep, M., Armstrong, J.E., and Van Baalen, C., 1975. Evidence for the occurrence of specific iron(III)-binding compounds in near-shore marine ecosystems. Appl. Microbiol., 30:186-188.

Heather, N.H., Olsen, R.A., and Jackson, L.L., 1984. Chemical identification of iron reductants exuded by plant roots. J. Plant Nutr., 7:667-676.

Jeffreys, R.A., Hale, V.Q., and Wallace, A., 1961. Uptake and translocation in plants of labeled iron and labeled chelating agents. Soil Sci., 92:268-273.

Jeffreys, R.A., and Wallace, A., 1968. Detection of iron-EDDHA in plant tissue. Agron. J., 60:613-616.

Kannan, S., 1971. Kinetics of iron absorption by excised rice roots. Planta, 96:262-270.

Kramer, D., Romheld, V., Landsberg, L., and Marschner, H., 1980. Induction of transfer cell formation by iron deficiency in the root system of *Helianthus annuus L.* Planta, 147:335-339.

Landsberg, E.C., 1984. Regulation of iron stress response by whole plant activity. J. Plant Nutr., 7:609-621.

Lankford, C.E., 1973. Bacterial assimilation of iron. Crit. Rev. Microbiol. 2:273-331.

Lewin, R., 1984. How microorganisms transport iron. Science, 225:401-402.

Lin, W., 1982. Isolation of NADH oxidation systems from the plasmalemma of corn root protoplast. Plant Physiol., 70:326-328.

Lindsay, W.L., 1979. Chemical Equilibria in Soils. Wiley-Interscience, New York.

Lindsay, W.L., and Schwab, A.P., 1982. The chemistry of iron in soils and its availability to plants. J. Plant Nutr., 5:821-840.

Lindsay, W.L., Hodgson, J.F., and Norvell, W.A., 1967. The physicochemical equilibrium of metal chelates in soils and their influence on the availability of micronutrient cations. Trans. Comm. II and IV Int. Soc. Soil Sci., Aberdeen, PP. 305-316.

Lindsay, W.L., and Norvell, W.A., 1969. Equilibrium relationships of Zn, Fe, Ca, and H with EDTA and DTPA in soils. Soil Sci. Soc. Am. Proc., 33:62-68.

Marschner, H., and Romheld, V., 1983. *In vivo* measurement of root-induced pH changes at the soil root interface: Effect of plant species and nitrogen source. Z. Pflanzenphysiol., 111:241-251.

Marschner, H., Romheld, V., and Azarabadi, S., 1978. Iron stress response of efficient and inefficient plant species. In: "Plant Analysis and Fertilizer Problems". Plant Nutrition Proc. 8th Int. Coll., (A.R. Ferguson, R.C. Bieleski and I.B. Ferguson, Eds.). Auckland, New Zealand.

Murphy, T.P., Lean, D.R.S., and Nalewajako, C., 1976. Blue-green algae: The excretion of iron-selective chelators enables them to dominate other algae. Science, 192:900-901.
Neilands,J.B., 1980. Microbial metabolism of iron. In: Iron in Biochemistry and Medicine, II., pp. 529-571. (A. Jacobs and M. Morwood, eds.), Academic Press, London.
Neilands, J.B., 1982. Microbial envelope proteins related to iron. Ann. Rev. Microbiol., 36:285-309.
Norvell, W.A., 1972. Equilibria of metal chelates in soils. In: "Micronutrients in Agriculture,", pp. 115-138. (J.J. Mortvedt, P.M. Giordano, and W.L. Lindsay, Eds.). Soil Sci. Soc. Am., Madison, WI.
O'Connor, G.A., Lindsay, W.L., and Olsen, S.R., 1970. Diffusion of iron chelates in soil. Soil Sci. Soc. Am. Proc., 34:407-410.
Odum, E.P., 1971. Fundamentals of Ecology. W.B. Saunders Co., Philadelphia.
Olsen, R.A. and Brown, J.C., 1980. Factors related to iron uptake by dicotyledonous and monocotyledonous plants. I. pH and Reductant. J. Plant Nutr., 2:629-645.
Olsen, R.A., Bennett, J.H., Blume, D., and Brown, J.C., 1981. Chemical aspects of the Fe stress response mechanism in tomatoes. J. Plant Nutr., 3:905-921.
Olsen, R.A., Brown. J.C., Bennett, J.H., and Blume, D., 1982. Reduction of Fe(III) as it relates to iron chlorosis. J. Plant Nutr. 5:433-447.
Orlando, J.A., and Neilands, J.B., 1982. Ferrichrome compounds as a source of iron for higher plants. In: "Chemistry and Biology of Hydroxamic Acids,". pp. 123-129. (H. Kehl, Ed.), S. Karger, Basel.
Pitman, M.G., 1976. Iron uptake by plant roots. In: Encyclopedia of Plant Physiol. Vol. 2. (U. Luttge and M.G. Pitman, Eds.), Springer-Verlag, New York.
Powell, P.E., Cline, G.R., Reid, C.P.P., and Szaniszlo, P.J., 1980. Occurrence of hydroxamate siderophores iron chelators in soils. Nature, 287:833-834.
Powell, P.E., Szaniszlo, P.J., Cline, G.R., and Reid, C.P.P., 1982. Hydroxamate siderophores in the iron nutrition of Plants. J. Plant Nutr., 5:653-673.
Powell, P.E., Szaniszlo, P.J., and Reid, C.P.P., 1983. Confirmation of hydroxamate siderophores in soil by a novel *Escherichia coli* bioassay. Appl. Environ, Microbiol., 46:1080-1083.
Raymond, K.N., Muller, G., and Matzanke, B.F., 1984. Complexation of iron by siderophores: A review of their solution and structural chemistry and biological function. Topics Curr. Chem., 123:49- 102.
Reid, C.P.P., Crowley, D.E., Powell, P.E., Kim, H.J., and Szaniszlo, P.J., 1984a. Utilization of iron by oat when supplied as ferrated hydroxamate siderophore or as ferrated synthetic chelate. J. Plant Nutr., 7:437-447.
Reid, R.K., Reid, C.P.P., Powell, P.E., and Szaniszlo, P.J., 1984b. Comparison of siderophore concentrations in aqueous extracts of rhizosphere and adjacent bulk soils. Pedobiologia, 26:263-266.
Romheld, V. and Marschner, H., 1981. Iron deficiency stress induced morphological and physiological changes in root tips of sunflower. Physiol., Plant., 53:354-360.
Romheld, V., and Marschner, 1983. Mechanism of iron uptake by peanut plants. Plant Physiol., 71:949-954.
Scholz, G., 1983. A report on iron uptake by higher plants. Biol. Zbl., 102:65-75.
Schwab, A.P., and Lindsay, W.L., 1983. Effects of redox on the solubility and availability of iron. Soil. Sci. Soc. Am. J., 47:201-205.
Sijoms, P.C., and Bienfait, H.F., 1984. Mechanism of iron reduction by roots of *Phaseolus vulgaris* L. J. Plant Nutr., 7:687-693.
Sommers, L.E., and Lindsay, W.L., 1979. Effect of pH and redox on predicted heavy metal equilibria in soils. Soil Sci. Soc. Am. J., 43:39- 47.

Spositio, G., Holtzclaw, K.M. and Le Vesque-Madore, C.S., 1981. Trace metal complexation by fulvic acid extracted from sewage sludge: I. Determination of stability constants and linear correlation analysis. Soil Sci. Soc. Am. J., 45:465-468.

Stutz, E., 1964. Aufnahne von Ferrioxamine B durch Tomatenpflanzen. Experimentia, 20:430-431.

Sugiura, Y. and Nomoto, K., 1984. Phytosiderophores. Structure and properties of mugineic acids and their metal complexes. Structure and Bonding, 58:107-135.

Szaniszlo, P.J., Powell, P.E., Reid, C.P.P., and Cline, G.R., 1981. Production of hydroxamate siderophore iron chelators by mycorrhizal fungi. Mycologia, 73:1158-1174.

Thomas, J.D., and Mathers, A.C., 1979. Manure and iron effects on sorghum growth on iron-deficient soil. Agron. J., 71:792-794.

Van Egmond, F., and Atkas, M., 1977. Iron nutritional aspects of the ionic balance of plants. Plant and Soil, 48:685-703.

Venkat Raju, K., and Marschner, H., 1972. Regulation of iron uptake from relatively insoluble iron compounds by sunflower plants. Z. Pflanzenernaehr, Bodenkd, 133:227-241.

Venkat Raju, K., and Marschner, H., 1981. Inhibition of iron-stress reactions in sunflower by bicarbonate. Z. Pflanzenernaehr, Bodenkd, 144:339-355.

Vose, P.B., 1982. Iron nutrition in plants: A world overview. J. Plant Nutr., 5:233-249.

Wallace, A. and Lunt, G.R., 1960. Iron chlorosis in horticultural plants. A review. Proc. Am. Soc. Hort. Sci. 75:819-841.

Wallace, A., 1962. A decade of synthetic chelates in plant iron nutrition. Arthur Wallace, Los Angeles, CA.

Wallace, A, 1982. Historical landmarks in progress relating to iron chlorosis in plants. J. Plant Nutr. 5:277-289.

Wiebe, C. and Winklemann, G., 1975. Kinetic studies on the specificity of chelate iron uptake in *Aspergillus*, J. Bacteriol., 123:837-842.

Winklemann, G. and Braun, V.,1981. Stereoselective recognition of ferrichrome by fungi and bacteria. FEMS Microbiol. Lett., 11:237-241.

THE EFFECT OF PSEUDOMONAS SIDEROPHORES ON IRON NUTRITION OF PLANTS

Y. Hadar[1], E. Jurkevitch[2], and Y. Chen[3]

Department of Plant Pathology and Microbiology[1] and
The Seagram Center for Soil and Water Sciences[2,3]
The Hebrew University of Jerusalem
Faculty of Agriculture, Rehovot, Israel

INTRODUCTION

Iron is a major constituent of the soil, however its availability to plants and microorganisms is limited at neutral or basic pH levels (O'Connor et al., 1971). Fluorescent pseudomonads are able to produce iron chelating compounds, siderophores, under iron stress conditions (Meyer and Abdallah, 1978; Neilands, 1981; Raymonds et al., 1984; Teintze et. al., 1981). The siderophores enable the bacteria to grow at low iron concentrations and compete with other microorganisms. Antagonism by soil pseudomonads towards certain soil microflora has been the subject of a number of recent reports. Kloepper et al., (1980) showed that either *Pseudomonas* strain B10 or its siderophore, pseudobactin can cause soils conducive to *Fusarium* to become suppressive.

The application of the same strain resulted in improved growth and increased yield of several crops. The mechanism suggested is iron competition with deleterious native microflora (Suslow, 1982). *Pseudomonas* strains isolated from *Fusarium* suppressive soils also exhibited an ability to reduce Fusarium wilt (Scher and Baker, 1982). Effects on other plant pathogens have been described by Misaghi et al.,(1982) and Vandenbert et al., (1983).

Hubbard et al., (1983) demonstrated that seed-colonizing pseudomonads are responsible for the failure of *Trichoderma hamatum* as a biocontrol agent in low iron soils. Addition of iron to these soils permitted *T. hamatum* to protect seeds, thus suggesting that pseudomonads inhibit *T. hamatum* through iron competition by siderophores production. Siderophores produced *in vitro* also inhibited *T. hamatum* mycelial growth, however, inhibition was abolished by the addition of iron (Hubbard et al., 1983).

Kloepper et al., (1980) suggested that the plant must be able to utilize iron from pseudomonads siderophores, otherwise, the plant would become chlorotic. However, none of the studies described above investigated the effect of pseudomonads or their siderophores on the iron nutrition of the host plant.

Powell et al., (1980) and Akers (1983) showed that hydroxamate siderophores concentration in soil were at levels sufficient to affect the absorption of Fe by plants. Cline et al., (1982; 1983) showed that siderophores are able to form stable Fe chelates over wide soil pH range. Recently, Cline et al., (1984) and Reid et al., (1984) showed the ability of FeDFOB siderophore produced by *Streptomyces pilosus*, to reduce chlorosis and enhance growth of sorghum and sunflower plants grown in iron deficient nutrient solutions.

The purpose of this study was to determine the effect of siderophores produced by fluorescent pseudomonads on iron chlorosis of peanuts grown under iron stress.

IRON EXTRACTION WITH SIDEROPHORES OF PSEUDOMONADS

Siderophore-producing bacteria were isolated on selective medium (Grant and Holt, 1977) from plants grown in calcareous soil, and were grown on succinate medium (Meyer and Abdallah, 1978). Siderophores solution produced by strain 3 was compared with DTPA for its ability to extract iron from soils (Table 1). DTPA concentration was 5mM (Lindsay and Norvell, 1978) and the siderophore apparent concentration 0.27 mM. Soil samples of 10 g were equilibrated with 20 ml aliquots for 2 h. The solutions were separated from the soil by centrifugation and iron was determined in the supernatant using 2380 Perkin Elmer atomic absorption spectrophotometer. The effect of the extraction time on extractable iron was measured in Mitzpeh Massua soil which was later used for the plant growth chamber experiments.

It was found that iron concentration in the extracts increased with time. The levels were 3.2, 4.4, 5.6, 6.2 and 6.2 mgFe/Kg soil for 1,2,3,5 and 24 h of extraction respectively. The pseudomonads siderophores extracted the same amount of iron as DTPA in the highly calcareous soil (63% $CaCO_3$), but lower iron levels were extracted in the other two soils. It should be noted that the DTPA concentration was much higher than that of the siderophore (Table 1).

GROWTH CHAMBER EXPERIMENTS

A bioassay pot test described by Barak and Chen (1982) was used to evaluate the effect of bacterial treatments on iron chlorosis. Peanuts

Table 1. Soil Iron Extraction by DTPA and Siderophores Solution

		Extractable iron (mg/Kg soil)	
Soil	$CaCO_3$ (%)	DTPA	*P. putida* siderophores
Light Clay	63	4.0 ± 0.21*	4.4 ± 0.07
Clay	29	12.8 ± 0.16	6.8 ± 0.11
Sandy Loam	5	5.4 ± 0.28	3.4 ± 0.22

* Each number represents average of 5 replicates

Table 2. The effect of *Pseudomonas putida* (strain 1) and its siderophore on peanuts chlorophyll content

Treatment	Chlorophyll (mg/g leaf)
Control	2.01B +
FeEDDHA	2.70A
Bacterial suspension	2.04B
Bacterial suspension plus Fe- siderophore	2.45AB
Free siderophore	2.08B
Fe-siderophore	2.42AB

+ Numbers followed by same letter are not significantly different (p-0.05)

were grown in a calcareous soil, containing 60% $CaCO_3$. Two types of control were included: (i) no fertilization, (ii) treatment with FeEDDHA (Sequestrene 138, Ciba-Geigy, 20 mg/Kg soil). *Pseudomonas putida* (strain 1) culture grown on succinate salts medium was applied to the soil three times during the growth period in the following ways:

(1) Bacterial suspension: the seeds were treated with $2.6.10^9$ CFU/seed. A bacterial suspension was applied with the irrigation.

(2) A bacterial suspension plus Fe-siderophore complex: watering solutions containing the bacteria in this growth medium plus iron (added in excess to 1 mM as $FeCl_3$) in order to form the Fe siderophore complex. Due to the high pH (8.5) the uncomplexed iron precipitated immediately.

(3) A siderophore treatment: the bacteria were removed by centrifugation from the growth medium which was then dialysed against water.

(4) A siderophore-iron complex: same as (3) but with addition of 1 mM $FeCl_3$ after dialysis.

Four weeks after planting, leaves were analysed for their chlorophyll content. Results are presented in Table 2. The pigment alone or the bacterial suspension did not prevent iron deficiency. However, treatments containing the Fe-siderophore complex resulted in an increase in chlorophyll content relative to the untreated control.

A second experiment was performed to study the effect of concentrated siderophore solution and to examine a new strain of *P. putida* (No. 3) isolated from peanut roots grown in the same soil. The treatments, besides the controls, were:

(1) Strain 3 as a bactericidal suspension plus Fe siderophore complex. This strain produced 3 times as much siderophore as the other strains (measured by optical density at 450 nm).

(2) Fe-siderophore complex of strain 1 concentrated by evaporation after dialysis against water.

(3) Strain 1 suspension plus its Fe-siderophore complex.

In this experiment the solutions were centrifuged to remove the uncomplexed iron. Treatments (1) and (3) were applied every week as

Table 3. Amount of iron added to soil and their effect on peanuts leaves chlorophyll, soil EC and pH

Treatments	Iron added mg/Kg soil	Chlorophyll content mg/g leaf	Soil electrical conductivity (dsim/m)	Soil pH
Control	0	1.9c[+]	0.28	7.9
FeEDDHA	1.5	3.8a	0.30	7.8
Bacterial suspension + Fe siderophore (strain 3)	4.35	4.2a	0.40	7.8
Bacterial suspension + Fe siderophore (strain 1)	2.0	1.9c	0.47	8.1
Fe siderophore (strain 1)	5.8	2.7b	0.42	7.9

* Each number is an average of 6 replicates
Numbers followed by the same letter are not significantly different ($p=0.05$)

irrigation while (2) was given at the 2nd and 3rd week of growth. Results are described in Table 3. It seems that strain 3 did not differ from the FeEDDHA treatment in preventing of lime-induced chlorosis. The complex of strain 1 also significantly increased chlorophyll content in comparison to the untreated control by 40%. Bacterial treatments only slightly changed the soil pH and electrical conductivity (Table 3).

CONCLUSIONS

Siderophore-producing *Pseudomonas putida* strains were isolated from rhizosphere of plants grown under iron stress. The strains differ significantly in siderophore production, and strain 3 was found to produce about three times as much as the other isolates.

Siderophores separated from growth media were able to extract between 50-100% of the DTPA extractable iron from 3 different soils. However, application of free siderophores or bacteria to soil did not result in an increase in chlorophyll content in peanuts. Application of either Fe-siderophore or siderophore bacterial suspension containing the complex were able to increase chlorophyll content by 30-100% respectively compared to the untreated control. The bacterial application may serve as a potential treatment for the remedy of lime-induced chlorosis. Fluorescent pseudomonads may play an active role in plant iron nutrition as well as in the control of soil borne plant pathogen.

Further research will be conducted in our laboratory in order to compare the properties of siderophores produced by different bacterial strains and to study the potential role of Pseudomonads Fe-siderophores in the iron nutrition of plant grown in calcareous soils.

ACKNOWLEDGEMENT

This study was supported by the Israel National Council for Research and Development.

REFERENCES

Akers, H.A., 1983. Multiple hydroxamic acid microbial iron chelators (siderophores) in soils. Soil Sci., 135:156-159.

Barak, P. and Chen, Y., 1982. The evaluation of iron deficiency using a bioassay-type test. Soil Sci. Soc. Am. J., 46:1019-1022.

Cline, G.R., Powell, P.E., Szaniszlo, P.J. and Reid, C.P.P., 1982. Comparison of the abilities of hydroxamic acid, synthetic and other natural organic acids to chelate iron in nutrient solution. Soil Sci. Soc. Am. J., 46:1158-1164.

Cline, G.R., Powell, P.E., Szaniszlo, P.J., and Reid, C.P.P., 1983. Comparison of the abilities of hydroxamic and other natural organic acids to chelate iron in soils. Soil Sci., 136:145-157.

Cline, G.R., Reid, C.P., Powell, P.E. and Szaniszlo, P.J., 1984. The effects of a hydroxamate siderophore on iron absorption by sunflower and sorghum. Plant Physiol., 76:36-39.

Grant, M.A. and Holt, J.G., 1977. Medium for the selective isolation of members of the genus *Pseudomonas* from natural habitats. Appl. Environ. Microbiol., 33:1222-1224.

Hubbard, J.P., Harman, G.E., and Hadar, Y., 1983. Effect of soil borne *Pseudomonas* spp. on the biological control agent *Trichoderma hamatum* on pea seeds. Phytopathology, 73:655-659.

Kloepper, J.W., Leong, J., Teintze, M. and Schroth, M.N., 1980. *Pseudomonas* siderophores: A mechanism explaining disease-suppressive soils. Curr. Microbiol., 4:317-320.

Lindsay, W.L. and Norvell, W.A., 1978. Development of a DTPA soil test for zinc, iron, manganese and copper. Soil Sci. Soc. Am. J., 42:421-428.

Meyer, J.M. and Abdallah, M.A., 1978. The fluorescent pigment of *Pseudomonas* fluorescens: biosynthesis, purification and physicochemical properties. J. Gen. Microbiol., 107:319-328.

Misaghi, I.L., Stowell, L.J., Grogan, R.G., and Spearman, L.C., 1982. Fungistatic activity of water soluble fluorescent pigments of fluorescent *Pseudomonas*. Phytopathology, 72:33-36.

Neilands, J.B., 1981. Microbial iron compounds. Ann. Rev. Biochem. 50: 715-731.

O'Connor, G.A., Lindsay, W.L., and Olsen, S.R., 1971. Diffusion of iron and iron chelates in soil. Soil Sci. Soc. Am. J., 35:407-410.

Powell, P.E., Cline, G.R., Reid, C.P.P. and Szaniszlo, P.J., 1980. Occurrence of hydroxamate siderophore iron chelators in soils. Nature, 287:833-834.

Raymonds, K.N., Muller, G., and Matzanke, B.F., 1984. Complexation of iron siderophores. A review of their solution and structural chemistry and biological function. Topics Curr. Chem., 123:49-102.

Reid, C.P.P., Crowley, D.E., Kim, H.J., Powell, P.E. and Szaniszlo, P.J., 1984. Utilization of iron by oat when supplied as ferrated synthetic chelate or as ferrated hydroxamage siderophore. J. Plant Nutr., 71: 437-447.

Scher, F.M., and Baker, R., 1982. Effect of *Pseudomonas putida* and a synthetic iron chelator on induction of soil suppressiveness to Fusarium wilt pathogens. Phytopathology, 72:1567-1573.

Suslow, T.V., 1982. Role of root colonizing bacteria in plant growth. In: "Phytopathogenic Prokeryotes," Vol. 1, 187-219. (M.S. Mount and G.H. Lacy, eds.), Academic Press, New York.

Teintze, M., Hussain, M.B., Barnes, C.L., Leong, J. and Van der Hielm, D., 1981. Structure of ferric pseudobactin. A siderophore from a plant growth promoting *Pseudomonas*. Biochemistry, 20:6446-6457.
Vandenberg, P.A., Gonzales, C.F., Wright, A.M. and Kunka, B.S., 1983. Iron chelating compounds produced by soil pseudomonads: correlation with fungal growth inhibition. Appl. Environ. Microbiol., 46:128-132.

THE FACILITATION OF IRON UPTAKE IN BACTERIA AND PLANTS BY SUBSTITUTED CATECHOLS

R.C. Hider

Department of Chemistry
Essex University
Colchester, Essex, U.K.

Although iron is one of the most abundant metals in uncontaminated soil, its extremely high affinity for hydroxide, silicate and phosphate renders the water soluble fraction precipitously low, $<10^{-17}$ M at neutral pH values. As a result, bacteria, fungi, phytoplankton and vascular plants have evolved a diverse range of methods for specifically accumulating iron. Aluminium is also an abundant component of soil and by virtue of a similar cationic charge density to that of iron(III) has some comparable coordination properties. The chemistry associated with iron accumulation mechanisms must be capable of distinguishing iron(III) and aluminium (III). Aerobic bacteria and fungi commonly produce siderophores which are low molecular weight compounds (Mr = 500-1000) possessing a high affinity for iron(III) ($K_f > 10^{30}$) the biosynthesis of which is induced by low iron levels (Hider, 1984; Neilands, 1981). These compounds are usually hexadentate ligands (I, II) containing catecholate, hydroxamate and carboxylate moieties. Siderophores coordinated to iron are typically accumulated by microorganisms using specific translocation mechanisms and the tightly bound iron is removed for utilisation by the cell (Hider, 1984). In contrast to microorganisms, no hydroxamate or catecholate containing siderophore has

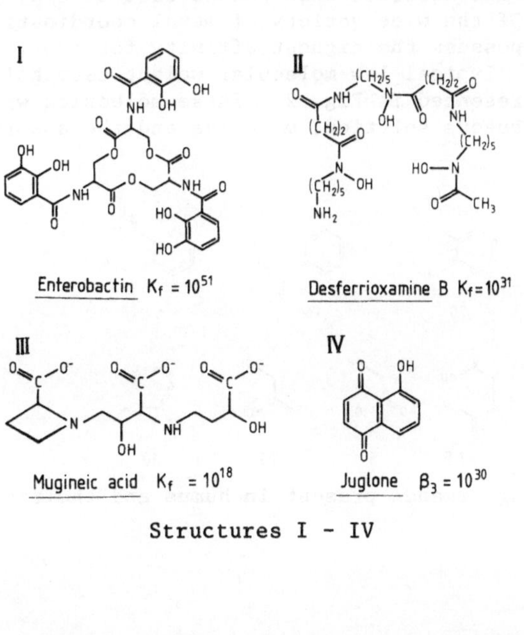

I — Enterobactin $K_f = 10^{51}$

II — Desferrioxamine B $K_f = 10^{31}$

III — Mugineic acid $K_f = 10^{18}$

IV — Juglone $\beta_3 = 10^{30}$

Structures I - IV

been isolated from the root secretions of higher plants. This is surprising and possibly an important distinction which will find beneficial application in agriculture.

Many vascular plants are capable of responding to low iron stress and possess wide ranging mechanisms for enhancing the solubility of iron(III): (i) secretion of protons by roots (Marschner et al., 1974); (ii) secretion of reducing compounds by roots (Marschner, 1974); (iii) secretion of phytosiderophores (Sugiura and Nomato, 1984). Although phytosiderophores are widely distributed in higher plants (Rippenger and Schreiber, 1982), they have only been isolated from the root washings of graminaceous plants (Sugiura and Nomato, 1984). Those secreted are amino hydroxylcarboxylates (III) with a high affinity for iron(III) (Kf = 10^{18}), although not as high as that possessed by bacterial and fungal siderophores.

The competition for iron between different siderophores, inorganic oxyanions and the soil humus must be critically dependent on conditions of moisture, temperature, pH and the presence of competing cations, for instance aluminium (III). This balance can be shifted by the introduction into the soil of high affinity iron(III) ligands. The antagonist effects of walnut (*Juglans nigra*) for instance, have been recorded on such diverse plants as pine trees, potatoes and cereals. *J. nigra* produces a precursor of juglone (IV) in its leaves and roots (Hether et al., 1984). On leaching from the plant, juglone is formed in the soil and by virtue of its high affinity for iron(III) renders the iron non-available to many plants. The chemical composition of humus therefore and its ability or otherwise to interact with iron is of critical importance, particularly in alkaline soils and aquatic environments.

Coordination of Iron to Soil Humus

The major insoluble portion of soil organic matter is humus which is traditionally divided into the acid insoluble humic and acid soluble fulvic fractions. Both fractions are of similar composition and differ only in that the humic fraction possesses a higher molecular weight range and a low carboxylic acid content (Buffle, 1984). Degradation products of lignin form the major component of humus and as such large proportions of phenolic and benzene carboxylic acids are present, together with fatty acids, peptides and oligosaccharides attached to phenolic functions. In soil and aquatic environments these organic molecules tend to form micelles (Ogner and Schnitzer, 1970) and are absorbed onto the surface of clay particles (Degens and Mopper, 1976). Of the wide variety of metal coordination sites present in humus, catechols possess the highest affinity for iron(III) over the pH range 6-10 (Fig. 1.) Typical low molecular weight catechol derivatives found in humus are presented in Fig. 2. These molecules will distribute themselves between aqueous solution, micelles and the absorbed organic layers on clay.

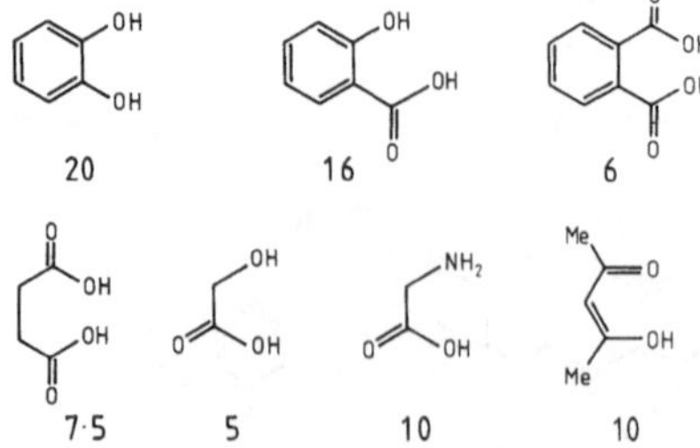

20 16 6

7·5 5 10 10

Fig. 1. Cation binding ligands present in humus and their K_1 values for iron(III)

Redox Activity of Iron-Catechol Complexes

Catechol is a powerful reducing agent possessing a redox potential similar to that of the iron(III)/iron(II) couple. The oxidation products, semiquinone and quinone are also capable of coordinating cations but generally with a reduced affinity (Pierpont and Buchanan, 1981). The electronic structure of the iron complex can exist in two isoelectric forms (Fig. 3). Chelate bonding by the semiquinone structure is analogous to acetylacetonate bonding and as such can be strong. Furthermore, chelation can protect semiquinones from further oxidation to the corresponding quinones (Eaton, 1964). Thus iron(III) catechol complexes are capable of undergoing intramolecular electron transfer reactions, the position of equilibrium of such reactions being influenced by the solution pH, the dielectric strength of the medium and presence of other ligands (Hider, 1984). Over the pH range 2.0 to 4.5 the green iron(II) semiquinone complex (V) is the dominant species in aqueous solution (Hider et al., 1983) (Fig. 4) as indicated by Mössbauer spectra (Fig. 5) (Hider et al., 1981). Over this pH range the coordinated iron(II) prevents the semiquinone undergoing disproportionation to catechol and benzoquinone. At pH values below 2.0 disproportionation does occur (Mentasti and Pelizzetti, 1973). Iron(II) species are also present over the pH range 4.5-6.0, the tris-catecholato-iron(III) complex (VI) dominating only in alkaline solutions (pH >9.0) (Fig. 4) (Hider et al., 1981). An important consequence of internal redox reactions involving iron(III) is that iron is converted into a kinetically labile iron(II) species.

The possible role of catechols (polyphenols) in the dissolution of soil iron(III) oxides and subsequent reduction to iron(II) was first reported by Bloomfield (1957; King and Bloomfield, 1966) and subsequently confirmed by others (Beres and Kiraly, 1958; Coulson et al., 1960; Hingston, 1962). Indeed, Mössbauer spectroscopy studies of humic (Goodman and Cheshire, 1979)

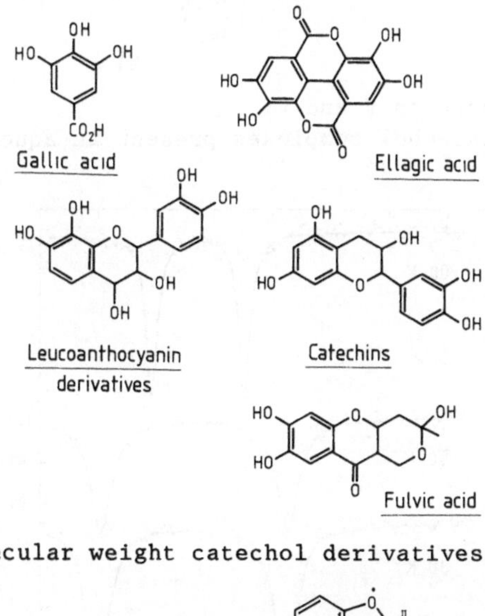

Fig. 2. Low molecular weight catechol derivatives present in humus

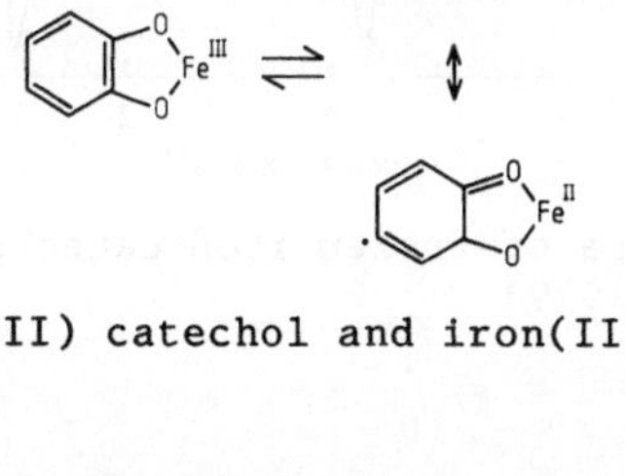

Fig. 3. Iron(III) catechol and iron(II) semiquinone

and fulvic acids (Goodman and Cheshire, 1984) have directly demonstrated the ability of these materials to reduce iron(III) under acid conditions. 40% of fulvic acid iron complex was found to be iron(II) at pH 2.0, 20% at pH 4.0 but only a trace at pH 5.0 (Goodman and Cheshire, 1984). However, as indicated above, minor humic acid-induced change in iron oxidation state

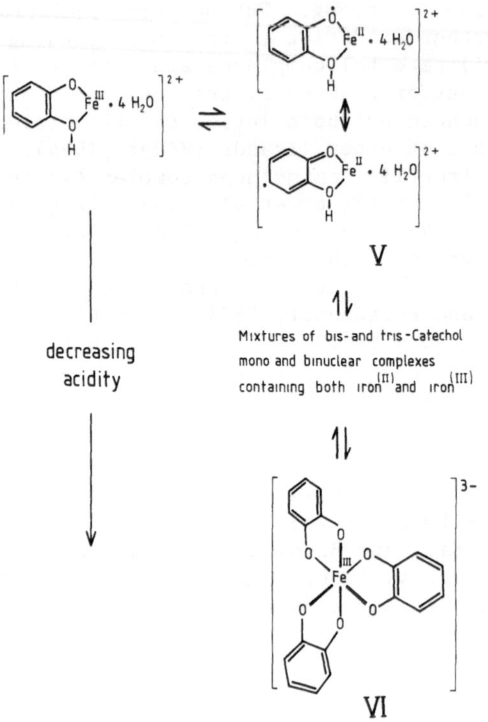

V

decreasing acidity

Mixtures of bis- and tris-Catechol mono and binuclear complexes containing both iron$^{(II)}$ and iron$^{(III)}$

VI

Fig. 4. / Structures V and VI
Iron catechol complexes present in aqueous media

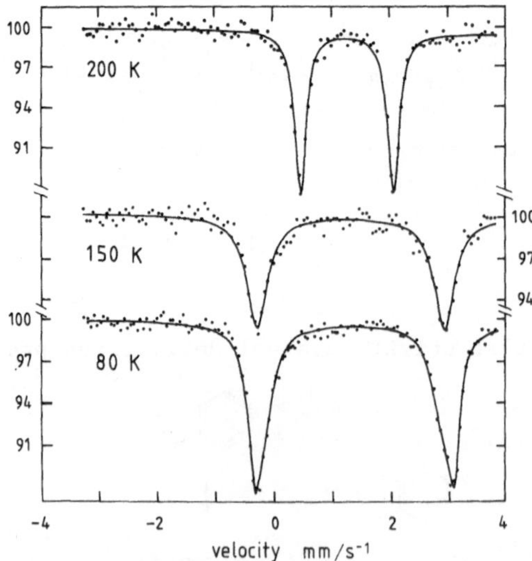

Fig. 5. Mössbauer spectra of frozen iron catechol solutions (pH 4.0) (Hider et al., 1979)

can have a considerable influence on the bioavailability of soil iron, by virtue of the kinetic lability of iron(II). Acid dependent release of iron offers an explanation for the extremely low iron availability observed with neutral and alkaline soils and why plant iron deficiency is widespread on such soils.

Iron Absorption by Vascular Plants

Some plant species are more susceptible towards low iron induced chlorosis than others. Given that the iron content per unit mass of plant is approximately constant irrespective of species, this finding implies that some root systems are more efficient at iron absorption than others. Sharp differences can even be found between different strains of the same species. It is clear therefore that plant roots can influence soil iron chemistry. The use of iron(II) specific coordination reagents has demonstrated that the reduced form of iron and not iron(III) is absorbed by soybean roots (Chaney et al., 1972). This finding has formed the basis of a generally applicable model for plant iron absorption (Fig. 6). For artificially added chelates such as EDTA the root surface offers no diffusion barrier and the chelates can enter the free space of the roots. However in the absence of such chelates the inorganic iron is present as an insoluble form and as such cannot enter the free space. Under these

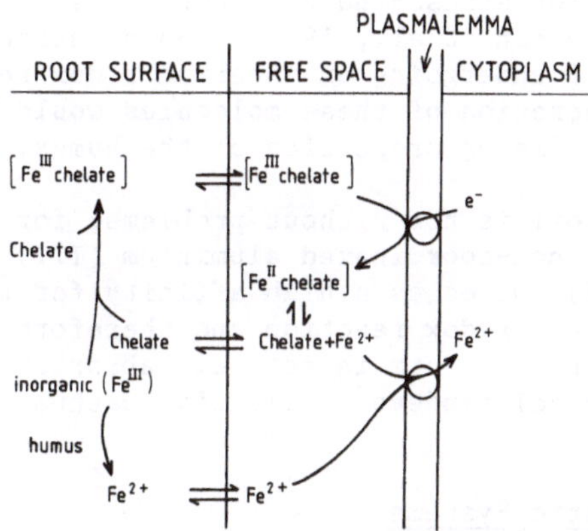

Fig. 6. Iron absorption by plant roots, based on a model proposed by Chaney et al., 1972)

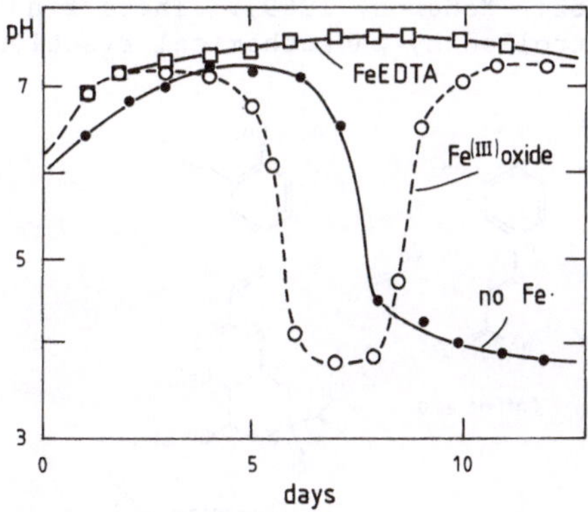

Fig. 7. Influence of iron supply on the pH of nutrient solution supporting the growth of Sunflower plants. Marschner et al., 1974

conditions the chemistry in the close vicinity of the root surface, the rhizosphere, is of critical importance. One method by which plants can influence the iron availability in the rhizosphere is by secretion of protons, thereby decreasing the pH. In 1961 Brown et al., reported that wild type soybean roots possess a greater ability for iron stress induced H^+ secretion than a chlorosis susceptible strain. Similar findings were subsequently reported for tomato (Brown and Ambler, 1974) and sunflower (Marschner et al., 1974). When sunflower plants are grown in the presence of iron EDTA, the pH of the nutrient solution remains constant, whereas when no iron is added the pH drops to below 4.0 (Fig. 7). In the presence of iron(III) oxide the pH drops to 4.0 for several days until the iron stress is relieved whereupon the pH returns to neutrality (Fig. 7). Similar observations have been made with the tomato (Olsen and Brown, 1980). At pH 4.0, the concentration of iron(III) is still low and yet plants grow well under these conditions (Kashirad and Marschner, 1974). A humic acid based internal redox reaction generating iron(II) is capable of accounting for this high growth rate. Significantly many graminaceous plants, some of which are known to produce siderophores (Sugiura and Nomato, 1984), lack this response (Olsen and Brown, 1980).

In addition to protons, some chlorosis-resistant plants also secrete organic molecules which are themselves capable of reducing iron(III) (Brown et al., 1971; Marschner et al., 1974; Olsen and Brown, 1980). A range of phenolic compounds including caffeic (VII) and chlorogenic (VIII) acids have been detected in iron stress-induced root secretions of the tomato (Hether et al., 1984; Olsen et al., 1981). Both catechol containing compounds are capable of undergoing an internal redox reaction with iron(III). Thus the secretion of these molecules would be predicted to supplement the iron mobilising properties of the humus.

Acidification of soil is not without problems, for instance an increase in the concentration of non-coordinated aluminium (III) will result. Although aluminium (III) possesses a high affinity for catechol, it is not susceptible to an internal redox reaction and therefore remains coordinated at pH 4.0. Aluminium in this form is not well absorbed by plants. Thus the involvement of catechol renders a selective pathway for the absorption of iron.

Iron Absorption in Aquatic Systems

As with terrestrial plants, work with phytoplankton indicates that iron is absorbed in the iron(II) state (Anderson and Morel, 1980; Anderson and Morel, 1982). The concentration of iron(II) in lake surface water follows a diurnal pattern and increases with increased light intensity (Anderson and Morel, 1982; McMahon, 1969). These findings suggest that iron(II) levels are controlled by photochemical reactions. Such conclusions

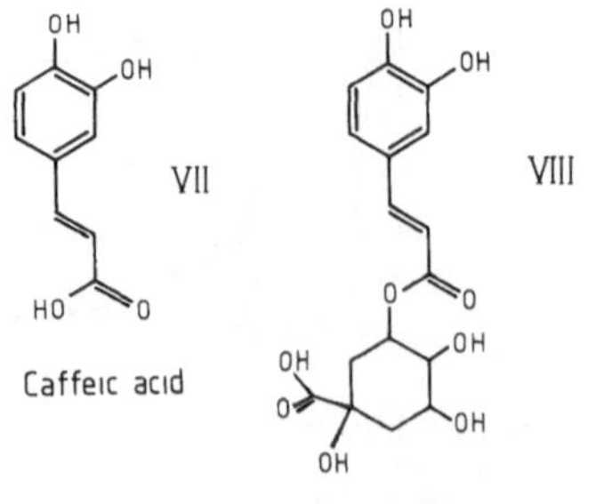

VII Caffeic acid

VIII Chlorogenic acid

Structures VII and VIII

have been extended to marine waters (Waite and Morel, 1984). Humic substances coprecipitated with iron(III) oxide enhance the efficiency of photochemical reduction processes, possibly by acting as a chromophore (Finden et al., 1984; Franko and Heath, 1982; Waite and Morel, 1984). Catechol containing components in humic material possess an important (David and David, 1976) advantage over many other potential ligands, in addition to their high affinity for iron(III). They possess a strong absorption spectrum in the visible region, much more intense than that corresponding to iron(III) dicarboxylate complexes. This is important as although electromagnetic radiation of wavelength below 400 nm is nondetectable at depths greater than 1 m, the diurnal variation of iron(II) can be observed at depths of 4-6 m (McMahon, 1969). Clearly then, the photoreduction process uses light of longer wavelengths, i.e. 500-600 nm, a region where iron catechol complexes possess a relatively high extinction coefficient. Two mechanisms are possible (Fig. 8). Scheme A involves photodissociation of an iron-oxygen bond generating a hydroxyl radical (David and David, 1976). Scheme B involves an intramolecular redox process in a charge transfer excited state (Balzani and Carassiti, 1970). Both mechanisms generate the relatively labile iron(II) cation which would rapidly dissociate from the complex.

Iron Absorption by Microorganisms

Many bacteria and fungi produce catechol or hydroxamate based siderophores that possess a high affinity for iron (Raymond et al., 1985) and hence are capable of scavenging iron(III) from the immediate environment of the microorganisms colony. Iron(III) is removed from both hydroxamate and catechol siderophores by enzyme catalysed reduction (Lodge et al., 1980; Lodge et al., 1982). The standard redox potential of iron hydroxamate complexes is in the region of -450 mV which is close to that of the NAD^+/NADH couple (-320 mV). Consequently a redox reaction can occur between the two couples under standard aqueous physiological conditions. With enterobactin (I) however the iron(III) complex possesses a redox potential close to -750 mV at neutral pH (Cooper et al., 1978), which is well below the range of physiological reducing agents and yet iron is removed via enzyme catalysed oxidation of FMN (Lodge et al., 1980). A possible mechanism for this reductive removal of iron involves an internal redox reaction between catechol and iron(III) at the active site of the enzyme (Hider, 1984; Hider et al., 1984; Hider et al., 1979).

An important difference between the mechanisms of iron absorption by bacteria and fungi on one hand and vascular plants and many phytoplankton on the other, is the former pair absorb iron(III), while the latter pair absorb iron(II). This differential preference for the redox state of iron may well have important ecological implications.

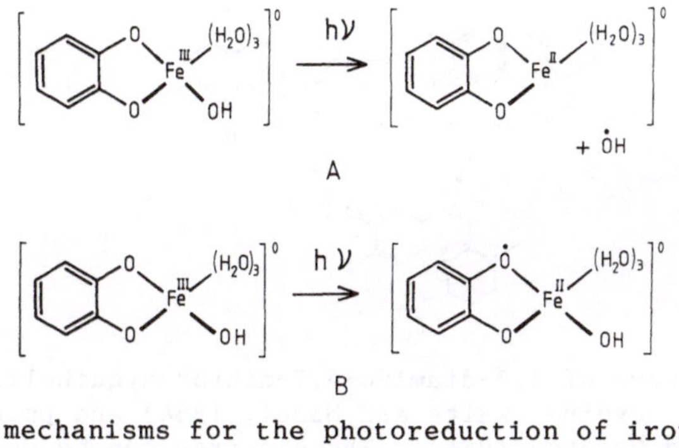

Fig. 8. Possible mechanisms for the photoreduction of iron catechol complexes; A, photodissociation; B, intra-molecular redox reaction

Interspecies Competition for Iron

Bacteria, fungi and vascular plants compete with each other for iron. Although higher plants absorb iron(II), this is usually derivatised from iron(III). In principle, therefore, the growth of the three classes will be adversely affected by the presence of high affinity iron(III) ligands, e.g. juglone and hydroxamate siderophores, unless the organism possesses a specific translocation mechanism for the iron-siderophore complex. Competition of this type is particularly clear in marine and lake environments. Desferrioxamine B(II), a hydroxamate siderophore inhibits the uptake of iron by plankton (Anderson and Morel, 1982), and the siderophore(s) excreted by blue-green algae inhibit the growth of eukaryotic plankton (Murphy et al., 1976).

The significance of the presence of low levels of hydroxamates (Powell et al., 1980) in soil is not clear. If they are tris hydroxamate it is unlikely that they will be able to efficiently donate iron to higher plants because even in a temporary acid environment, such as that found in the rhizosphere, hydroxamate siderophores will bind iron(III) preferentially to the bidentate catechols found in humus. It is likely that these molecules are of fungal origin and are secreted in order to scavenge iron which then becomes specifically directed to the secreting fungal colony.

In contrast to the concept outlined above, bacteria of the genus *Pseudomonas* secrete siderophores (Teintze et al., 1981) into the soil which suppress certain plant diseases (Kloepper et al., 1980; Schroth and Hancock, 1982). Evidence indicates that they do so by efficiently scavenging iron and thereby inhibit the growth of deleterious rhizoplane fungi and bacteria (Schroth and Hancock, 1982; Teintze et al., 1981). Pyoverdin Pα, one of these siderophores, possesses a high affinity for iron(III), $K = 10^{3}$ (Meyer and Abdallah, 1978). This value is higher than the association constant of many fungal hydroxamates and thus, for instance, pyoverdin would be predicted to outcompete ferrichrome for available iron.

How this class of siderophore limits iron absorption by fungi and bacteria without decreasing the iron supply to plant roots has not been established. A likely explanation lies in the unique structure of the *Pseudomonas* siderophores which contain 2,3-diamino-6,7-dihydroxyquinoline (Teintze et al., 1981; Wendenbaum et al., 1983). This moiety, like catechols, would be predicted to undergo an internal redox reaction with iron(III) forming a semiquinone (Fig. 9, A) and iron(II). The presence of the two positively charged nitrogen atoms will favour this redox reaction, by decreasing the pKa of the phenolic functions and by stabilising the semiquinone structure. The positive charge associated with the nitrogen atoms exerts a strong electron withdrawing effect. This will decrease the

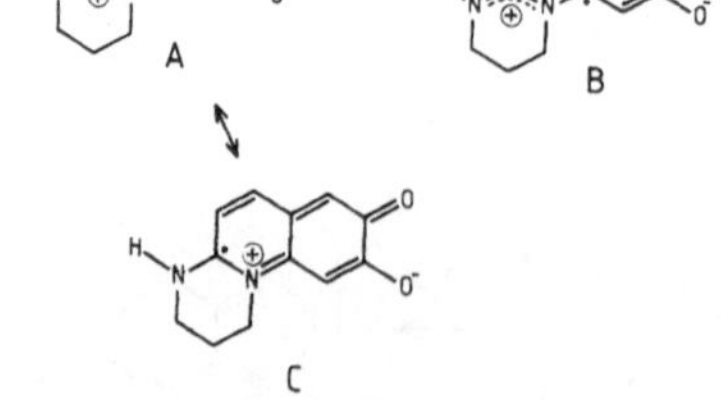

Fig. 9. The semiquinone of 2,3-diamino-6,7-dihydroxyquinoline, the chromophore of pyoverdine (Waite and Morel, 1984) and pseudobactin (Sugiura and Nomato, 1984). The electron can be delocalised over the quinoline ring system.

electron density on both of the iron coordinating oxygen anions, thus decreasing the affinity of the substituted quinoline for iron(III) to below that of catechol. The semiquinone (A), resulting from an internal redox reaction, would be stabilised by the delocalisation of the electron to sites adjacent to the electron deficient nitrogens (B and C). That this substituted quinoline can undergo facile redox reactions involving semiquinones is demonstrated by the conversion of pseudobactin A to pseudobactin (Teintze and Leong, 1981). Thus when compared with simple catechol ligands 2,3-diamino-6,7-dihydroxyquinoline will have reduced affinity for iron(III) and enhanced affinity for iron(II), that is, conversion from iron(III) to iron(II) will be favoured. Under acid conditions therefore, 2,3-diamino-6,7-dihydroxyquinoline containing siderophores would be predicted to release iron(II) more readily than catechol containing siderophores. As hexadentate hydroxamates have a relatively low affinity for iron(II), this iron would become available for absorption by plant roots (Fig. 10).

Internal redox reactions centred on iron would appear to have considerable relevance to the generation of disease resistant soil (Fig. 10).

Acknowledgement

The author would like to thank the British Technology Group for their continuing support of iron biochemistry at Essex University

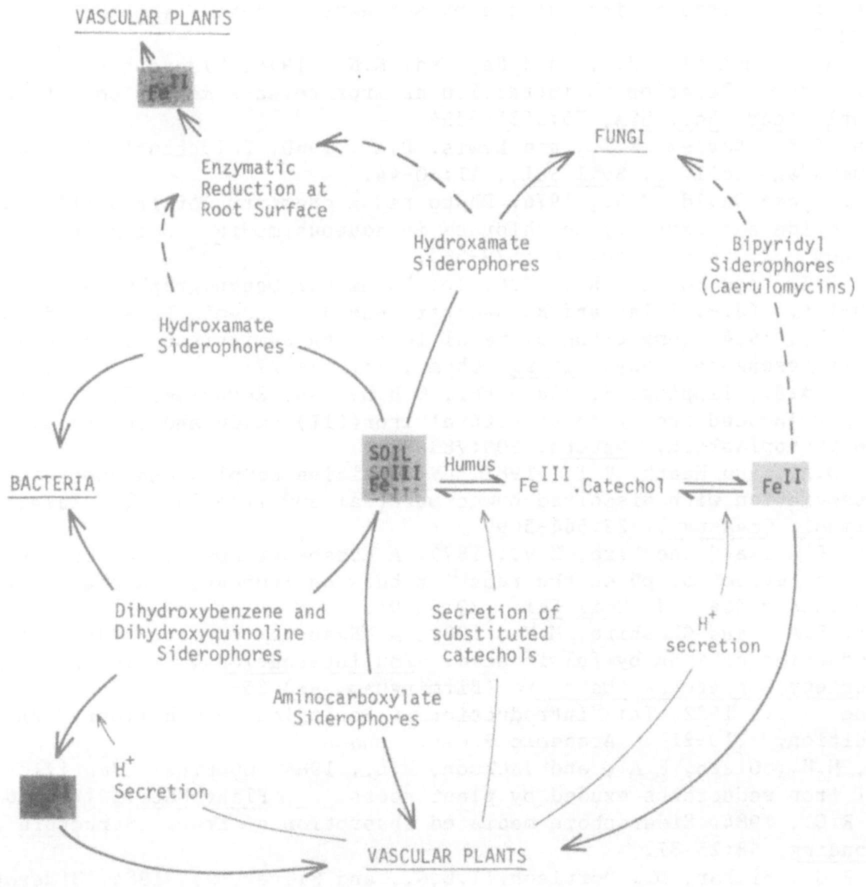

Fig. 10. Iron mobilisation in soil. The relative contributions of routes A, B and C will depend on the plant species and the microbiological flora of the soil.

REFERENCES

Anderson, M.A., and Morel, F.M., 1980. Uptake of Fe(II) by a diatom in oxic culture medium. Mar. Biol. Lett., 1:263-268.

Anderson, M.A., and Morel, F.M., 1982. The influence of aqueous iron chemistry on the uptake of iron by the coastal diatom *(Thalassiosira weissflogii)*. Limnol. Oceanogr., 27:789-813.

Balzani, V., and Carassiti, V., 1970. In: "Photochemistry of coordination compounds". Academic Press, London. pp.74-75.

Beres, T., and Kiraly, I., 1958. Investigation of the reducing effect of peat fulvic acids on ferric ions. Agrokem. Talajtan., 7:281-286.

Bloomfield, C., 1957. The possible significance of polyphenols in soil formation. J. Sci. Food Agric., 8, 389-392.

Brown, J.C., and Ambler, J.E., 1974. Iron-stress response in tomato *(Lycopersicon esculentum)*. Physiol. Plant Pathol., 31:221-224.

Brown, J.C., Chaney, R.L., and Ambler, J.E., 1971. A new tomato mutant inefficient in the transport of iron. Physiol. Plant Pathol., 25: 48-53.

Brown, J.C., Holmes, R.S., and Tiffin, L.O., 1961. Iron chlorosis in soybeans as related to the genotype of root stalks. Soil Sci., 91: 127-132.

Buffle, J., 1984. Natural organic matter and metal-organic interactions in aquatic systems. In: "Circulation of metals in the environment". 165-222. (H. Sigel, Ed.) Dekker, New York.

Chaney, R.L., Brown, J.C., and Tiffin, L.O., 1972. Obligatory reduction of ferric chelates in iron uptake by soybeans. Plant Physiol., 50 : 208-213.

Cooper, S.R., McArdle, J.V., and Raymond, K.N., 1978. Siderophore electrochemistry: Relation to intracellular iron release mechanism. Proc. Natl. Acad. Sci. USA, 75:3551-3554.

Coulson, C.B., Davies, R.I., and Lewis, D.A., 1960. Polyphenols in plant, humus and soil. J. Soil Sci., 11:30-44.

David, F., and David, P.G., 1976. Photo redox chemistry of iron(III) chloride and iron(II) perchlorate in aqueous media. A comparative study. J. Phys. Chem., 80:579-583.

Degens, E.T., and Mopper, K., 1976. In: "Chemical Oceanography Vol.6". 60-114. (J.P. Riley and R. Chester, Eds.) Academic Press, London.

Eaton, D.R., 1964. Complexing of metal ions with semiquinones. An electron spin resonance study. Inorg. Chem., 3:1268-1271.

Finden, D.A.S., Tipping, E., Jaworski, G.H.M., and Reynolds, C.S., 1984. Light induced reduction of natural iron(III) oxide and its relevance to phytoplankton. Nature, 309:783-784.

Franko, D.A., and Heath, R.T., 1982. UV-sensitive complex phosphorus: association with dissolved humic material and iron in a bog lake. Limnol. Oceanogr., 27:564-569.

Goodman, B.A., and Cheshire, M.V., 1979. A Mössbauer spectroscopic study of the effect of pH on the reaction between iron and humic acid in aqueous media. J. Soil Sci., 30:85-91.

Goodman, B.A., and Cheshire, M.V., 1984. A Mössbauer-effect study of the reduction of iron by fulvic acid. 2nd International Humic Substances Society Conference Abstracts (Birmingham, UK) 25-27.

Harborne, J.B., 1982. In: "Introduction to Ecological Biochemistry" 2nd Edition. 210-211. Academic Press, London.

Hether, N.H., Olsen, R.A., and Jackson, L.L., 1984. Chemical identification of iron reductants exuded by plant roots. J. Plant Nutr., 7:667-676.

Hider, R.C., 1984. Siderophore mediated absorption of iron. Structure and Bonding, 58:25-87.

Hider, R.C., Bickar, D., Morrison, I.E.G., and Silver, J., 1984. Siderophore iron-release mechanisms. J. Am. Chem. Soc., 106:6983-6987.

Hider, R.C., Howlin, B., Miller, J.R., Mohd-Nor, R., and Silver, J., 1983. Solution chemistry and Mössbauer studies on iron(II) and iron(III) catechol complexes. Inorg. Chim. Acta, 80:51-56.

Hider, R.C., Mohd-Nor, R., Silver, J., Morrison, I.E.G., and Rees, L.V.C., 1981. Model compounds for microbial iron-transport compounds. J. Chem. Soc. Dalton, 609-622.

Hider, R.C., Silver, J., Neilands, J.B., Morrison, I.E.G., and Rees, L.V.C., 1979. Identification of iron(II) enterobactin and its possible role in *E. coli* iron transport. FEBS Lett., 102:325-328.

Hingston, F.J., 1962. Activity of polyphenolic constituents of leaves of *Eucalyptus* and other species in complexing and dissolving iron oxide. Aust. J. Soil Res., 1:64-73.

Kashirad, A., and Marschner, H., 1974. Effect of pH and phosphate on iron nutrition of sunflower and corn plants. Agrochimica., 18:497-508.

King, H.G.C., and Bloomfield, C., 1966. The reaction between water-soluble tree leaf constitutents and ferric oxide in relation to podzolisation. J. Sci. Food Agric., 17:39-43.

Kloepper, J.W., Leong, J., Teintze, M., and Schroth, M.N., 1980. Enhanced plant growth by siderophores produced by plant growth-promoting rhizobacteria. Nature, 286:885-886.

Lodge, J.S., Gaines, C.G., Arceneaux, J.E.L., and Byers, B.R., 1980. Non-hydrolytic release of iron from ferrienterobactin analogs by extracts of *Bacillus subtilis.* Biochem. Biophys. Res. Commun., 97:1291-1295.

Lodge, J.S., Gaines, C.G., Arceneaux, J.E.L., and Byers, B.R., 1982. Ferrisiderophore reductase activity in *Agrobacterium tumefaciens.* J. Bacteriol., 149:771-774.

Marschner, H., Kalisch, A., and Römheld, V., 1974. Mechanism of iron uptake in different plant species. Proc. 7th Int. Coll. Plant Analysis Fertilizer Problems, 273-281.

McMahon, J.W., 1969. The annual and diurnal variation in the vertical distribution of acid-soluble ferrous and total iron in a small dimictic lake. Limnol. Oceanogr., 14:357-367.

Mentasti, E., and Pelizzetti, E., 1973. Reactions between iron(III) and catechol; Equilibria and kinetics of complex formation in aqueous acid solution. J. Chem. Soc. Dalton, 2605-2608.

Meyer, J.M., and Abdallah, M.A., 1978. The fluorescent pigment of *Pseudomonas fluorescens*: Biosynthesis, purification and physico-chemical properties. J. Gen. Microbiol., 107, 319-328.

Murphy, T.P., Lean, D.R.S., and Nalewajko, C., 1976. Blue-green algae: Their excretion of iron-selective chelators enables them to dominate other algae. Science, 192:900-902.

Neilands, J.B., 1981. Microbial iron compounds. Annu. Rev. Biochem., 50: 715-731.

Ogner, G., and Schnitzer, M., 1970. Humic substances: Fulvic acid - dialkyl phthalate complexes and their role in pollution. Science, 170:317-318.

Olsen, R.A., and Brown, J.C., 1980. Factors related to iron uptake by dicotyledonous and monocotyledonous plants. J. Plant Nutr., 2: 629-645.

Olsen, R.A., Clark, R.B., and Bennett, J.H., 1981. The enhancement of soil fertility by plant roots. Am. Sci., 69:378-384.

Pierpont, C.G., and Buchanan, R.M., 1981. Transition metal complexes of o-benzoquinone, o-semiquinone and catecholate ligands. Coord. Chem. Rev., 38:45-87.

Powell, P.E., Cline, G.R., Reid, C.P.P., and Szaniszlo, P.J., 1980. Occurrence of hydroxamate siderophore iron chelators in soils. Nature, 287: 833-834.

Raymond, K.N., Müller, G., Matzanke, B.F., 1985. Complexation of iron by siderophores a review of their solution and structural chemistry and biological function. In: "Topics in Current Chemistry". 49-102. Springer Verlag.

Rippenger, H., and Schreiber, K., 1982. Nicotianamine and analogous amino acids, enologenous iron carriers in higher plants. Heterocycles, 17: 447-461.
Schroth, M.N., and Hancock, J.G., 1982. Disease-suppressive soil and root-colonizing bacteria. Science, 216:1376-1381.
Suguira, Y., and Nomato, K., 1984. Phytosiderophores. Structure and Bonding, 58:107-135.
Teintze, M., Hossain, M.B., Barnes, C.L., Leong, J., and Van Der Helm, D., 1981. Structure of ferric pseudobactin, a siderophore from a plant growth promoting *Pseudomonas*. Biochemistry, 20:6446-6457.
Teintze, M., and Leong, J., 1981. Structure of pseudobactin A, a second siderophore from plant growth promoting *Pseudomonas* B10. Biochemistry, 20:6457-6462.
Waite, T.D., and Morel, F.M.M., 1984. Photoreductive dissolution of colloidal iron oxides in natural waters. Environ. Sci. Tech., 18:860-868.
Wendenbaum, S., Demange, P., Dell, A., Meyer, J.M., and Abdallah, M.A., 1983. The structure of pyoverdine Pα, the siderophore of *Pseudomonas aeruginosa*. Tetrahedron Lett., 24:4877-4880.

DIVERSE EFFECTS OF SOME BACTERIAL SIDEROPHORES ON THE UPTAKE OF IRON BY PLANTS

J.O. Becker[1], R.W. Hedges[1] and E. Messens[2]

Plant Genetic Systems, Plateaustraat 22,[1]
B-9000 Gent, Belgium

Laboratorium voor Histologie, Rijksuniversiteit Gent,[2]
Ledeganckstraat 35, B-9000 Gent, Belgium

Around the roots of higher plants, supported at least in part by nutrients derived from those roots, a characteristic ecosystem (termed the rhizosphere) develops. This consists primarily of bacteria and fungi (Rovira and McDougall, 1967). The predominant constituents of the bacterial flora, which may attain a concentration greater than 10^8 per gram of soil (Lochhead and Thexton, 1947) are members of a relatively few genera including pseudomonads, agrobacteria and enterobacteria.

Plant roots must carry out several physiological functions if the plant is to flourish and most of these involve interaction with the soil around the root surface. It is well known that the physical and chemical activities of roots have a major influence upon the nature and composition of soils and recent studies are beginning to make it clear that microorganisms of the rhizosphere are involved in these interactions playing a range of roles some beneficial to plant growth whilst others are deleterious.

From an evolutionary viewpoint it is obvious that plants must have adapted so as to carry out their essential functions in the presence of high concentrations of rhizosphere organisms (and their metabolic products) whilst these microorganisms will undergo powerful selection so as to exploit to the full the ecological opportunities presented by the activities of the roots.

Several examples of such interactions are known and the physiological effects are probably best understood in the case of fungi whose hyphae can penetrate root tissues with results as diverse as overt plant disease when the fungus overwhelms the plant's resistance, a parasitic relationship whereby achlorophyllous orchids can only survive by exploiting the fungus (Black, 1980) and apparently mutually advantageous (mycorrhizal) associations (Harley and Smith, 1983). These last seem to be rather precariously balanced relationships in which either partner may gain dominance at the other's expense and it seems likely that some of the other rhizosphere symbioses may be similarly poised so that their nature may be determined by physiological and environmental influences.

Interactions of bacteria with plant roots are less well characterised. The best studied are those between rhizobia and leguminous plants. This

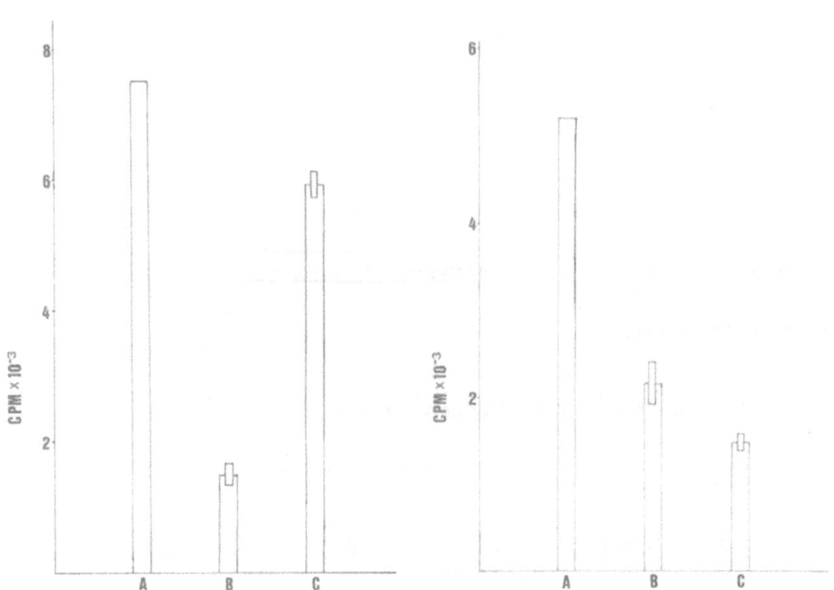

Fig. 1. The effect of exogenous siderophores on the uptake of ^{59}Fe by pea plants. Fig. 1a shows the effect of pseudobactin (6 days growth) and Fig. 1b the effect of agrobactin (4 days of growth). In each case column A shows the original level of ^{59}Fe in the growth medium, column B shows the level of ^{59}Fe in medium after growth of the plant without siderophore and column C the level after growth in the presence of siderophore. (Redrawn from Becker et al., 1985 and Becker et al., 1985).

Fig. 2. Autoradiograms of bean shoots grown for four days in low iron medium containing ^{59}Fe. Plant A grew in medium free of exogenous siderophore, whilst plant B grew in medium containing 5 μM agrobactin. (Reproduced, with permission from Becker et al., 1985).

relationship has been characterized as a beneficial disease, since it can be advantageous to the host when the soil lacks available nitrogenous compounds but seems to be basically parasitic when there is no such shortage.

Relationships between plants and other rhizosphere bacteria are even more obscure. Thus, tumorigenic agrobacteria appear as overt parasites, although it has been suggested that the roots induced by *A. rhizogenes* can assist the plant's survival by increasing the water absorptive powers (Strobel and Nachmias, 1985). The large majority of agrobacteria found in rhizosphere soil are, however, non-pathogenic (Schroth et al., 1971). Their role in rhizosphere ecology is obscure since we do not know whether they should be regarded as populations distinct from the tumorigenic (pTi^{+}) lines or as potential recruits to the set of tumor-inducing organisms (by plasmid transfer). The interactions of plant roots with other bacteria such as pseudomonads and enterobacteria are even less well understood.

Mutually beneficial interactions between plants and bacteria do not always involve special anatomic structures (such as nodules or tumors). Sulfur bacteria (*Beggiatoaceae*) in the rhizosphere of rice plants can oxidize phytotoxic sulfides, whilst catalase, secreted by the roots, protects the bacteria from the hydrogen peroxide which they produce as a respiratory end product (Heritage and Foster, 1984; Joshi and Hollis, 1977; Pitts et al., 1972). As with the mycorrhizal symbioses, the interactions of the typical rhizosphere bacteria with plant roots are complex and it is likely that they will vary with varying ecological circumstances. For example, an isolate of *Pseudomonas cepacia*, which is a pathogen of onions, can colonize the roots of that plant and protect it against fungal infection (Kawamoto and Lorbeer, 1976).

We approached this biological complex by focusing upon a single essential function which must be carried out by plant roots and asking how the bacterial compounds of the rhizosphere population might influence this process. We chose to consider the uptake of iron because this has been shown to be a physiologically demanding process (except under unusually acid soil conditions) (Lindsay, 1974). The physiological adaptions of the root to acquiring iron from soil are beginning to be understood (Marschner, 1983; Mino et al., 1983; Scholz et al., 1985) and molecular genetic analysis of the iron uptake machinery of bacteria is well advanced (Lewin,

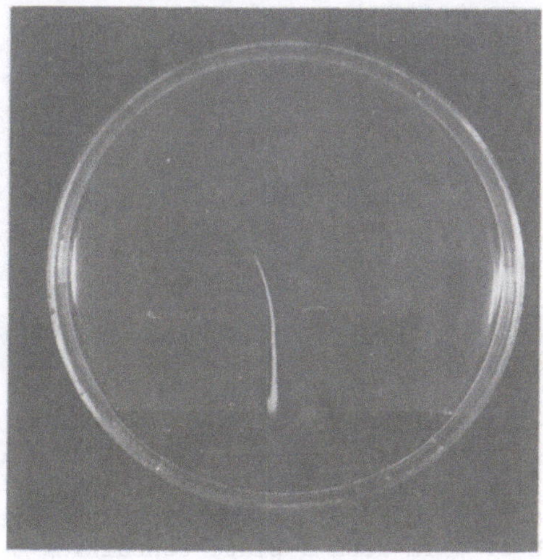

Fig. 3. Radish seedlings growing on low iron medium containing $^{59}Fe^{3+}$

1984). There are also precedents for the interactions of iron acquisition mechanisms of unrelated organisms. Thus it is well known that *E. coli* (Hartmann and Braun, 1980) can exploit the iron-chelating hydroxamates synthesised by fungi and more recently it has been shown that these compounds can supply iron to higher plants (Orlando and Neilands, 1982; Powell et al., 1982). Such observations suggest ways in which fungi (e.g. mycorrhizae) might be able to stimulate the growth of plants and rhizosphere bacteria.

On the other hand, it has been shown that the siderophore, pseudobactin, produced by fluorescent pseudomonads is able to compete for iron with the uptake systems of plant pathogenic bacteria and fungi and by denying them this metal, to reduce their ability to parasitize the plants (Becker and Cook, 1984; Scher and Baker, 1982; Schroth and Hancock, 1982; Suslow and Schroth, 1982). Thus, both negative and positive interactions are possible consequences of the production of a siderophore (not to mention the provision of iron to the producer organism).

Our tests so far, have been upon the effects of purified siderophores on the uptake of ferric iron by plant roots (Becker et al., 1985; Becker et al., 1985). The first problem was the appropriate concentration to use. We decided to use a concentration of 5 or 10 μM equivalent to that formed

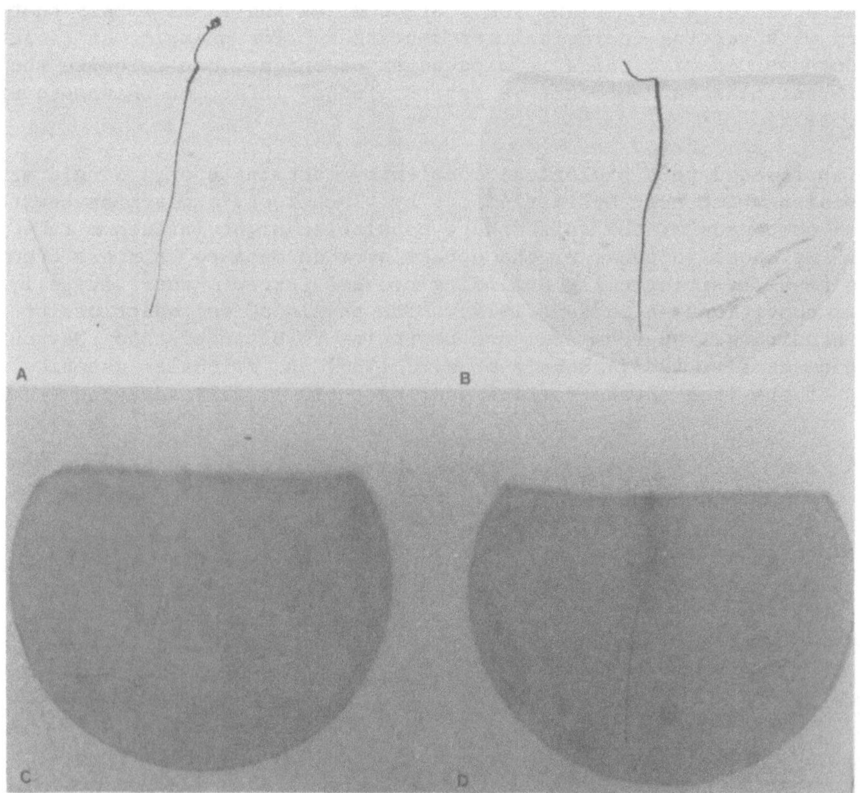

Fig. 4. Photographs and autoradiograms of agar films showing the effect of agrobactin on the uptake of iron by the roots of radish seedlings.
A - photograph of film from a plate with no exogenous siderophore
B - photograph of film from a plate containing 5 μM agrobactin
C - autoradiogram of the film A
D - autoradiogram of the film B

by a pure culture of the bacterium grown in succinate medium. This is about 1/10th to 1/5th of the maximum rate of siderophore production. In a natural rhizosphere one would not expect a pure culture of a single bacterium, so, the concentration used is around the maximum level rather than one typical of the root environment. Thus, it could be feared that in our experiments unnatural concentrations of siderophores might damage the roots. So, where high siderophore concentrations could lead to inhibition of iron uptake, we performed control experiments at lower concentrations to show that this was not the case.

Another sort of artefactual result could be produced by the precipitation or adsorption of iron onto the surface of the roots. We have therefore sought to show that the iron taken up in our experiments was physiologically available to the plant, by preparation of autoradiograms of the shoots, showing preferential transport to developing leaves and by studies of the rate of synthesis of chlorophylls, a process known to require iron (Brown, 1956).

Our first experiments were performed using pseudobactin (Becker et al., 1985). This siderophore is produced by fluorescent pseudomonads in quantities capable of reducing the concentration of iron in the rhizosphere to levels insufficient to support growth of some pathogenic fungi. Since this effect is the basis of a system of biological control, we sought to learn whether the siderophore affected the uptake of iron by plants. In fact, we found that the presence of pseudobactin had an inhibitory effect on the uptake of ferric iron apparently due to competition for iron between the siderophore and the iron uptake system of the plants (Fig. 1) (Becker et al., 1985). The ecological significance of this effect is not well understood but the encouraging results obtained when plants are inoculated with pseudobactin-producing bacteria (Kloepper et al., 1980) suggests that the

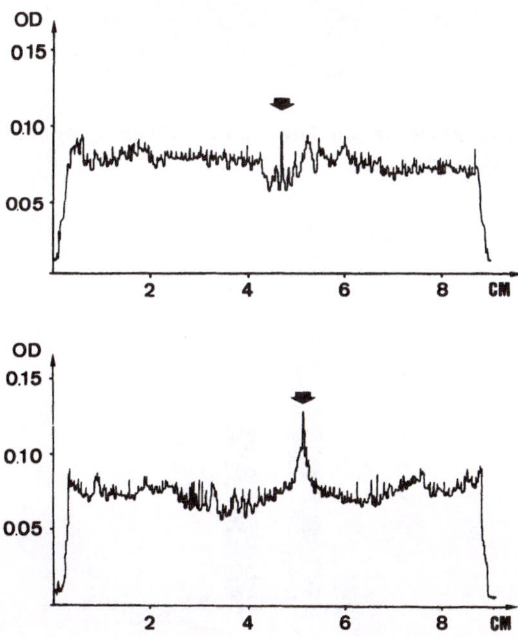

Fig. 5. Tracings of optical density measurements across autoradiograms of agar films taken from plates upon which radish seedlings had grown. Fig. 5A shows a plate without added siderophore whilst 5B shows a plate containing 5 μM agrobactin. Position of the root is indicated

inhibition is not serious for the plant, perhaps because some of its roots are relatively free of the bacteria.

It has been reported that *Fusarium oxysporum* f.sp. *lycopersici* has a higher requirement for certain micronutrients (including iron) than its host, tomato (Jones and Woltz, 1981). Thus, a lowering of the concentration of such nutrients below a physiological optimum could reduce the pathogenic potential of the fungus and on balance, benefit the host plant. This does not, however, seem to be an invariable characteristic of interactions of pathogenic fungi with plant roots. Reis et al., (1981, 1982) found that

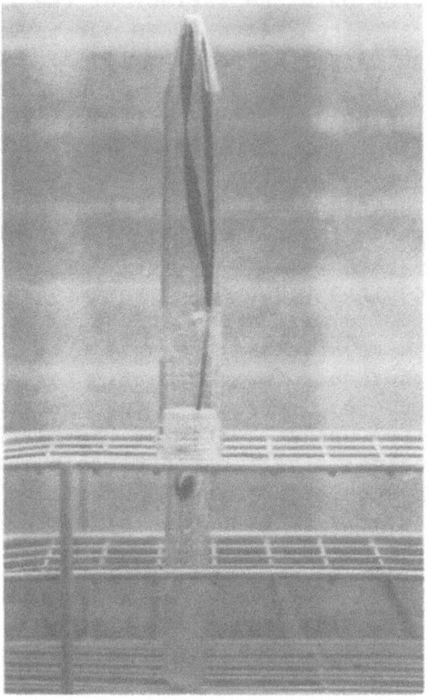

Fig. 6. Wheat seedling growing on low iron medium (containing $^{59}Fe^{3+}$) solidified with agar.

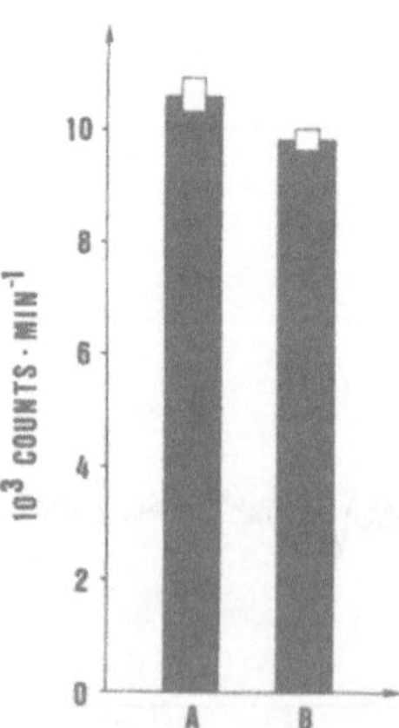

Fig. 7. Histogram of uptake of ^{59}Fe by wheat plants grown as shown in Fig. 6. Column A shows the uptake into plants grown in the presence of 5 µM agrobactin. Column B shows the uptake in the absence of siderophore.

decreasing the supply of iron to wheat plants challenged with *Gaeumannomyces graminis* var *tritici* tended to increase the incidence of infection and the severity of the disease. Thus, no general conclusions can be drawn from the available data on relative iron requirements of plants and pathogens.

It is not clear whether the bacterium gains any advantage from reducing the supply of iron to the plant. One of the modes of response of the plant to iron deficiency is the secretion of organic acids (e.g. caffeic acid) which can reduce ferric to ferrous iron and so increase the solubility of the metal ions (Julian et al., 1983; Marschner, 1983; Olsen et al., 1981). These organic acids are considered to have antifungal properties (Kuc, 1982) and can act as carbon sources for at least some fluorescent pseudomonads (unpublished results). Thus, by stimulating the production of these compounds the pseudomonads may be able to inhibit competing organisms and to provide themselves with nutrients.

Quite different results were obtained when we studied the effect of agrobactin on plant roots. By a variety of techniques, using a range of species we have consistently found a channelling of ferric iron into plants by the action of agrobactin. Figs. 1 and 2 show this effect. Direct measurements of the effects of agrobactin upon the ability of pea and bean plants to remove ^{59}Fe from low iron plant growth medium showed that stimulation of greater than 50% was possible (Becker et al., 1985). Particularly clear cut results were achieved when radish (*Raphanus sativus*) seeds were surface-sterilized, germinated on water agar for two days and then transferred to plates of 2% agar, based on a low (20 µM) iron plant growth medium containing $^{59}FeCl_3$ (Becker et al., in press). The experimental arrangement is shown in Fig. 5. Autoradiograms are shown in Fig. 3. Furthermore we observed that roots of control plants, but not of those provided with agrobactin solubilize the chalk presumably by secretion of organic acids. Production of these acids must be metabolically expensive for the plant and the acidification of the rhizosphere may be physiologically unfavourable to growth of both plant roots and bacteria but conducive to growth of fungi including phytopathogens (Anonymous, 1981; Costilow, 1981; Deacon, 1984). In some cases tracings of optical densities of autoradiograms taken from the agar films showed a depletion of ^{59}Fe in the neighbourhood of the control roots (Fig. 4). This was never observed on tracings from autoradiograms of agrobactin containing agar films (Fig. 5). We suppose that the siderophore solubilized the iron and permitted its less restricted distribution.

Not only dicotyledonous plants but also monocotyledons were stimulated in their ability to take up iron by the presence of agrobactin. Wheat seedlings *(Triticum sativum* cv. Vuka) were grown in individual containers (Fig. 6). They were shown to be able to remove ^{59}Fe from a low iron medium solidified with agar (Fig. 7). The effect of agrobactin is, thus, not correlated with the host range of the pathogenic agrobacteria which is not surprising since the iron uptake mechanism is not required for pathogenicity (Leong and Neilands, 1981) (and *vice versa*, non-pathogenic agrobacteria have normal iron uptake ability).

Since it is known that plants can make use of iron bound to hydroxamate siderophores produced by fungi it seemed interesting to test the ability of a bacterial hydroxamate to influence iron uptake by plants. We therefore tested aerobactin (Gibson and Magrath, 1969) a siderophore produced by klebsiellae, bacteria which have frequently been reported as rhizosphere constituents. We found that these compounds had little or no effect on iron uptake. This is probably due to the relatively low affinity for iron shown by these compounds (Neilands, 1981). Probably the plant uptake system is quite simply able to outcompete the exogenous siderophore and take up the ferric iron despite its presence.

All too obviously, our experiments are at a very preliminary stage, yet we can conclude that the different siderophores produced by various rhizosphere bacteria play diverse roles in the ecology of this ecosystem influencing the host plant and its associated fungi. We suggest that this represents an evolutionary progression, the earliest siderophores simply solubilizing ferric iron in a form capable of being reduced (and thereby released from the siderophore) by most organisms. Subsequent evolution has led to the development of systems capable of delivering iron specifically to the producing organism and even to the ability to inhibit growth of competing organisms (Becker et al., 1985; Neilands, 1981; Scher and Baker, 1982; Schroth and Hancock, 1982). If this scheme is confirmed it will represent the first example of the evolutionary origin of antibiotic agents from metabolically functional products.

ACKNOWLEDGEMENT

We thank Karin Tenning for her patient and skilful work in preparing the successive versions of this script.

REFERENCES

Anonymous, 1981. Lime and liming. Eighth edition, Reference book 35, M.A.F.F., Her Majesty's Stationery Office, London.

Becker, J.O., and Cook, R.J., 1984. *Pythium* control by siderophore-producing bacteria on roots of wheat. Phytopathology 74:806.

Becker, J.O., Hedges, R.W., and Messens, E., 1985. Inhibitory effect of pseudobactin on the uptake of iron by higher plants. Applied Environ. Microbiol., 49:1090-1093.

Becker, J.O., Messens, E., and Hedges, R.W., 1985. Influence of agrobactin on the uptake of ferric iron by plants. FEMS Micobiol. Ecol., (in press).

Becker, J.O., Messens, E., and Hedges, R.W. A convenient autoradiographic technique for study of uptake of minerals by plant roots and the effect of environmental factors upon the process. Plant Soil. (Submitted for publication).

Black, R., 1980. The role of mycorrhizal symbiosis in the nutrition of tropical plants. In: "Tropical Mycorrhizal Research". 191-202. (Ed. P. Mikola). Clarendon Press, Oxford, U.K.

Brown, J.C., 1956. Iron chlorosis. Annu. Rev. Plant Physiol., 7:171-190.

Costilow, R.N., 1981. Biophysical factors in growth. In: "Manual of Methods for General Bacteriology". (Eds. P. Gerhardt, R.G.E. Murray, R.N. Costilow, E.W. Nester, W.A. Wood, N.R. Krieg and G.B. Phillips) American Society for Microbiology, Washington, D.C.

Deacon, J.W., 1984. Introduction to modern mycology. Second edition, Blackwell, Oxford.

Gibson, F., and Magrath, D.I., 1969. The isolation and characterization of a hydroxamic acid (aerobactin) formed by *Aerobacter aerogenes* 62-1. Biochim. Biophys. Acta, 192:175-184.

Harley, J.L., and Smith, S.E., 1983. "Mycorrhizal symbiosis". Academic Press, London.

Hartmann, A., and Braun, V., 1980. Iron transport in *Escherichia coli*: uptake and modification of ferrichrome. J. Bacteriol., 143:246-255.

Heritage, A.D., and Foster, R.C., 1984. Catalase and sulfur in the rice rhizosphere: an ultrastructural histochemical demonstration of a symbiotic relationship. Microbiol. Ecol., 10:115-121.

Jones, J.P., and Woltz, S.S., 1981. *Fusarium* - incited diseases of tomato and potato and their control. In: "*Fusarium*: diseases, biology and taxonomy". 157-168. (Eds. P.E. Nelson, T.A. Tousson and R.J. Cook). Pennsylvania State University Press.

Joshi, M.M., and Hollis, J.P., 1977. Interaction of *Beggiatoa* and rice plant: detoxification of hydrogen sulfide in the rice rhizosphere. Science, 195:179-180.

Julian, G., Cameron, H.J., and Olsen, R.A., 1983. Role of chelation by ortho dihydroxy phenols in iron absorbtion by plant roots. J. Plant Nutr., 6:163-175.

Kawamoto, S.O., and Lorbeer, J.W., 1976. Protection of onion seedlings from *Fusarium oxysporum* f. sp. *cepae* by seed and soil infestation with *Pseudomonas cepacia*. Plant Dis. Rep., 60:189-191.

Kloepper, J.W., Leong, J., Teintze, M., and Schroth, M.N., 1980. Enhanced plant growth by siderophores produced by plant growth-promoting rhizobacteria. Nature, 286:885-886.

Kuc, J., 1982. Phytoalexins from the solanaceae. In: "Phytoalexins". 81-105. (Eds. J.A. Mailey and J.W. Mansfield) Blackie, Glasgow and London.

Leong, S.A., and Neilands, J.B., 1981. Relationship of siderophore-mediated iron assimilation to virulence in crown gall disease. J. Bacteriol., 147:482-491.

Lewin, R., 1984. How microorganisms transport iron. Science, 225:401-402.

Lindsay, W.L., 1974. Role of chelation in micronutrient availability. In: "The Plant Root and its Environment". 507-524. (Ed. E.W. Carson) University Press of Virginia, Charlottesville.

Lochhead, A.G., and Thexton, R.H., 1947. Qualitative studies of soil micro-organisms. VIII. The 'rhizosphere effect' in relation to the amino acid nutrition of bacteria. Can. J. Res., Section C, 25:20-26.

Marschner, H., 1983. General introduction to the mineral nutrition of plants. In: "Encyclopedia of Plant Physiology". Vol. 15A. 5-60. (Eds. A. Läuchli and R.L. Bieleski). Springer Verlag, Berlin.

Mino, Y., Ishida, T., Ota, N., Inoue, M., Nomoto, K., Takemoto, T., Tanaka, H., and Suguira, Y., 1983. Mugineic acid-iron(III) complex and its structurally analogous cobalt(III) complex: characterization and implication for absorption and transport of iron in graminaceous plants. J. Am. Chem. Soc., 105:4671-4676.

Neilands, J.B., 1981. Microbial iron compounds. Annu. Rev. Biochem., 50:715-731.

Olsen, R.A., Clark, R.B., and Bennett, J.H., 1981. The enhancement of soil fertility by plant roots. Am. Sci., 69:378-384.

Orlando, J.A., and Neilands, J.B., 1982. Ferrichrome compounds as a source of iron for higher plants. In: " Chemistry and Biology of Hydroxamic Acids". 123-129. (Ed. H. Kehl). S. Karger, Basel.

Pitts, G., Allam, A.I., and Hollis, J.P., 1972. *Beggiatoa*: occurrence in the rice rhizosphere. Science, 178:990-992.

Powell, P.E., Szaniszlo, P.J., Cline, G.R., and Reid, C.P.P., 1982. Hydroxamate siderophores in the iron nutrition of plants. J. Plant Nutr., 5:653-673.

Reis, E.M., Cook, R.J., and McNeal, B.L., 1981. Effect of plant nutrients on take-all of wheat. Phytopathology 71:108.

Reis, E.M., Cook, R.J., and McNeal, B.L. 1982. Effect of mineral nutrition on take-all of wheat. Phytopathology, 72:224-229.

Rovira, A.D., and McDougall, B.M., 1967. Microbiological and biochemical aspects of the rhizosphere. In: "Soil Biochemistry". 417-463. (Eds. A.D. McLaren and G.H. Petersen). M. Dekker Inc., New York.

Scher, F.M., and Baker, R., 1982. Effect of *Pseudomonas putida* and a synthetic iron chelator on induction of soil suppressiveness to *Fusarium* wilt pathogens. Phytopathology, 72:1567-1573.

Scholz, G., Schlesier, G., and Seifert, K., 1985. Effect of nicotianamine on iron uptake by the tomato mutant 'chloronerva'. Physiol. Plant., 63:99-104.

Schroth, M.N., and Hancock, J.G., 1982. Disease suppressive soil and root colonizing bacteria. Science, 216:1376-1381.

Schroth, M.N., Weinhold, A.R., McCain, A.H., Hildebrand, D.C., and Ross, N., 1971. Biology and control of *Agrobacterium tumefaciens*. Hilgardia, 40:537-552.

Strobel, G.A., and Nachmias, A., 1985. *Agrobacterium rhizogenes* promotes the initial growth of bare root stock almond. J. Gen. Microbiol., 131:1245-1249.

Suslow, T.V., and Schroth, M.N., 1982. Role of deleterious rhizobacteria as minor pathogens in reducing crop growth. Phytopathology, 72:111-115.

CHARACTERIZATION AND STRUCTURAL ANALYSIS OF THE SIDEROPHORE PRODUCED BY THE PGPR *PSEUDOMONAS PUTIDA* STRAIN WCS358

G.A.J.M. van der Hofstad, Joey D. Marugg, G.M.G.M. Verjans and P.J. Weisbeek

Department of Molecular Cell Biology
University of Utrecht
Utrecht
The Netherlands

INTRODUCTION

Under iron limitation, the plant-growth-promoting *Pseudomonas putida* strain WCS358 produces a yellow-green fluorescent siderophore (Geels and Schippers, 1983; Geels and Schippers, 1983; Marugg et al., in press). The absorption spectra of WCS358 culture medium before and after addition of $FeCl_3$ and at different pHs are shown in an accompanying paper by our group (Weisbeek et al., in press). They strongly resemble those of the pseudobactin/pyoverdine class of siderophores (Hider, 1984).

We describe here the isolation and charactization of the siderophore from WCS358 together with a preliminary structure assignment.

Isolation of the ferric siderophore

A modification of the procedures described by Meyer and Abdallah (see Hider, 1984) was used, comprising the following steps:

- 10-fold concentration of culture medium of a stationary WCS358 culture supplemented with 0.6 g/l $FeCl_3$
- 100% $(NH_4)_2SO_4$ precipitation
- extraction with phenol/chloroform
- precipitation of the brown ferric pseudobactin from the organic phase with ether
- extraction with small volumes of water
- re-extraction of the aqueous phase with chloroform to remove phenol
- lyophilisation

Yield per litre of medium: 200 mg.

The purity was checked by means of HPLC gel filtration analysis. One major peak comprising more than 95% of A_{280}-absorbing material was found.

Amino Acid Analysis

Analysis of a 6 N HCl hydrolysate is shown in Fig. 1. The composition data indicates the purity of the isolated siderophore.

The following amino acids and their molar ratios were found: Lys(2) Orn (<½) Ala (1) Glu (<½) Ser (1) Thr (2) Asp (1) and, very probably, HO-asp (1). By comparison with the reactions of known pseudobactin-like siderophores, we presume that Orn and Glu were formed from HO-Orn during acid hydrolysis (Fig. 2). This gives the following amino acid composition for the siderophore of WCS358:
Lys (2) HO-Orn (1) Ala (1) Ser (1) Thr (2) Asp (1) and HO-Asp (1).

MW determination

Fast atom bombardment mass spectrometry (FAB) was used for MW determination.

Contrary to expectation from the amino acid composition, no $[M+H]^+$ peak was detected in the range up to 2000 Daltons. Using HPLC gel filtration chromatography with an exclusion range of 16kD (Fig. 3), most of the ferric WCS358 siderophore was found in a high MW form of around 9 kD

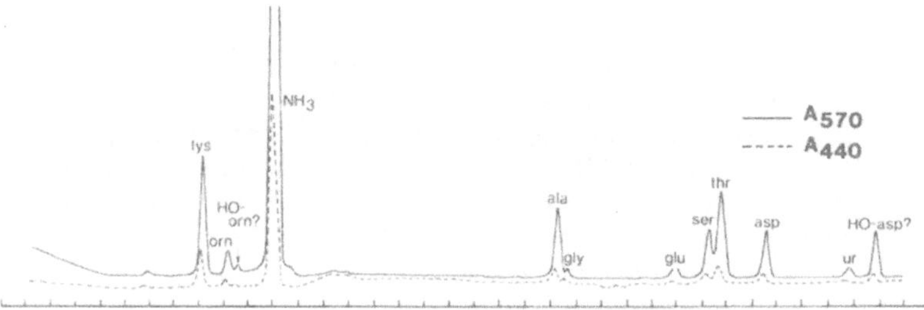

Fig. 1. Amino acid analysis of a 6N HCl hydrolysate from the ferric siderophore of WCS358 using a LKB 3201 Amino Acid Analyser with a programme for biological samples.

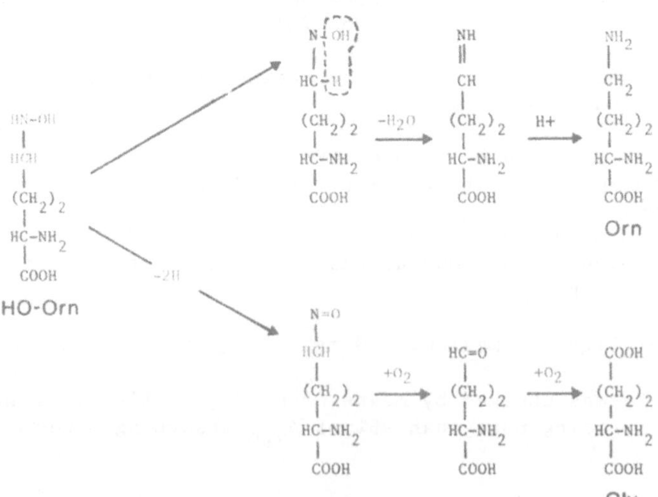

Fig. 2. Presumable formation of Orn and Glu from HO-Orn during acid hydrolysis.

(presumably a hexamer). Penta-(7.5), tri-(4.5), di-(3) and monomer (1.5 kD) forms were also present, the last in only very small amounts.

Amino Acid Sequence

Manual Edman degradations were carried out according to Chang (1983) with combined use of PITC and DABITC. The aminoacyl-DABTH derivates were analyzed on polyamide sheets. The first degradation step did not result in an identifiable amino acid. Step 2 showed unambiguously Ser and its byproducts. Steps 3, 4, 5 and 6 respectively revealed Asp, Thr, Ala and again Thr. In step 7, an as yet unidentified amino acid was found. Step 8 revealed Lys and its byproducts and step 9 a component near Arg, which

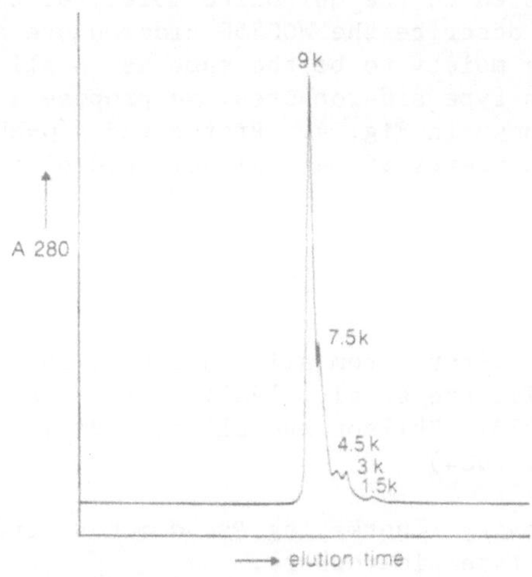

Fig. 3. HPLC gel filtration chromatography (exclusion 16 kD) in 0.1 M phosphate pH 6.8

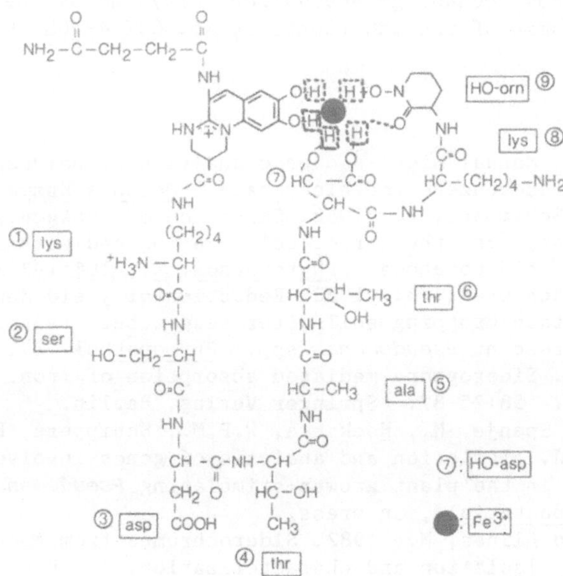

Fig. 4. Tentative structure of pseudobactin 358

is possibly HO-Orn. It can be concluded that the amino acid from step 1 has a free ε-NH_2 group. Since the second amino acid is Ser, the first member of the polypeptide chain must be NH-bound Lys. Presumably, its amino acid-DABTH derivative did not move in the second direction because of the presence of the fluorescing group of the siderophore. The unidentified amino acid from step 7 is presumably HO-Asp. Accordingly, the following tentative amino acid sequence can be deduced:
ε-NH_2-Lys . Ser . Asp . Thr . Ala . Thr . HO-Asp . Lys . HO-Orn.

Structure

The absorption spectra (Weisbeek et al., in press) and fluorimetric data of the siderophore from WCS358 are very similar to those from pseudobactin/pyoverdin type siderophores. This suggests the presence of a fluorescing group related to the quinoline moiety of that class of siderophores. We therefore describe the WCS358 siderophore as pseudobactin 358. Assuming the quinoline moiety to be the same as in all other published pseudobactin-pyoverdin type siderophores, we propose a structure for pseudobactin 358 as shown in Fig. 4. Proton and ^{13}C-NMR techniques, as well as FAB mass spectrometry of partial acid hydrolysates, will be used to test this proposal.

CONCLUSIONS

Pseudobactin 358 differs from all other published siderophores in its class (Hider, 1984; Teintze et al., 1981; Teintze and Leong, 1981; Philson and Llinas, 1982; Philson and Llinas, 1983; Wendenbaum et al., 1983; Yang and Leong, 1984).

It appears that every fluorescing *Pseudomonas* strain produces a specific pseudobactin-type siderophore.

ACKNOWLEDGEMENTS

The research described was performed in part with financial support from The Netherlands Technology Foundation (STW) and by the Biomolecular Engineering Programme of the EEC (Contract No. GBI-4-108-NL).

REFERENCES

Chang, J-Y., 1983. Manual micro-sequence analysis of polypeptides using dimethylaminoazobenzene-isothiocyanate. Methods Enzymol., 91:455-466.

Geels, F.P., and Schippers, B., 1983. Selection of antagonistic fluorescens *Pseudomonas* spp. and their root colonization and persistence following treatment of seed potatoes. Phytopathol. Z., 108:193-206.

Geels, F.P., and Schippers, B., 1983. Reduction of yield depressions in high frequency potato cropping soil after seed tuber treatments with antagonistic fluorescent *Pseudomonas* spp. Phytopathol. Z., 108:207-214.

Hider, R.C., 1984. Siderophore mediated absorption of iron. In: "Structure and Bonding". 58:25-87. Springer Verlag, Berlin.

Marugg, J.D., van Spanje, M., Hoekstra, W.P.M., Schippers, B., and Weisbeek, P.J. Isolation and analysis of genes involved in siderophore biosynthesis in the plant growth-stimulating *Pseudomonas putida* strain WCS358. J. Bacteriol., in press.

Philson, S.B., and Llinas, M., 1982. Siderochromes from *Pseudomonas* fluorescens. Isolation and characterization. J. Biol. Chem., 257: 8081-8085.

Philson, S.B., and Llinas, M., 1983. Siderochromes from *Pseudomonas* fluorescens. Structural homology as revealed by NMR spectroscopy. J. Biol. Chem., 257:8086-8090.

Teintze, M., Hossain, M.B., Barnes, C.L., Leong, J., and Van der Helm, D., 1981. Structure of ferric pseudobactin, a siderophore from a plant growth promoting *Pseudomonas*. Biochemistry, 20:6446-6457.

Teintze, M., and Leong, J., 1981. Structure of pseudobactin A, a second siderophore from plant growth promoting *Pseudomonas* B10. Biochemistry, 20:6457-6462.

Weisbeek, P.J., van der Hofstad, G.A.J.M., Schippers, B., and Marugg, J.D. Genetic analysis of the iron uptake system of two plant growth-promoting *Pseudomonas* strains. In press.

Wendenbaum, S., Demange, P., Dell, A., Meyer, J.M., and Abdallah, M.A., 1983. The structure of pyoverdine $_{Pa}$, the siderophore of *Pseudomonas aeruginosa*. Tetrahedron Lett., 24:4877-4880.

Yang, C.C., and Leong, J., 1984. Structure of pseudobactin 7SR1, a siderophore from a plant-deleterious *Pseudomonas*. Biochemistry, 23:3534-3540.

PHYSICAL, BIOLOGICAL AND HOST FACTORS IN IRON COMPETITION IN SOILS

R. Baker[1], Y. Elad[2], and B. Sneh[3]

Department of Plant Pathology and Weed Science[1]
Colorado State University, Fort Collins
Colorado, USA 80523

Department of Plant Pathology and Microbiology[2]
Faculty of Agriuculture, The Hewbrew University of
Jerusalem, Rehovot, 76100 Israel

The George S. Wise Faculty of Life Sciences[3]
Tel Aviv University, Ramat-Avid, Tel Aviv 69978 Israel

INTRODUCTION

A Metz fine sandy loam soil, found in the Salinas Valley of California was suppressive to Fusarium wilt diseases (Baker and Chet, 1982; Smith, 1977; Smith and Snyder, 1972). A factor associated with biological control appeared to be operative because the suppressiveness could be transferred to conducive soil; that is, only 600 g/m^2 of suppressive soil added to conducive soil in a commercial greenhouse decreased the incidence of Fusarium wilt of carnation over a 2-yr period (Baker, 1980). Also suppressiveness was nullified by heat-treatment of soil at 48 C (Scher and Baker, 1980).

Sneh and Baker (1980) introduced mycelial mats of *Fusarium oxysporum* f. sp. *lini* into conducive or suppressive soil to act as "bait" for potential antagonists. Bacteria isolated from conducive soil by this method did not reduce incidence of flax wilt when introduced into conducive or steamed soil; however, *Pseudomonas* spp. isolated from the Metz fine sandy loam induced suppressiveness. Kloepper et al., (1980) also induced suppressiveness by addition of fluorescent pseudomonads to conducive soil and suggested that these antagonists produced siderophores that competed with Fusarium wilt pathogens for available iron (Fe). This reduced efficiency for host colonization by the wilt pathogens. Here we review subsequent evidence that supports the hypothesis that biological factors, whose activity is enhanced by the physical/chemical environment found in the Metz soil, are responsible, at least in part for Fusarium suppressiveness.

BASIC MECHANISMS OF COMPETITION FOR IRON

In a narrow sense, competition is "active demand in excess of immediate supply of material or condition on the part of two or more organisms" (Clark, 1963). In soil, microorganisms compete exclusively for substrate and/or essential elements (Baker, 1981). The basic mechanism involved in biological control by competition is that the pathogen must be deprived of a nutrient essential for pathogenesis. Factors found to be potentially essential are nitrogen, carbon (Griffin, 1981) and Fe (Scher and Baker, 1982). Thus by proper manipulation of one or more of these factors in soil, they could be limiting for pathogenesis. The mechanism by which fluorescent pseudomonads appear to function in biocontrol is by competition for Fe.

The Fe available in soil for plants and microorganisms is defined by the equation:

$$Fe(OH)_3 + H^+ \rightleftarrows Fe^{3+} + 3H_2O$$

The principal factor mediating the equilibrium of this equation is the pH of the soil. Thus, the influence of pH may be considerable in determining whether Fe^{3+} is available to the pathogen, and ultimately, whether a soil is suppressive or conducive since Fe^{3+} is necessary for germ-tube elongation from spores or formae specialis of *Fusarium oxysporum* (Scher and Baker, 1982). Indeed, when pH of the suppressive Metz soil (native pH ca 8.0) was reduced to pH 7.0, about half of the suppressiveness was lost. At pH 6.0 the soil was conducive (Scher and Baker, 1980).

When *Pseudomonas putida,* ethylenediaminedi-O-hydroxyphenyl-acetic acid (EDDHA) or the ferrated form (FeEDDHA) was added to conducive soil, the soil became suppressive to Fusarium wilt of flax, cucumber, or radish (Dupler and Baker, 1984; Elad and Baker, 1985; Scher and Baker, 1980; Scher and Baker, 1982). Suppressiveness was not induced by Fe-ethylenediaminetetra-acetic acid (FeEDTA), or Fe-diethylenediaminetetraacetic acid (FeDTPA). These results may be explained by the mechanism of competition for Fe illustrated in Fig. 1.

The three ligands bind Fe^{3+} at various stability constants: EDDHA, $\log_{10} K - 33.9 >$ DTPA, $\log_{10} K - 27.3 >$ EDTA, $\log_{10} K = 25$ (Lindsey, 1974). Fe-binding siderophores were detected for most microorganisms examined (Neilands, 1973), and the causal agents of Fusarium wilt produced siderophores of the hydroxamate class under Fe-deficient conditions. These have stability constants in the proximity of $\log_{10} K = 29$. This constant is many powers of ten less than most bacterial catechol siderophores. *P. putida* produces a mixed hydroxamate catechol siderophore (pseudobactin as reported by Teintze et al., (1980).

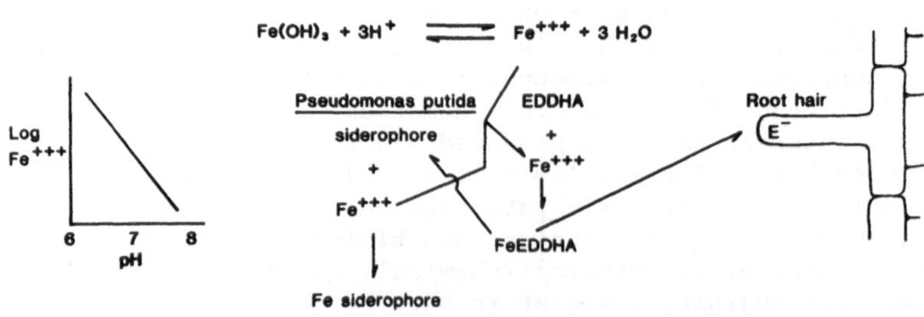

Fig. 1. Mechanisms and pathways involved in soil suppressive to Fusarium wilt pathogens through competition for iron (Fe). (by permission, R. Baker).

Theory expansion suggests that adequate Fe^{3+} is available to Fusarium wilt pathogens for germination and penetration through root tips of the host in conducive soil. Since many pseudomonads have high rhizosphere competence (Schmidt, 1979), there should be intense competition for Fe^{3+} at the rhizoplane (infection court) when *P. putida* produces siderophores, especially in alkaline soils. Therefore, Fe^{3+} is bound in such a way that it is unavailable to *F. oxysporum*.

The ligand EDDHA has a higher stability constant than the siderophore produced by the pathogen. Fe^{3+} complexed by this ligand in the rhizosphere would be unavailable to the pathogen; however, EDDHA-bound Fe can be utilized by the host. This frees EDDHA to bind more Fe^{3+} at the root surface (Lindsey, 1974) and the soil is suppressive. This is a dynamic system in which the siderophores produced by pseudomonads, the EDDHA, and the root combine to render Fe^{3+} limiting for pathogenesis by Fusarium wilt pathogens.

EVIDENCE FOR THE MECHANISM OF COMPETITION FOR FE

Sneh et al., (1984) isolated over 700 bacteria from the suppressive Metz soil. The isolates were capable of producing fungal cell-wall degrading enzymes and/or lysing living or dead germ tubes of chlamydospores and/or producing fluorescent compounds (siderophores) in Fe-deficient nutrient medium. Representative groups with or without these attributes were added to conducive soil and evaluated for their influence or chlamydospore germination of *F. oxysporum* f. sp. *cucumerinum* in rhizosphere and nonrhizosphere soil, and for their ability to induce suppressiveness to Fusarium wilt of cucumber. With only one exception (*Enterobacter cloacae),* nonfluorescent bacteria induced little or no inhibition of chlamydospore germination or disease suppression. There was, however, a direct correlation ($\underline{r}$ = 0.99) between the amount of siderophore production (*in vitro)* by various fluorescent pseudomonads and their inhibition of chlamydospore germination in soil. There was also a direct correlation between siderophore production by the various strains and suppression of Fusarium wilt of cucumber ($\underline{r}$ = 0.747 [Sneh et al., 1984], radish ($\underline{r}$ = 0.825) or pea ($\underline{r}$ = 0.872 [Elad and Baker, 1985]).

Other possible mechanisms for suppression were tested. As noted above, ability of potential antagonists to induce lysis and/or enzymatic disintegration of germ tubes was not related to suppressiveness (Sneh et al., 1984). An alternative mechanism could be cross protection. In unpublished research, carnation cuttings took up (through the wounded tissue) a suspension of the Metz soil for 2 hr, were rooted, and eventually challenge inoculated with *F. oxysporum* f. sp. *dianthi*. There was no evidence of decrease in incidence of disease by such treatment suggesting that cross protection was not involved in suppressiveness.

Antibiosis was considered as another candidate mechanism for suppression by pseudomonads. Wong and Baker (1984) found, however, no correlation between *in vitro* antibiosis in nutrient agar and suppression of *Gaeumannomyces graminis* var. *tritici* or var. *avenae* although disease suppression was induced by siderophore producing strains.

Chlamydospore germination in rhizosphere soil was inhibited slightly by addition of a nonsiderophore-producing mutant (Elad and Baker, 1985). Available Fe added to the system did not influence inhibition suggesting that another essential element, probably carbon (C), was limiting. This inhibition was not substantial enough, however, to account for suppressiveness. Indeed, such studies provided no evidence for the induction of suppressiveness by increases in biomass that should limit C availability. Growth rates and, thus, biomass of pseudomonads added initially in similar concentrations to soil were not significantly different. Yet, their ability to inhibit chlamydospore germination in soil (Sneh et al., 1984), in the rhizosphere or to decrease disease (Elad and Baker, 1985) differed. Such responses were correlated directly with production of siderophores.

Thus, no evidence for participation of pseudomonads in exploitation, antibiosis, cross protection or competition for C in the induction of suppressiveness was obtained. So far, the only mechanism clearly identified that induces substantial suppressiveness in soil is competition for Fe.

INFLUENCE OF THE CHEMICAL AND PHYSICAL ENVIRONMENT

The profound influence of soil reaction on suppressiveness and Fe availability (Scher and Baker, 1982) was reviewed above. It is well known that siderophore production also occurs only in low Fe environments (Neilands et al., 1980).

This phenomenon was illustrated when fluorescent pseudomonads or partially purified siderophore preparations were applied to soils with pH 5.0 - 7.0 and germination of chlamydospores observed (Elad and Baker, 1985). Predictably, more inhibition of chlamydospore germination occurred in soil at pH 6.0 - 7.0 than at 5.0 - 5.5 by either treatment. However, the applied siderophore preparation reduced chlamydospore germination significantly at pH 5 - 5.5, whereas, the bacteria did not, even though present at population levels of 5.7×10^8 cuf/g soil.

Baker (1968) proposed that competition for a particular element can be determined by adding the candidate limiting factor to the system and observing whether biological control was nullified. Thus, Sneh et al., (1984) added available Fe to soil and observed that it counteracted the inhibitory effect of fluorescent pseudomonads on chlamydospore germination. Counteraction also was observed, however, when soil was amended with Zn^{2+}, Co^{2+}, Mn^{2+} and MoO_4^{2-}. In their investigations, Elad and Baker, (1985) suggested two lines of evidence to explain these phenomena: (i) added micronutrients delayed multiplication of pseudomonads in soil thereby extending the lag phase. Thus, it appears that high population densities and/or activity of the biocontrol agent are necessary for inducing inhibition in the initial stages of chlamydospore germination. (ii) Siderophores produced by pseudomonads apparently lack specificity for binding Fe^{3+} and may incorporate or substitute other cations on binding sites. However, when Fe^{3+} was added to siderophore preparations originally provided with these other cations, Fe gradually replaced these previously bound microelements. This suggests that the affinity of the siderophore-binding-site for Fe^{3+} is higher than the other cations tested.

HOST INFLUENCES

The ligand, EDDHA, has a higher stability constant than hydroxamate siderophores produced by Fusarium pathogens and, indeed, induced suppressiveness when added to conducive soil (Scher and Baker, 1982). However, quantities of Fe limiting to germination of chlamydospores should only occur on the host rhizoplane where there is intense competition for this element by both the root and rhizobacteria, like fluorescent pseudomonads, which utilize Fe in FeEDDHA (Scher et al., 1984). Once stripped of its Fe, the ligand binds more Fe from the available Fe^{3+} pool (Lindsey, 1974). Since there are no roots or active rhizosphere bacteria to compete for this trace element in soil distant from the root, the amount of Fe^{3+} would be determined by soil pH regardless of the addition of FeEDDHA. Thus, FeEDDHA was only effective in reducing chlamydospore germination in the rhizosphere but not in soil (Elad and Baker, 1985; Scher and Baker, 1984). These phenomena suggest that there is an increment of control in suppressive soils attributable to the host.

There is probably more to this interaction than a simple qualitative relationship. Elad and Baker (1985) observed that siderophore production was reduced proportionally more than rate of bacterial cell division when fluorescent pseudomonads were grown in dilution series of synthetic medium containing low Fe. Therefore, in plants producing lower amounts of exudates in the rhizosphere, pseudomonads should produce lower amounts of siderophores. Evidence for this hypothesis was obtained by observing the level of chlamydospore germination of many isolates of *Fusarium* spp. germinating in rhizospheres of their respective hosts in soil infested or not infested with a siderophore-producing strain of *P. putida*. There was a significant, direct correlation between the germination of chlamydospores in soil to which the pseudomonad was not added and the degree inhibition induced when the strain was introduced. This suggests that the more nutrients exuded in the rhizosphere (as reflected by higher levels of chlamydospore germination) the more inhibition induced by the pseudomonad (Fig. 2). Therefore, increased inhibition results in the rhizosphere when more nutrients (except Fe) are available for the production of siderophores.

In the past, experimenters have applied one application of organic nitrogen and/or carbon sources to induce chlamydospore germination of *Fusarium* spp. in soil (e.g. Smith, 1977; Smith and Snyder, 1972). This, of course, does not reflect the situation found in soils where root exudates triggering germination are continuously introduced into certain areas of the rhizosphere. For a closer simulation of this situation, nutrients were supplied in pulses and striking differences were observed in germination inhibition in soil induced by fluorescent pseudomonads between this procedure and a one-time application of nutrients at the beginning of the experiment (Elad and Baker, 1985). Except at extremely low or high concentrations (where no differences were observed), inhibition of germination was less when nutrients were pulsed than when applied in total at one time. This provides at least a partial explanation of the results of Sneh et al., (1984) who observed less inhibition induced by

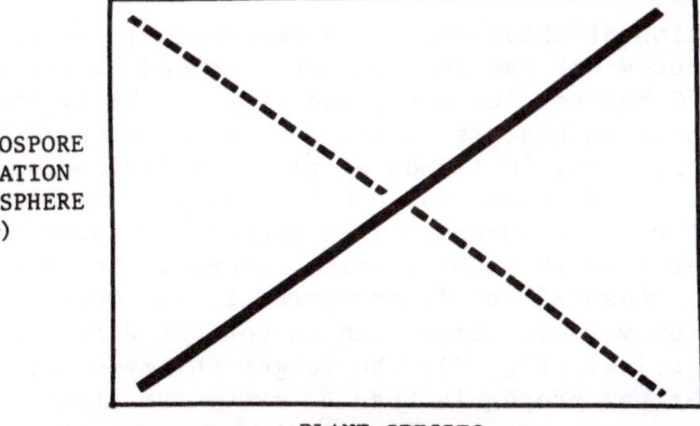

Fig. 2. Diagramatic presentation of the evidence accumulated by Elad and Baker (1985) suggesting that the more nutrients supplied by root exudates of various plants (as reflected by higher levels of chlamydospore germination of many *Fusarium* spp.) the more inhibition induced by siderophore-producing pseudomonads.

CHLAMYDOSPORE CHARACTERISTICS	FUSARIUM OXYSPORUM F. SP. CUCUMERINUM (SOIL IS SUPPRESSIVE)	FUSARIUM SOLANI F. SP. PHASEOLI (SOIL IS NOT SUPPRESSIVE)
SIZE	○	 ◯
VOLUME	1	4.99
WEIGHT	1	4.60
IRON CONTENT	1	3.00

Fig. 3. Ratios of some properties of chlamydospores of *Fusarium oxysporum* f. sp. *cucumerinum* and *F. solani* f. sp. *phaseoli* (Elad and Baker, 1985).

fluorescent pseudomonads in the rhizosphere as compared to a test in which soils were supplied with one application of nutrients.

SPECIFIC SUPPRESSION OF *FUSARIUM* SPP.

Soil was suppressive in the Salinas Valley to *formae specialis* of *F. oxysporum* but not to other *Fusarium* spp. (Smith, 1977; Smith and Snyder, 1972). A hypothesis explaining this phenomenon was advanced by Elad and Baker (1985).

Low levels of inhibition of chlamydospore germination in rhizospheres induced by addition of fluorescent pseudomonads was noticed in isolates of *F. solani* attributable to C but not Fe. Also, addition of the ligand, EDDHA, (that binds Fe in soil so that it is not available to the hydroxamate siderophores of *Fusarium* spp. [Neilands et al., 1980]) suppressed germination of isolates of *F. oxysporum* but not *F. solani* or *F. Graminearum*. What is the explanation for these phenomena? In general, chlamydospores of *F. oxysporum* are smaller than in other species. Indeed, when chlamydospores of *F. solani* f. sp. *phaseoli* and *F. oxysporum* f. sp. *cucumerinum* were compared, the ratios of volume, weight and Fe-content were three to five times greater for *F. solani* (Fig. 3); the larger chlamydospores of *F. solani* contained more Fe per propagule than *F. oxysporum*. Therefore, they did not require an exogenous supply of this element even in low Fe environments.

CONCLUSIONS

The research, reviewed above, contributes to our basic understanding of the influence of pseudomonads and their siderophores in soil suppressive to Fusarium wilt pathogens. Further, such studies identify the important environmental parameters, such as Fe activity, soil pH, and nutritional status in the rhizosphere, that are critical in inducing suppressiveness. Recently, the critical Fe level associated with biocontrol was below 10^{-22} M in an analog system where chlamydospore germination was observed in the presence of *P. putida* (Simeoni et al., 1985). Thus the key strategy in biocontrol by use of this system, is management of Fe-activity.

REFERENCES

Baker, R., 1968. Mechanisms of biological control of soil-borne pathogens. Ann. Rev. Phytopath. 6:263-294.

Baker, R., 1980. Measures to control Fusarium and Phialophora wilt pathogens of carnations. Plant Dis. 64:743-749.

Baker, R., 1981. Ecology of the fungus, *Fusarium:* Competition, In: *Fusarium:* diseases, biology, and taxonomy. (P.E. Nelson, T.A. Toussoun, and R.J. Cook, Eds.) Pennsylvania State University Press, University Park.

Baker, R., 1983. State of the art: plant diseases, IN: Proceedings of the National Interdisciplinary Biological Control Conference, 15-17 Feb., 1983, Las Vegas, NV,(S.L. Battenfield, Ed.) CSRS /USDA Washington, D.C.

Baker, R., and Chet, I., 1982. Induction of suppressiveness. In:"Suppressive soils and plant disease," (R.W. Schneider, Ed.), The Americal Phytopathological Society, Saint Paul.

Clark, F.E., 1963. The concept of competition in microbial ecology. In: "Ecology of Soil-borne plant pathogens," (K.F. Baker and W.C. Snyder, Eds.), University of California Press, Berkeley.

Dupler, M., and Baker, R., 1984. Survival of *Pseudomonas putida*, a biological control agent in soil. Phytopathology, 74:195-200.

Elad, Y., and Baker, R., 1985. Influence of trace amounts of cations and siderophore-producing pseudomonads on germination of *Fusarium oxysporum* chlamydospores. Phytopathology (in press).

Elad, Y., and Baker, R., 1985. The role of competition for iron and carbon in suppression of chlamydospore germination of *Fusarium* spp. by *Pseudomonas* spp. Phytopathology (in press).

Griffin, G.J., 1981. Physiology of conidium and chlamydospore germination in *Fusarium.* In: "Diseases, biology and taxonomy," (P.E. Nelson, T.A. Toussoun, and R.J. Cook, Eds.), Pennsylvania State University Press, University Park.

Kloepper, J.W., Leong, J., Teintze, M., and Schroth, M.N., 1980. *Pseudomonas* siderophores: a mechanism explaining disease-suppressive soils. Curr. Microbiol. 4:327-320.

Lindsey, W.L., 1974. Role of chelation in micronutrient availability. In: "The plant root and its environment," (R.W. Carson, Ed.) University Press Virginia, Charlottesville.

Neilands, J.B., 1973. Microbial iron transport compounds (siderophores). In: "Inorganic Biochemistry," (I.G.L. Eickhorn, Ed.) Elsevier, Amsterdam.

Neilands, J.B., Peterson, T., and Leong, J., 1980. High affinity iron transport in microorganisms. Iron III coordination compounds of the siderophores agrobactin and parabactin. ACS Symp. Ser. 140:263-278.

Scher, F.M., and Baker, R., 1980. Mechanism of biological control in a Fusarium-suppressive soil. Phytopathology, 70:412-417.

Scher, F.M., and Baker, R., 1982. Effect of *Pseudomonas putida* and a synthetic iron chelator on induction of soil suppressiveness to Fusarium wilt pathogens. Phytopathology, 72:1567-1573.

Scher, F.M., and Baker, R., 1984. A fluorescent microscopic technique for viewing fungi in soil and its application to studies of a Fusarium-suppressive soil. Soil Biol. Biochem. 15:715-718.

Scher, F.M., Dupler, M., and Baker, R., 1984. Effect of synthetic iron chelates on population densities of *Fusarium oxysporum* and the biological control agent *Pseudomonas putida* in soil. Can. J. Microbiol. 30:1271-1275.

Schmidt, E.L., 1979. Initiation of plant root microbe interactions. Ann. Rev. Microbiol., 33:355-379.

Simeoni, L.A., Lindsay, W.L., and Baker, R., 1985. Critical iron level associated with biological control of Fusarium wilt. Proceedings of the 3rd International Symposium (in press).

Smith, S.N., 1977. Comparison of germination of pathogenic *Fusarium oxysporum* chlamydospores in host rhizosphere soils conducive and suppressive to wilts. Phytopathology, 67:502-510.

Smith, S.N., and Snyder, W.C., 1972. Germination of *Fusarium oxysporum* chlamydospores in soils favorable and unfavorable to wilt establishment. Phytopathology, 62:273-277.

Sneh, B., Dupler, M., Elad, Y., and Baker, R., 1984. Chlamydospore germination of *Fusarium oxysporum* f. sp. *cucumerinum* as affected by fluorescent and lytic bacteria from a Fusarium-suppressive soil. Phytopathology, 74:1115-1124.

Teintze, M., Hassain, M.B., Baines, C.L., Leong, J., and van der Helm, D., 1980. Structure of ferric pseudobactin a siderophore from a plant growth promoting *Pseudomonas*. Biochemistry, 20:6446-6457.

Wong, P.T.W., and Baker, R., 1984. Suppression of wheat take-all and Ophiobolus patch by fluorescent pseudomonads from a Fusarium-suppressive soil. Soil Biol. Biochem., 16:397-403.

IMPORTANCE OF SIDEROPHORES IN MICROBIAL INTERACTIONS IN THE RHIZOSPHERE

J.E. Loper and M.N. Schroth

Former Graduate Research Assistant and Professor
Department of Plant Pathology, University of California
Berkeley, 94720.

Present address of senior author: Soilborne Diseases Laboratory, Plant Protection Institute, Agricultural Research Service, United States Department of Agriculture, Beltsville, Maryland, 20705

INTRODUCTION

Plants, animals, and microorganisms require iron for oxidation-reduction and other cellular reactions and have evolved mechanisms to acquire it from the environment. With animal pathogens, siderophore synthesis plays a direct role in the infection process (Weinberg, this book). Growth of numerous bacterial genera is stimulated by excess iron in body fluids and tissues of vertebrate hosts. Conversely, host defense mechanisms include withholding iron from the pathogen. It is not surprising, therefore, that the activities of microorganisms affecting plant health also would be influenced by the availability of iron in the environment and their competitive abilities to sequester it.

The pioneering efforts of J.B. Neilands on the role of iron in microbial physiology and the close proximity of our laboratories at Berkeley resulted in a discussion in the late 1970's among Biochemistry and Plant Pathology graduate students concerning the ecological significance of siderophores produced by rhizosphere bacteria. At that time we were studying mechanisms by which certain pseudomonads enhanced plant growth and increased crop yields when inoculated onto seed or propagative plant material. These bacteria established large population sizes on plant roots which were maintained throughout the growing season. (T. Burr et al., 1978; J. Kloepper et al., 1980). A question was raised regarding the role of siderophore production by pseudomonads in their competitive abilities at the root surface. This question was of broad interest since pseudomonads are important as saprophytes, foliar epiphytes, phytopathogens, seed inoculants to enhance plant growth (Burr and Caesar, 1984; Schroth and Hancock, 1981; Schroth and Hancock, 1982; Suslow, 1982) and as potential biocontrol agents of phytopathogens (Howell and Stipanovic, 1979; Howell and Stipanovic, 1980; Scher and Baker, 1982; Sneh et al., 1984).

ROLE OF IRON IN MICROBIAL INTERACTIONS

In Vitro Antiobiosis and Siderophores

The fluorescent pseudomonads produce an extracellular water-soluble, yellow-green pigment which fluoresces under ultraviolet irradiation. The fluorescent pigments produced by these bacteria function as siderophores, as characterized by their synthesis only under iron-limiting conditions, (Lenhoff, 1963) specific and high affinities for Fe(III), (Meyer and Abdallah, 1978; Teintze et al., 1981) and their roles in transport of Fe^{III} into the bacterial cell (Meyer and Hornsperger, 1978). *Pseudomonas* strains that produce siderophores exhibit *in vitro* antibiosis against indicator bacterial or fungal strains (Hemming et al., 1982; Kloepper et al., 1980a: Kloepper and Schroth, 1981a; Misaghi, et al., 1982) which is nullified by the addition of iron (10^{-3} M $FeCl_3$) to the culture medium.

Plant Growth Promotion and Disease Protection

Several lines of evidence suggest that beneficial pseudomonads enhance plant growth by producing extracellular siderophores which sequester the iron in the root environment making it less available to competing deleterious microflora. This hypothesis had its genesis with the finding that mutants of plant growth promoting rhizobacteria (PGPR) which did not exhibit *in vitro* antibiosis were deficient in siderophore production (Sid^-) and no longer caused plant growth promotion. (Kloepper and Schroth, 1981a). These Sid^- mutants also failed to cause a detectable alteration in the composition of the root microflora as was observed with parental strains. Sixteen Sid^- mutants of five strains were obtained by UV irradiation, or treatment with N-methyl-N-nitro-N-nitrosoguanidinine or ethyl methansulphonate and were tested in greenhouse or field experiments. These mutants were phototrophic and exhibited no detectable differences in growth rates in culture from their parental strains. However, the single site natures of these chemically-induced mutations were not definitive. Furthermore, genetic analysis of fluorescent pigment production of *Pseudomonas syringae* (Loper et al., 1984b) and *Pseudomonas fluorescens* (Moores et al., 1984) has demonstrated the involvement of several genes. Some of these genes may also influence the production of other anti-fungal secondary metabolites. For example, single Tn5 insertions in *P. syringae* pv. *syringae* negated the production of both the fluorescent pigment/siderophore and syringomycin (Loper, unpublished). The availability of genetic tools for analysis of the pseudomonads now facilitates the derivation of well characterized single site mutations. Confirmation of the above experiments with single site insertion or deletion mutants and mutants restored to fluorescence by complementation with cloned wild type DNA fragments will provide more definitive evidence for the role of siderophores in plant growth promotion.

Evidence for the importance of siderophore production in microbial interactions in the rhizosphere was greatly strengthened by experiments examining the activity of pseudobactin, the purified siderophore of *Pseudomonas* strain B10, in plant growth promotion and disease suppression. (Kloepper et al., 1980a, Kloepper et al., 1980b). The activity of strain B10 in plant growth promotion and in suppression of the diseases Fusarium wilt of flax and take-all of wheat caused by *Fusarium oxysporum* f. sp. *lini* or *Gaeumannomyces graminis* var. *tritici,* respectively, was mimicked by the addition of pseudobactin directly to the potting soil. Accordingly, disease conducive soils could temporarily be made disease suppressive by the incorporation of either strain B10 or its siderophore. The effect of pseudobactin was nullified by the addition of Fe^{III} EDTA and was not observed when used in chelated form (ferric pseudobactin). The effect of pseudobactin on disease inhibition was attributed to an iron-starvation of *F. oxysporum* f. sp. *lini,* and *G. graminis* var. *tritici.* In the case of general plant growth promotion, it was similarly proposed that pseudobactin deprived various deleterious microorganisms of iron. Additional data indicating that plant growth promotion could result from an altered composition of the root microflora came from tests done under gnotobiotic conditions. (Kloepper and Schroth, 1981b). Plant growth was observed in raw field soil but not in sterilized soil, suggesting that the pseudomonads were suppressing deleterious microorganisms in the rhizosphere.

Siderophores and Germination of Fungal Spores

Siderophores produced by *Pseudomonas* spp. have been implicated in the suppression of several wilt diseases caused by *Fusarium* spp. Iron competition, by elaboration of siderophores, was suggested as a mechanism by which pseudomonads limited chlamydospore germination (Sneh et al., 1984) germ tube development from microconidia (Scher and Baker, 1982) and conferred disease suppressiveness,(Kloepper et al., 1980b;Scher and Baker, 1982). Recently an *Alcaligenes* sp. strain MFA1 was identified which inhibited colonization of carnation root surfaces by *F. oxysporum* f. sp. *dianthi,* delayed systemic establishment of the pathogen, and delayed wilt symptom development, (Yuen and Schroth, 1986a; Yuen et al., 1986). Although microconidial germination, germ tube elongation and mycelial growth of *F. oxysporum* f. sp. *dianthi* were limited by MFA1 in culture, the most striking effect of MFA1 in the soil was on chlamydospore germination. Chlamydospore germination in the spermospheres and rhizospheres of control plants was 28 and 43%, respectively, and 15 and 14%, respectively, when seeds and roots were treated with cell densities of MFA1 as low as 10^3 cfu/g soil. Siderophore(s) were implicated in the suppressive effect of MFA1 since the addition of iron (10^{-3} M $FeCl_3$) to soil nullified the inhibitory effect of the bacterium. Sid^- mutants of MFA1, obtained by ultraviolet or ethyl methanesulphonate mutagenesis, established rhizosphere population sizes equivalent to that of the parental strain but did not affect chlamydospore germination in soil. Since chlamydospores rather than microconidia or mycelium are the principal survival propagule of *F. oxysporum* f. sp. *dianthi* in the soil, their germination is critical to the infection process. It is worthwhile to note that commonly used screening procedures evaluating the influence of bacterial strains on mycelial growth or spore germination in culture may have failed to identify this effective bacterial antagonist. Chlamydospores of *F. oxysporum* f.sp. *dianthi* require an exogenous source of nutrients to germinate in the soil but will germinate readily on water agar culture. The difference in the nutritional status of chlamydospores in culture versus in the soil illustrates the importance of focussing experiments designed to evaluate the role of siderophores in rhizosphere ecology at the root surface where environmental conditions most closely approximate those in nature.

FACTORS AFFECTING THE EFFICACY OF MICROBIAL ANTAGONISTS

Variation in Soil Microflora and Soil Chemical-Physical Properties

Although reports are widespread describing the successful use of beneficial bacteria for yield enhancement, (Burr et al., 1978; Kloepper et al., 1980c; Suslow, 1982; Suslow and Schroth, 1982b) or biological control in the field (Burr and Caeser, 1984; Schroth and Hancock, 1981; Schroth and Hancock, 1982; Weller and Cook, 1983) this approach has yet to be exploited on a commercial scale. One limitation to the agronomic application of fluorescent pseudomonads is the variation in efficacy observed among field experiments. For example, a collection of fluorescent pseudomonads characterized as plant growth promoting rhizobacteria by scientists from the University of California at Berkeley is described in Table 1. Statistically significant increases in yield were observed only

Table 1. Description of *Pseudomonas* strains with demonstrated field efficacy

Strain	Source	Significant Trials/total[a]	Reported Field Results
A1	potato periderm	4/5	Yield increases of potato and sugar beet (Kloepper et al., 1980; Suslow and Schroth, 1982b)
B4	sugar beet rhizosphere Shafter, CA	3/7	Yield increases of sugar beet (Suslow and Schroth, 1982b)
B10	potato periderm Tulelake, CA	2/4	Yield increases of potato (Kloepper et al., 1980)
BK-1	potato periderm Shafter, CA	2/8	Yield increases of potato (Burr et al., 1978; Kloepper et al., 1980)
E6	celery rhizosphere Moss Landing, CA	2/3	Yield increases of potato and sugar beet (Kloepper et al., 1980; Suslow and Schroth, 1982b)
RV3	sugar beet rhizosphere	4/7	Yield increases of potato and sugar beet (Suslow and Schroth, 1982b)
SH5	sugar beet rhizosphere	5/9	Yield increases of potato and sugar beet (Suslow and Schroth, 1982b)
TL3	potato periderm	4/9	Yield increases of potato (Burr et al., 1978)

[a]From Suslow (1982). The number of successful trials (where significant increases in final yield were observed) over the number of trials in which each strain was tested from 1978 to 1982.

in a subset of the field tests carried out by the Berkeley group from 1978-1983 (Suslow, 1982). Since the mode-of-action of plant growth promotion is, in part, due to the suppression of deleterious microorganisms that are a part of the native rhizosphere microflora, (Suslow and Schroth, 1982a, Yuen and Schroth, 1986b) the variation in growth enhancement among fields may be explained by differences in the levels of deleterious microorganisms in those fields. Unfortunately, field test sites for the strains listed in Table 1 were, for the most part, not well characterized with respect to their levels of major and minor pathogens. The enormous diversity of soil-borne organisms which may impact crop yields makes such characterization impractical. In the case of sugar beet, several deleterious organisms have been identified and characterized (Suslow and Schroth, 1982a). "Minor" pathogens impacting yields have not been well characterized in many other crop plants.

Although a thorough characterization of minor pathogens has not typically accompanied field studies investigating the efficacy of beneficial organisms, the importance of cropping history on yield responses of potato observed with treatment with fluorescent pseudomonads was emphasized by Geels and Schippers (1983). These scientists observed differential yield responses of potatoes which were planted in fields with different cropping patterns following seed piece treatment with fluorescent pseudomonads. While bacterial seed piece treatment resulted in yield increases in soil from a field continuously cropped with potato (potato soil), no increase was observed in soil from a field cropped with wheat (wheat soil). Yields of non-treated controls in the potato soil were significantly less than those in the wheat soil. Unidentified deleterious microorganisms were responsible for the yield depression observed in soils continuously cropped with potatoes (Hoekstra, 1981: Maenhout and Hoekstra, 1980). Geels and Schippers' (1983) results suggest that the yeild enhancement effect of fluorescent pseudomonads depends on the yield depression caused by the continuous cropping of potato in a given field. The variation in efficacy of seed piece treatments observed among field soils was attributed to differences in cropping history.

In addition to variations in cropping history, soil chemical and physical factors which vary among field sites may influence the reproducibility of results among field experiments. Presumably, any factor influencing either the growth of or production of a critical metabolite such as a siderophore by a bacterial antagonist would greatly influence the efficacy of that antagonist in growth promotion or suppression of a specific disease. Accordingly, an examination was made of the potential effects of soil temperature and of sugars, amino acids, and organic acids commonly found in root exudates on the growth and fluorescent pigment/siderophore production of the fluorescent pseudomonads described in Table 1.

Influence of Temperature on Fluorescent Pigment Siderophore Production of *Pseudomonas* Strains

Temperature influences the siderophore production of *E. coli* (Kochan, 1977), *Salmonella typhimurium* (Garibaldi, 1972), *Pasteurella multocida* (Kluger and Rothenburg, 1979) and a *Pseudomonas* sp. (Garibaldi, 1971). The iron requirement for maximal cell yields of these bacteria increases as the temperature of incubation is increased. In animal systems, the decreased production of siderophores by bacterial pathogens and the decreased plasma iron concentrations at elevated body temperatures (fever) may be a coordinated host defense mechanism. Growth of bacterial pathogens may be inhibited at these elevated temperatures due to the iron-depletion imposed by these coordinated events. (Garibaldi, 1972; Kluger and Rothenburg, 1979).

Garibaldi (1971) first reported the influence of temperature on the iron metabolism of the fluorescent pseudomonads. The biosynthesis of hydroxamate iron transport compounds by one *Pseudomonas* strain occurred only at temperatures of 28 C or lower. At temperatures exceeding 28 C, this bacterial strain did not grow in an iron-limited basal medium. This growth inhibition was reversed upon addition to the medium of iron or hydroxamate iron transport compounds.

Although soil temperatures of agricultural fields are generally lower than 28 C, temperatures of plant surfaces or upper soil layers commonly exceed 28 C in some agricultural regions. For example, soil temperatures exceeding 28 C were observed at soil depths of 15 or 30 cm in several fields in the California central valley from 1980-1982 (Stapleton and DeVay, 1984). These temperatures were not conducive for iron-independent growth of the *Pseudomonas* strain studied by Garibaldi.

The growth of plant growth-promoting *Pseudomonas* strains under iron-limiting conditions was also influenced by temperature (Table 2). Strains B4, B10 and E6 did not grow at temperatures less than 12C on King's B medium (KBM) supplemented with the iron-chelating compound, ethylenediamine-di (o-hydroxyphenylacetic acid)(EDDHA, 1000 µg/ml). However, these bacteria grew on KBM alone or on KBM-EDDHA supplemented with 10^{-4}M $FeCl_3$. The plant

Table 2. Influence of temperature on iron-limited growth of bacterial strains

	Growth on KBM-EDDHA[a] Strains		
Temperature (C)	B4	B10	E6
5	-	-	-
8	-	-	-
12	+	+	+
21	+	+	+
24	+	+	+
27	+	+	+
30	+	+	+
33	-	+	+

[a]The ability of each strain to grow on Kings B medium (KBM) supplemented with EDDHA (KBM-EDDHA) was tested under various growth temperatures. The EDDHA solution was prepared as described by Ong et al., (1979) after removing iron as described by Rogers (1973). KBM-EDDHA was allowed to stand for at least 24 h at 4 C prior to use to allow slow chelation of iron. Bacterial suspensions were adjusted to a uniform density of 0.1 O.D. at 640 nm and streaked with a calibrated 1 µl loop on KBM-EDDHA medium. Only four suspensions were streaked on each plate to negate the influence of cross-feeding. Iron-limited growth inhibition was reversed by addition of a freshly prepared, filter-sterilized solution of 10^{-2} M $FeCl_3$ to the KBM-EDDHA medium to a final concentration of 10^{-4} M $FeCl_3$. All strains grew on KBM and on KBM-EDDHA supplemented with 10^{-4} M $FeCl_3$ at all temperatures tested. Abbreviations: + = visible growth, - = no visible growth.

growth-promoting pseudomonads varied with respect to their iron-limited growth at 33 C (Table 3). While five strains were fluorescent on KBM and grew on KBM-EDDHA at both 27C and 33C, strains B4 and RV3 did so only at 27C. The growth inhibition of these strains on KBM-EDDHA at 33C was reversed by the addition of purified pseudobactin, indicating that siderophore biosynthesis but not utilization was inhibited at elevated temperatures. These results support and extend those of Garibaldi (1971) which showed that the iron-limited growth of a *Pseudomonas* strain at elevated temperatures was reversed by the addition of hydroxamate iron-transport compounds which were produced by that strain at lower temperatures. In this case, we have observed that *Pseudomonas* strains vary in their temperature ranges conducive to siderophore production and iron-dependent growth.

Influence of Root Exudates on the Fluorescent Pigment/Siderophore Production of Pseudomonas Strains

The term "rhizosphere," originally coined by Hiltner (1904) refers to the zone of soil in which microbial activity is affected by the roots of any plant species. The increased microbial activity in the rhizosphere

Table 3. Influence of temperature on growth, fluorescence and pseudobactin utilization of bacterial strains.[a]

Strain	Fluorescence on KBM		Growth on KBM-EDDHA		Pseudobactin utilization [b]	
	27 C	33 C	27 C	33 C	27 C	33 C
A1	+	+	+	+	+	+
B4	+	-	+	-	+	+
B10	+	+	+	+	+	+
Bk-1	+	+	+	+	+	+
E6	+	+	+	+	+	+
RV3	+	-	+	-	+	+
SH5	+	+	+	+	+	+

[a]Experimental conditions were as described in Table 2. All strains grew at 27 or 33 C on KBM-EDDHA supplemented with 10^{-4} M $FeCl_3$.

[b]Pseudobactin utilization was tested by incorporating 0.1 ml of a bacterial suspension (0.4 O.D. at 640 nm) into 20 ml KBM-EDDHA, 10 µl of purified pseudobactin (2 mM) was spotted onto the center of a Petri dish containing the solidified agar. A ring of growth around the pseudobactin spot indicated the ability of a strain to utilize ferric-pseudobactin as a source of iron. Purified pseudobactin was a gift from J. Leong.

Abbreviations: + - fluorescence, visible growth or ability to utilize pseudobactin as an iron source, for columns 1, 2 and 3 respectively. - = lack of these visible reactions.

in contrast to the bulk soil is generally attributed to the release by plant roots of organic materials which serve as growth substrates for microorganisms. A wide range of organic materials are released by roots, including carbohydrates, amino acids, organic acids, enzymes, and vitamins (Rovira, 1969). The ability of a bacterial strain to utilize such substrates determines, in part, its population size in the rhizosphere (Bowen, 1979). Particular genera, such as *Pseudomonas*, predominate in the rhizosphere of some soils (Rouatt and Katznelson, 1961), presumably because of their abilities to utilize organic substrates exuded by plant roots.

Organic substrates influence both the growth rate and the metabolite production of microorganisms. Fluorescent pigment production of the pseudomonads is influenced by minerals (King et al., 1948), amino acids (De Ley, 1964), peptones (King et al., 1954) and carbon sources (Gouda and Chodat, 1963; Vidaver, 1967). However, no systematic study has been previously reported which determined the influence of organic compounds commonly associated with root exudates on the fluorescence of rhizosphere-associated *Pseudomonas* strains.

Sugars, amino acids and organic acids commonly found in root exudates generally supported growth and fluorescence of seven rhizosphere pseudomonads (Table 4).With the exception of citrate, the organic substrates which supported growth also supported fluorescence of most strains. However, fluorescence was masked on some carbon sources when a poorly buffered minimal medium was used. For example, no fluorescence was observed on glucose, fructose or mannose as sole carbon sources when the HEPES buffer was omitted from the minimal medium described in Table 4. The pH's of the agar surfaces of these unbuffered media were approximately 5.5 from two to seven days after inoculation. The absorption maximum of pseudobactin, the fluorescent pigment of *P. fluorescens* strain B10, shifts from 400 nm to slightly longer wavelengths and the extinction coefficient increases with an increase in pH from 3 to 9 (Teintze et al., 1981). In our experience, the pigmentation of culture filtrates of *Pseudomonas* strains was more intense as pH increased from 5 to 9. An increased absorbance at 400 nm was also observed as exemplified by *P. fluorescens* strain BK1. Clearly, the lack of fluorescence observed on unbuffered media containing glucose, fructose or mannose, is not due to the inability of the pseudomonads to produce the fluorescent pigment/siderophore on these media, but may be due to a visible change in fluorescence with changes in pH.

Generally, the seven *Pseudomonas* strains were remarkably versatile with respect to their abilities to grow and fluoresce on a variety of organic substrates. Although some strain differences were observed with respect to growth and fluorescence on a given substrate, these differences were minor in the context of the wide variety of organic substrates utilized. Citric acid supported growth but not fluorescence of several strains. However, it was demonstrated by Meyer and Abdallah (1978) that fluorescent pigment/siderophore production on citrate media was observed when iron-limiting conditions were imposed by pre-treatment of the growth medium with 8-hydroxyquinoline. Since citric acid itself may serve as an iron-transport compound (Neilands, 1977), the internal iron concentration of cells grown on a citrate medium may be greater than that of cells grown without the presence of citrate (Meyer and Abdallah, 1978).

DISCUSSION

Siderophore production by specific bacterial antagonists is increasingly implicated as a key mechanism in plant growth promotion and the control of certain diseases. The complementary studies by Baker, Kloepper, Schippers, Scher, Weller, as reviewed in this book argue strongly for the premise that the availability of iron in the root environment and the abilities of various microorganisms to compete for it are factors governing microbial ecology in the rhizosphere.

Although pseudomonads produce copious quantities of fluorescent pigment/siderophores *in vitro,* there has been only indirect evidence for their production in the rhizosphere. We conclude from our nutritional studies that sugars, amino acids and organic acids present in root exudates can support both the growth and fluorescent pigment/siderophore production of several beneficial *Pseudomonas* strains. The wide variety of substrates supporting fluorescent pigment/siderophore production suggests that shifts in exudate composition are not likely to prevent siderophore production. Even organic substrates such as citric acid, on which pigment production is not normally observed, will support pigment production under iron-limiting conditions. Therefore, the level of available iron in the rhizosphere and the quantity rather than the composition of organic substrates will determine the siderophore production of rhizosphere pseudomonads.

Soil temperatures of most agricultural regions are permissive for fluorescent pigment production of the seven rhizosphere pseudomonads described here. However, temperatures exceeding those permissive for fluorescent pigment/siderophore production may occur in upper soil layers of some regions. Since the lack of pigment production observed with some strains at temperatures greater than 30C was not reversed under iron-limiting conditions, non permissive temperatures may limit siderophore

Table 4. Growth and fluorescence of beneficial pseudomonads on organic substrates commonly found in root exudates.

Organic substrate and sources of identification in root exudates[a]	Strain						
	B4	RV3	SH5	A1	E6	BK-1	B10
Amino Acids:							
α-alanine (2,4,5,6,21,37,44, 47,48,50,54,56,63,64,65,66)	+	+	+	+	+	+	+
-aminobutyric acid (2,6,48,50,63,64,65,66)	+	+	+	+	+	+	+
L- or D-arginine (2,5,44,54,64,65)	+	+	+	+	+	+	+
L-glutamine (2,4,6,21,47,48, 50,54,56,63,65,66,67)	+	+	+	+	+	+	+
L-histidine (4,5,44,54)	+	+	+	+	+	ng	+
DL-homoserine (2,5,47,56)	ng	ng	ng	ng	ng	ng	ng
DL-leucine (2,4,5,6,44,47, 48,50,54,63,64,65)	+	+	+	+	+	+	+
L-lysine (2,4,5,44,47,50,54 56,64,65,66)	+	+	+	+	+	+	+
D-methionine (2,4,44,48,50, 53,63,64)	ng	ng	ng	ng	ng	ng	ng
L-phenylalanine (4,5,44,48, 50,54,56,63,64,65,66)	ng	ng	ng	+	+	ng	ng
DL-proline (4,5,21,37,44, 50,54,63,64,65,66	+	+	+	+	+	+	+
DL-serine (2,4,5,6,21,37, 44,47,48,50,54,63,64,65,66)	ng	+	ng	ng	ng	ng	ng
L-tyrosine (4,5,21,44,50,54, 56,63,66)	+	+	+	+	+	+	+
L-valine (2,4,5,6,44,48,50, 54,64,65)	+	+	+	+	+	+	+
Organic acids:							
L-aspartic acid (2,4,5,6,37,44, 47,48,50,54,63,64,65,66)	+	+	+	+	+	+	+
citric acid (4,50,66)	+	-	NT	+	-	-	-
D- or L-glutamic acid (2,5,6 21,37,44,48,50,54,63,64, 65,66)	+	+	+	+	+	+	+
glycolic acid (4,50,63,65,66)	ng	ng	ng	ng	ng	ng	ng
p-hydroxybenzoic acid (31)	+	+	+	+	+	+	+
L(+) lactic acid (4,65,66)	+	+	+	+	+	+	+
D-malic acid (4,37,50,63,65,66)	ng	+	ng	+	+	ng	+
DL-pipecolic acid (10%) (66)	+	+	+	+	+	ng	+

Organic substrate and sources of identification in root exudates[a]	Strain						
	B4	RV3	SH5	A1	E6	BK-1	B10
propionic acid (10%) (50)	+	+	+	+	+	ng	+
Succinic acid (4,50,63,65,66)	+	+	+	+	+	+	+
syringic acid (31)	ng	ng	ng	ng	ng	ng	ng
n-valeric acid (50)	+	+	+	+	+	ng	+
Sugars:							
L(+)arabinose (15,50,55,63, 64,65,66)	ng	+	NT	ng	+	+	+
D(-) fructose (4,15,50,54,55, 56,63,64,65,66)	+	+	-	+	+	-	-
galactose (50,55,63,64,65,66)	ng	+	NT	ng	+	+	+
D(+) glucose (4,15,50,54,55, 63,64,65,66)	+	+	+	+	+	+	+
D(+) mannose (15,55)	+	+	NT	ng	+	+	+
D(-) ribose (15,50,55,63,64, 65,66)	+	+	NT	+	+	+	+

Organic substrates were added to minimal medium after autoclaving to a final concentration of 1 g per liter, except where noted above. Minimal medium consisted of per liter $(NH_4)_2SO_4$, 1.32 g; $MgSo_4 \cdot 7H_2O$, 0.25 g: HEPES buffer (pH 7.5), 0.05 M; phosphate buffer (pH 7.5),).005M. Plates were spotted with 10 µl of a turbid bacterial suspension, incubated at 25 C for 7 days, and observed daily for growth and fluorescence. The pH of the agar surface remained at 7.5 + 0.2 over the duration of the experiment.

Abbreviations: ng = no visible growth, NT = not tested, + = fluorescence observed from bacterial growth, - = no fluorescence observed although bacterial growth was present. Where D or L isomers are listed, both isomers were tested with similar results. With few exceptions, isomers of substrates identified in root exudates were not specified in cited literature.

production independent of the level of available iron in the rhizosphere.

The influence of temperature on siderophores could have strong implications on the ability of a particular *Pseudomonas* strain to effect biological control or yield enhancement under certain field conditions. Soil temperature influences the rhizosphere population size of inoculated *Pseudomonas* strains (Loper et al., 1985). The population sizes of strains B4 and B10 are more stable at 12 C than at 24 C although growth rates in the potato rhizosphere are greater at 24 than at 12 C. With further study, soil temperature effects may explain differences in field results observed in different agricultural areas. For example, strain B4, which was deficient in fluorescing pigment production at temperatures greater than 30 C, has only been effective in early spring plantings in California and in trials in Idaho where temperatures at the beginning of season were cooler than in most other California trials. In contrast, strain SH5 has been effective in several California trials but not in Idaho (Suslow, 1982). Certainly, differences in efficacy patterns such as those described for strains B4 and SH5 must be investigated before microbial inoculants can effectively be used on a commercial basis. As more is learned about the involvement of chemical and environmental factors which influence rhizosphere colonization and metabolite production by beneficial pseudomonads, we may be able to understand the differences in efficacy observed in natural soils. Furthermore, the knowledge that *Pseudomonas* strains differ in such factors as temperature ranges optimal for fluorescent pigment/siderophore production may give workers a predictive tool to use in determining which beneficial strain should be most effective under a given set of field conditions.

The role of siderophores in microbial interactions in the rhizosphere cannot be considered in isolation of other factors that influence the colonization and other activities of bacterial antagonists. Rhizosphere population sizes of bacterial antagonists determine the qualities of metabolites, such as siderophores, produced at the root surface. In pursuing a comprehensive approach towards an understanding of rhizosphere ecology, microbial characteristics of potential importance to rhizosphere colonization have been examined. For example, antibiotic production by *Erwinia carotovora* subsp. *betavasculorum* influenced potato tuber wound colonization by sensitive *E. carotovora* strains (Axelrood and Schroth, in press). Osmotolerance of *Pseudomonas* strains in culture was correlated with their abilities to establish stable populations in the rhizosphere of potato. (Loper et al., 1985). The variation in rhizosphere bacterial populations among root systems of several crop plants was both large and lognormally distributed (Loper et al., 1984a). Further, individual roots within a root system vary significantly in the rhizosphere population sizes of inoculated antagonistic bacteria (Bahme and Schroth, 1984). Clearly, there are exciting opportunities for the enhancement of bacterial antagonists. Our knowledge of factors limiting siderophore production and rhizosphere colonization by bacterial antagonists in the soil have identified some potentially important phenotypic characters which may be genetically manipulated. The ultimate success of biological control may rest in the success of such efforts towards the enhancement of these bacterial antagonists.

REFERENCES

Axelrood P.E., and Schroth, M.N. 198 . Antibiotic production by *Erwinia carotovora* subsp. *betavasculorum:* partial characterization and evidence for activity in potato tubers. In: Proc. 6th Conf. of Plant Pathogens, College Park, Maryland (In press).

Ayers, W.A., and Thornton, R.H., 1968. Exudation of amino acids by intact and damaged roots of wheat and peas. Plant Soil, 28:193.

Bahme, J.B., and Schroth, M.N., 1984. Colonization dynamics of a rhizobacterium on potato. Phytopathology, 74:806 (abst.).

Balasubramanian, A., and Rangaswami, G., 1969. Studies on the influence of foliar nutrient sprays on the root exudation pattern in four crop plants. Plant Soil. 30:210.

Boulter, D., Jeremy, J.J., and Wilding, M., 1966. Amino acids liberated into the culture medium by pea seedling roots. Plant Soil, 24:121.

Bowen, G.D., 1969. Nutrient status effects on loss of amides and amino acids from pine roots. Plant Soil, 30:139.

Bowen, G.D., 1979. Integrated and experimental approaches to the study of growth of organisms around roots. In: "Soil-borne Plant Pathogens," (B. Schippers and W. Gams, Eds.), Academic Press, Inc., London.

Burr, T.J., Schroth, M.N., and Suslow, T., 1978. Increased potato yields by treatment of seedpieces with specific strains of *Pseudomonas fluorescens* and *P. putida*, Phytopathology, 68:1377.

Burr, T.J., and Caesar, A., 1984. Beneficial plant bacteria, CRC Crit. Rev. Plant Sci., 2:1.

De Lay, J., 1964, *Pseudomonas* and related genera. Ann. Rev. Microbiol., 18:17.

Garibaldi, J.A., 1971. Influence of temperature on the iron metabolism of a fluorescent pseudomonad. J. Bacteriol. 105:1036.

Garibaldi, J.A., 1972. Influence of temperature on the biosynthesis of iron transport compounds by *Salmonella typhimurium*. J. Bacteriol. 110:262.

Geels, F.P., and Schippers, B., 1983. Reduction of yield depressions in high frequency potato cropping soil after seed tuber treatments with antagonistic fluorescent *Pseudomonas* spp. Phytopathol. Z. 108:207.

Gouda, P.S., and Chodat, F., 1963. Glyoxylate et succinate, facteurs déterminant respectivement l'hypochromie et l'hyperchromie des cultures de *Pseudomonas fluorescens*. Pathol. Microbiol., 26:655.

Hamlen, R.A., Lukezic, F.L., and Bloom, J.R., 1972. Influence of age and stage of development on the neutral carbohydrate components in root exudates from alfalfa plants grown in a gnotobiotic environment. Can. J. Plant Sci., 52:633.

Hemming, B.C., Orser, C., Jacobs, D.L., Sands, D.C., and Strobel, G.A., 1982. The effects of iron on microbial antagonism by fluorescent pseudomonads. J. Plant Nutr., 5:683.

Hiltner, L., 1904. Über neuere Erfahnrungen und Probleme auf dem Gebiet der Bodenbakteriologie und unter besonderer Berücksichtigung der Gründüngung und Brache, Arb. Deut. Landw. Ges., 98:59.

Hoekstra, O. , 1981. 15 jaar "De Schreef." Resultaten van 15 jaar vruchtwisselingsonderzoek op het bouwplannenproefveld "De Schreef," Publickatie PAGV 11:1.

Howell, C.R., and Stipanovic, R.D., 1979. Control of *Rhizoctonia solani* on cotton seedlings with *Pseudomonas fluorescens* and with an antibiotic produced by the bacterium. Phytopathology, 69:480.

Howell, C.R., and Stipanovic, R.D., 1980. Suppression of *Pythium ultimum*-induced damping-off of cotton seedlings by *Pseudomonas fluorescens* and its antibiotic, pyoluteorin. Phytopathology, 70:712.

Husain, S.S., and McKeen, W.E., 1963. Interactions between strawberry roots and *Rhizoctonia fragariae*, Phytopathology, 53:541.

King, J.V., Campbell. J.J.R., and Eagles, B.A., 1948. The mineral requirements for fluorescin production. Can. J. Res., 26C:514.

King, E.O., Ward, M.K., and Raney, D.E., 1954. Two simple media for the demonstration of pyocyanin and fluorescin. J. Lab. Clin. Med. 44:301.

Kleopper, J.W., Leong, J., Teintze, M., and Schroth, M.N., 1980a. Enhanced plant growth by siderophores produced by plant growth-promoting rhizobacteria. Nature, 286:885.

Kloepper, J.W., Leong, J., Teintze, M., and Schroth, M.N., 1980b. *Pseudomonas* siderophores: a mechanism explaining disease-suppressive soils. Curr. Microbiol., 4:317.

Kloepper, J.W., Schroth, M.N., and Miller, T.D., 1980. Effects of rhizosphere colonization by plant growth-promoting rhizobacteria on potato plant development and yield. Phytopathology, 70:1078.

Kloepper, J.W., and Schroth, M.N., 1981a. Relationship of an *in vitro* antibiosis of plant growth-promoting rhizobacteria to plant growth and the displacement of root microflora. Phytopathology, 71:1020.

Kloepper, J.W., and Schroth, M.N., 1981b. Plant growth-promoting rhizobacteria and plant growth under gnotobiotic conditions. Phytopathology, 71:642.

Kluger, M.J., and Rothenburg, B.A., 1979. Fever and reduced iron: their interaction as a host defense response to bacterial infection. Science, 203:374.

Kochan, I., 1977. Role of siderophores in nutrition/immunity and bacterial parasitism. In: "Microorganisms and Minerals," (E.D. Weinberg, Ed.) Marcel Dekker, New York.

Kovacs, Jr., M.F., 1971. Identification of aliphatic and aromatic acids in root and seed exudates of peas, cotton and barley. Plant Soil, 34:441.

Lenhoff, H.M., 1963. An inverse relationship of the effects of oxygen and iron on the production of fluorescin and cytochrome c by *Pseudomonas fluorescens*. Nature, 199:601.

Loper, J.E., Suslow, T.V., and Schroth, M.N., 1984a.Lognormal distribution of bacterial populations in the rhizosphere. Phytopathology, 74:1454.

Loper, J.E., Orser, C.S., Panapoulos, N.J., and Schroth, M.N., 1984b. Genetic analysis of fluorescent pigment production in *Pseudomonas syringae* pv. *syringae*, J. Gen. Microbiol., 130:1507.

Loper, J.E., Haack, C., and Schroth, M.N., 1985. Population dynamics of soil pseudomonads in the rhizosphere of potato (*Solanum tuberosum* L.), Appl. Evinron. Microbiol., 49:416.

Maenhout, C.A.A., and Hoekstra, O., 1980. Bodemvruchtbaarheid en bodemgezondheid in relatie tot vruchtwisseling en bouwplan. Bedrijfsontwikkeling, 11:587.

Magyorosy, A.C., and Hancock, J.G., 1974. Association of virus-induced changes in laimosphere microflora and hypocotyl exudation with protection to Fusarium stem rot. Phytopathology, 64:994.

Meyer, J.M., and Abdallah, M.A., 1978. The fluorescent pigment of *Pseudomonas fluorescens:* biosynthesis, purification and physicochemical properties. J. Gen. Microbiol. 107:319.

Meyer, J.M., and Hornsperger, J.M., 1978. Role of pyoverdine$_{pf}$, the iron-binding fluorescent pigment of *Pseudomonas fluorescens,* in iton transport. J. Gen. Microbiol. 107:329.

Misaghi, I.J., Stowell, L.J., Grogan, R.G., and Spearman, L.C., 1982. Fungistatic activity of water-soluble fluorescent pigments of fluorescent pseudomonads. Phytopathology, 72:33.

Moores, J.C., Magazin, M., Ditta, G.S., and Leong, J., 1984. Cloning of genes involved in the biosynthesis of pseudobactin, a high-affinity iron transport agent of a plant growth-promoting *Pseudomonas* strain. J. Bacteriol., 157:53.

Neilands, J.B., 1977. Siderophores: diverse roles in microbial and human physiology, In: "Iron Metabolism". Elsevier Press.

Ong, S.A., Peterson, T., and Neilands, J.B., 1979. Agrobactin, a siderophore from *Agrobacterium tumefaciens,* J. Biol. Chem., 254:1860.

Richter, M., Wilms, W., and Scheffer, F., 1968. Determination of root exudates in a sterile continuous flow culture II. Short-term and long-term variations of exudation intensity. Plant Physiol., 43:1747.

Rogers, H.J., 1973. Iron-binding catechols and virulence in *Escherichia coli,* Infec. Immun., 7:445.

Rouatt, J.W., and Katznelson, H., 1961. A study of the bacteria on the root surface and in the rhizosphere soil of crop plants. J. Appl. Bacteriol., 24:164.

Rovira, A.D., 1956. Plant root excretions in relation to the rhizosphere effect. I. The nature of root exudate from oats and peas. Plant Soil, 7:178.

Rovira, A.D., 1959. Root excretions in relation to the rhizosphere effect. IV. Influence of plant species, age of plant, light, temperature and calcium nutrition on exudation, Plant Soil, 11:53.

Rovira, 1969. Plant root exudates, Bot. Rev., 35:35.

Rovira, A.D., and McDougall, B.M., 1967. Microbiological and biochemical aspects of the Rhizosphere. In: "Soil Biochemistry,"(A.D. McLaren and G.F. Petersen, Eds.) Marcell Dekker, New York.

Scher, F.M., and Baker, R., 1982. Effect of *Pseudo onas putida* and a synthetic iron chelator on induction of soil suppressiveness to Fusarium wilt pathogens, Phytopathology, 72:1567.

Schroth, M.N., and Hancock, J.G., 1981. Selected topics in biological control. Ann. Rev. Microbiol., 35:453.

Schroth, M.N., and Hancock, J.G., 1982. Disease-suppressive soil and root-colonizing bacteria. Science, 216:1376.

Schroth, M.N., Toussoun, T.A., and Snyder, W.C., 1963. Effect of certain constituents of bean exudate on termination of chlamydospores of *Fusarium solani* f. *phaseoli* in soil. Phytopathology, 53:809.

Shay, F.J., and Hale, M.G., 1973. Effect of low levels of calcium on exudation of sugars and sugar derivatives from intact peanut roots under axenic conditions. Plant Physiol., 51:1061.

Smith, W.H., 1972. Influence of artificial defoliation on exudates of sugar maple. Soil Biol. Biochem., 4:111.

Sneh, B., Dupler, M., Elad, Y., and Baker, R., 1984. Chlamydospore germination of *Fusarium oxysporum* f. sp. *cucumerinum* as affected by fluorescent and lytic bacteria from Fusarium-suppressive soil. Phytopathology, 74:1115.

Stapleton, J.J., and DeVay, J.E., 1984. Thermal components of soil solarization as related to changes in soil and root microflora and increased plant growth response. Phytopathology, 74:255.

Suslow, T.V., 1982. Role of root-colonizing bacteria in plant growth. In: "Phytopathogenic Prokaryotes," Vol. 1., (M.S. Mount and G.H. Lacy, Eds.) Academic Press, New York.

Suslow, T.V. and Schroth, M.N., 1982a. Role of deleterious rhizobacteria as minor pathogens in reducing crop growth, Phytopathology, 72:111.

Suslow, T.V. and Schroth, M.N., 1982b. Rhizobacteria of sugar beets: effects of seed application and root colonization on yield. Phytopathology, 72:199.

Teintze, M., Hossain, M.B., Barnes, C.L., Leong, J. and van der Helm,D., 1981. Structure of ferric pseudobactin, a siderophore from a plant growth-promoting *Pseudomonas* . Biochemistry, 20:6446.

Vančura, V., 1964. Root exudates of plants. I. Analysis of root exudates of barley and wheat in their initial phases of growth. Plant Soil, 21:231.

Vančura, V., 1967. Root exudates of plants. III. Effect of temperature and "cold shock" on the exudation of various compounds from seeds and seedlings of maize and cucumber. Plant Soil, 27:319.

Vančura, V., and Hanzlikova, A., 1972. Root exudates of plants. IV. Differences in chemical composition of seed and seedlings exudates. Plant Soil, 36:271.

Vančura, V., and Hovadik,A., 1965. Root exudates of plants. II. Composition of root exudates of some vegetables. Plant Soil, 22:21.

Vidaver, A.K., 1967. Synthetic and complex media for the rapid detection of fluorescence of phytopathogenic pseudomonads: effect of the carbon source. Appl. Microbiol., 15:1523.

Weller, D.M., and Cook. R.J., 1983. Suppression of take-all of wheat by seed treatments with fluorescent pseudomonads, Phytopathology, 73:463.

Yuen, G.Y., and Schroth, M.N., 1986a. Inhibition of *Fusarium oxysporum* f. sp. *dianthi* by iron competition with an *Alcaligenes* sp., Phytopathology (in press).

Yuen, G.Y., and Schroth, M.N. 1986b. Interactions of *Pseudomonas fluorescens* strain E6 with ornamental plants and its effect on the composition of root-colonizing microflora. Phytopathology (in press).

Yuen, G.Y., Schroth, M.N., and McCain, A.H., 1986. Reduction in Fusarium wilt of carnations with suppressive soils and antagonistic bacteria. Plant Disease, 69:(in press).

SUPPRESSION OF ROOT DISEASES OF WHEAT BY FLUORESCENT PSEUDOMONADS AND MECHANISMS OF ACTION

David M. Weller and R. James Cook

Research Plant Pathologists, USDA-ARS
Root Disease and Biological Control Research Unit
367 Johnson Hall, WSU, Pullman, Washington, 99164-6430

INTRODUCTION

Take-all and Pythium root rot, incited respectively by *Gaeumannomyces graminis* (Sacc.) von Arx and Olivier var.*tritici* Walker and several *Pythium* species are major constraints to wheat production in the pacific Northwest of the U.S.A. Take-all occurs mainly where wheat is grown under irrigation such as the Columbia Basin and Snake River plains and in the high rainfall region of western Washington and Oregon. The take-all fungus infects the roots, crowns, and lower stems of wheat; diseased plants are chlorotic and stunted with roots that are blackened. In the field, diseased plants may die prematurely in irregular patches due to the interruption of the transport of water to the tops by infection of the roots and crowns.

Pythium root rot occurs throughout all wheat growing areas of the Northwest and ten species or varieties of *Pythium* have been shown to occur on wheat (Chamswarng and Cook (1985). *Pythium* spp. attack mainly root tips, rootlets, and root hairs of wheat, diminishing the uptake of water and nutrients and probably also the supply of plant growth hormones for transport to the tops, resulting in generally smaller plants that appear nutrient deficient. The adverse effects of Pythium root rot are not always apparent unless a healthy check (e.g. wheat growing in soil fumigated with methyl bromide, chloropicrin or dichloropropene plus chloropicrin is available for comparison (Cook and Haglund, 1982). Wheat with take-all and/or Pythium root rot may yield only 50-90% of its potential.

The damage caused by both diseases, especially take-all, can be minimized by using crop rotation and conventional tillage practices. However, in the Pacific Northwest it has now become common to grow several consecutive crops of wheat before introducing a break crop or fallow. Further, the practice of direct drilling wheat into minimally tilled soil is being adopted to control erosion; unfortunately, the severity of both diseases is greater with reduced or no tillage (Cook et. al., 1980; Moore and Cook, 1984).

Fumigation of the soil will control both diseases but the cost is prohibitive for commercial wheat production. Triadimenol as a seed treatment (Bockus, 1983) or benomyl as an in-furrow treatment (Bateman, 1981) will provide some control of take-all, but under Pacific Northwest conditions triadimenol is often phytotoxic and the use of benomyl would not be economically feasible. Metalaxyl is registered for use as a seed treatment against Pythium root rot, but control is often variable, owing possibly to the fact that some of the *Pythium* species are relatively insensitive to metalaxyl (Cook and Zhang, 1985). Adequate resistance to either take-all or Pythium root rot does not exist.

The challenge is to find new methods of disease control. This paper reviews research on the application of fluorescent pseudomonads for suppression of take-all and Pythium root rot of wheat in the Pacific Northwest of the U.S.A. and the mechanisms by which suppression may occur.

BIOLOGICAL CONTROL OF TAKE-ALL WITH FLUORESCENT PSEUDOMONADS

Selection of Effective Strains

The decision to pursue the use of fluorescent pseudomonads instead of other bacteria or fungi for biological control of take-all in the Pacific Northwest was based primarily on the results of prior studies that implicated pseudomonads (Cook and Baker, 1983; Cook and Rovira, 1976; Kloepper et al., 1980; Smiley, 1978; Smiley, 1979; Weller and Cook, 1981) as agents responsible for take-all decline (Shipton, 1975). Smiley (1979) and Weller and Cook (1981) both reported that the proportion of fluorescent pseudomonads inhibitory to *G. graminis* var. *tritici in vitro* was higher on wheat roots grown in take-all suppressive soils than on roots grown in conducive soils. Further, a significantly greater proportion of the pseudomonads from suppressive soils controlled take-all *in vivo* (Weller et al., 1985) as compared to those from conducive soils.

Besides their association with take-all decline several other characteristics of fluorescent pseudomonads make them ideal for biological control of take-all: they naturally occur in the rhizosphere; they are nutritionally versatile and are adapted to utilizing root exudates for growth; they grow rapidly in the rhizosphere; they can be introduced onto roots; and they produce siderophores and antibiotics (Weller, 1985).

The first step in the selection procedure is to isolate candidate fluorescent pseudomonads from roots growing in a take-all suppressive soil in the presence of the take-all fungus since the chances of finding effective strains are better in these soils. The ability of the candidate strains to inhibit *G. graminis* var. *tritici in vitro* is a second important characteristic. Although antibiosis cannot be used as a sole determinant in the selection procedure, antibiotics or siderophores have been shown to be important mechanisms of suppression in several examples of biological control of root diseases by fluorescent pseudomonads (Colyer and Mount, 1984; Howell and Stipanovic, 1979; Howell and Stipanovic, 1980; Kloepper et al., 1980; Kloepper and Schroth, 1981; Scher and Baker, 1982; Schroth and Hancock, 1981; 1982; Sneh et al.,1984; Wong and Baker, 1984; Xu, 1984.

An *in vivo* tube assay was developed to further facilitate selection of the most effective strains (Weller et. al., 1985). The assay uses tapered plastic tubes (2.5 cm diam x 16.5 cm long) each with a hole in the bottom and supported in a hanging position in plastic racks, 200 tubes per rack. Each tube is filled with a 6.5 cm-thick column of sterile vermiculite followed by 5.0 g of test soil. *G. graminis* var. *tritici* is added to the soil as colonized oat grains that are pulverized in a Waring Blendor and then sieved into fractions of known particle sizes (Wilkinson et. al., 1985). Particles of 0.25-0.5 mm are added to the soil at 0.15% and 0.45% w/w. Two bacteria-treated seeds (10^8 colony forming units (CFU)/seed) are placed on the soil and then covered with vermiculite. The tubes are incubated at 15-18 C and after 3-4 weeks the seedlings are evaluated for disease severity. At the inoculum potential used, strains with a maximum level of suppression can be identified while those that are only weakly suppressive can be eliminated.

Application of Effective Strains

Strains of *Pseudomonas fluorescens* and *Pseudomonas putida* applied as seed or soil treatments have been reported by several workers to suppress take-all in greenhouse and field tests (Bednářová-Civínová, et. al., 1981; Cook and Rovira, 1976; Sivasithamparam and Parker, 1978; Smiley, 1978; Vraný et. al., 1981; Wong and Baker, 1984).In studies conducted in Washington State since 1979, seed treatments of fluorescent pseudomonads have been tested in experimental plots at locations that are representative of areas where take-all can be a problem: the high rainfall areas of western Washington, the irrigated Columbia basin of central Washington, and the moderate rainfall region of eastern Washington. To treat seeds, the bacteria are grown on agar plates of King medium B (KMB) for 2-3 days, scraped into a 1% solution of methylcellulose and mixed with seed. The seed when dried yields about 10^8 CFU/seed (Weller and Cook, 1983). In field plots where *G. graminis* var. *tritici* was added to the seed furrow, spring or winter wheat treated with *P. fluorescens* strain 2-79 (NRRL B-15132) or 2-79 + *P. fluorescens* strain 13-79 (NRRL B-15134) had significantly less take-all and 5-27% greater yield than non-treated wheat in 6 of 8 tests; treatments were effective at all locations. The combination of strains 2-79 and 13-79 has performed better than each strain used individually in about half of the tests. Strain 2-79 also suppressed take-all in experimental plots in South Australia (D.M. Weller and A.D. Rovira, unpublished) and Holland (B. Schippers, personal communication). In four of five tests conducted in commercial fields where *G. graminis* var. *tritici* was naturally present and where commercial planting techniques were used, the bacterial treatment suppressed take-all and increased yield from 5-21% (R.J. Cook, E.N. Bassett and D.M. Weller, unpublished).

The seed treatments are more effective when tested against *G. graminis* var. *tritici* that is reintroduced into a fumigated (methyl bromide) soil than in a natural soil. For example, in a 1980 spring wheat plot, 2-79+13-79 increased yield 147% above the inoculated but untreated check (Weller and Cook, 1983). This is not unexpected since, although take-all will be more severe in fumigated soil, the pseudomonads proliferate in treated soil (Ridge, 1976) and they have less competition from indigenous bacteria during root colonization. The use of soil fumigation is a useful research tool since it can be used to demonstrate that the enhanced growth of *Pseudomonas*-treated wheat is due to the control of take-all rather than the control of other root pathogens or to direct stimulation of the plant.

Colonization of Roots

One characteristic important to suppression of take-all and other root diseases by fluorescent pseudomonads is the ability of the bacteria to establish and maintain a significant population on and around the root (Loper et. al., 1985; Suslow, 1982). Studies on root colonization by beneficial pseudomonads generally have been possible because of the use of antibiotic-resistant mutants (Kloepper et. al., 1980; Loper et. al., 1985; Loper et. al., 1984; Suslow and Schroth, 1982; Weller, 1983; Weller, 1984); drug markers permit recovery of an introduced strain against a high background of indigenous rhizosphere bacteria. *P. fluorescens* strain 2-79RN$_{10}$, resistant to rifampin and nalidixic acid, has been used as a model to study colonization, multiplication, and survival of introduced suppressive bacteria on roots of both winter and spring wheat. When applied to seed, 2-79RN$_{10}$ is subsequently detected along the entire length of seminal roots.

A population gradient develops along the root with the population highest near the seed and declining linearly towards the root tip. By studying population changes over time on individual sections of the root, introduced bacteria were shown to multiply and increase up to 100-fold on the root with apparent doubling times ranging from 15-67 hours (Weller, 1984). Evidence provided by Howie (1985) indicates that the bacteria are probably carried passively from the seed with the elongating root, and that bacterial motility has no primary role in initial root colonization. Howie (1985) also demonstrated that, depending on the soil, root colonization by these bacteria is maximal between -0.3 and -0.7 bars matric potential and at a rhizosphere pH between 6.0 and 6.5.

In another study, the population of $2\text{-}79RN_{10}$ was monitored on the roots of winter wheat throughout one growing season (Weller, 1983). Strain $2\text{-}79RN_{10}$ was detected at over 10^6 CFU/0.1 g of root (2.5-cm length of root) for one month after planting in October and at that time it comprised over 93% and 12%, respectively, of the total populations of fluorescent pseudomonads and aerobic bacteria. The introduced bacteria survived throughout the winter but declined to 2.8×10^3 CFU/0.1 g of root. With the onset of spring, the population of $2\text{-}79RN_{10}$ on the roots of plants with take-all increased nearly tenfold and subsequently remained fairly steady until harvest. In contrast, the same strain on roots of healthy plants (not exposed to *G. graminis* var. *tritici*) increased only slightly and then declined. Bacteria proliferate on roots infected by *G. graminis* var. *tritici* (Brown, 1981; Rovira and Wildermuth, 1981; Vojinović, 1972a; Vojinović, 1972b; Vojinović, 1973; Weller, 1983) possibly because of a greater availability of nutrients from take-all lesions. Of significance is that the ratio of the populations of strain $2\text{-}79RN_{10}$ on diseased and healthy roots is consistently greater than that of the indigenous fluorescent pseudomonads or total bacteria, suggesting that growth of the introduced bacteria is somehow stimulated selectively by the presence of the take-all fungus and/or lesions caused by the fungus.

Evidence for the Role of Antibiotics and Siderophores in Take-all Suppression

A shano silt loam collected from a field where take-all had declined (suppressive soil) and Puget and Ritzville silt loams from fields where crops other than wheat were grown (conducive soils), were diluted 1:10 with a fumigated soil (to minimize differences in soil physical/chemical factors among the soils) (Cook and Rovira, 1976; Shipton et. al., 1973), infested with 1.0% oat grain inoculum of *G. graminis* var. *tritici* and then sown to wheat. After two croppings, wheat in the suppressive soil mixture was protected against take-all, whereas, wheat in the mixtures with conducive soils was severely diseased. The roots of the wheat in the suppressive soil had about 10-fold more fluorescent pseudomonads than roots in the conducive soils and a significantly greater proportion of the pseudomonads from the suppressive soil were inhibitory to *G. graminis* var. *tritica in vitro.* Nine strains (highly inhibitory on both KMB and potato dextrose agar, PDA) from the suppressive soil and 18 strains (little or no inhibition on either KMB or PDA) from the conducive soils were compared in tube assay (Weller et. al., 1985) for ability to suppress take-all. The soil used in the assay was fumigated Shano silt loam (pH 6.3) amended with oat grain inoculum. The strains from the suppressive soil, as a group, were significantly ($p=0.05$) more suppressive of take-all than those from the conducive soils. Thus, with these strains at least, *in vitro* inhibition of *G. graminis* var. *tritici* correlates with *in vivo* suppression of take-all.

P. fluorescens strains Rla-80 and 2-79 inhibit the growth of *G. graminis* var. *tritici* on PDA and KMB and suppress take all *in vivo* (Weller and Cook, 1983; Weller et.al., 1985). On KMB, fluorescent pigment production by Rla-80 and 2-79 and inhibition of the fungus by Rla-80 both are eliminated by the addition of $FeCl_3$; for 2-79, 10μM $FeCl_3$ significantly reduces its inhibition of the fungus on KMB, but higher concentrations of iron increase inhibition probably because of initiation of antibiotic production. Five N-methyl-N -nitro-N-nitrosoguanidine mutants from 2-79 and six from Rla-80 were selected that showed reduced or no inhibition of *G. graminis* var. *tritica* on PDA and/or KMB. The mutants colonized wheat roots to the same extent as the respective parents but all were significantly less suppressive of take-all than the parents (Weller et. al., 1985).

To further determine the relative role of siderophores and antibiotics in take-all suppression, the bacteria were tested in a fumigated Shano silt loam, pH 6.3, amended with either FeEDTA or EDDHA. *P. pudita* strain L30b-80 was included along with the two strains described above since it suppressed take-all but inhibited *G. graminis* var. *tritica in vitro* only by production of a siderophore. In other studies (Kloepper et.al., 1980a; Kloepper et. al., 1980b, Scher and Baker, 1982; Wong and Baker, 1984) elimination of disease suppression by the addition of iron and induction of suppression with iron chelators have been used as evidence of a role for siderophores in biological control by fluorescent pseudomonads. Take-all was highly suppressed when EDDHA was added to soil at 1000 μg/g of soil indicating that *G. graminis* var. *tritici* was sensitive to iron deprivation in the soil. The suppressiveness of L30b-80 was totally nullified, that of Rla-80 was unchanged, and that of 2-79 was reduced in half by the addition of FeEDTA to the soil. Ferrous sulfate also eliminated the suppressiveness of strain L30b-80 but had no effect on suppressiveness of the other two strains. The results of these studies indicate that Rla-80 suppresses take-all mainly by antibiotic production, strain L30b-80 by siderophore production, and strain 2-79 by both antibiotic and siderophore production.

S. Guruiddaiah, D.M. Weller and R.J. Cook (unpublished) isolated a greenish-yellow antibiotic from strain 2-79 grown in potato dextrose broth. The compound belongs to the phenazine group of antibiotics, but appears to have a unique structure. It is active against a wide spectrum of fungi including root pathogens of wheat such as *Pythium* spp., *Rhizoctonia solani,* and *Fusarium culmorum.* Of the fungi tested, the antibiotic is most active against *G. graminis* var. *tritici* with complete inhibition of growth at less than 1 μg/ml. The antibiotic also suppresses the severity of take-all of wheat as a seed treatment. Strain 2-79 produces at least one other antibiotic but the phenazine-like compound is thought to have a primary role in suppression of take-all.

G. graminis var. *tritici* initially attacks the seminal roots of wheat. Brown, thick-walled macrohyphae (runner hyphae) grew along the surface of the roots, and from these thin-walled microhyphae (infection hyphae) penetrate the roots. The pattern of root colonization by $2\text{-}79RN_{10}$ closely follows the growth of the take-all fungus on the roots and the ability of the bacteria to become established and multiply along the length of seminal roots is important to its success in biological control. The siderophore of strain $2\text{-}79RN_{10}$ may have a primary role in disease suppression by competing with the take-all fungus for iron early in the parasitic phase, while runner hyphae are growing on the root surface. The fungus would be especially vulnerable to iron deprivation during this portion of the parasitic phase since nutrient reserves available in the inoculum particle would be low and competition for nutrients on the rhizoplane would be intense. The net result would be a lower inoculum potential of the fungus and possibly fewer initial infections of the root. In contrast, the phenazine-like antibiotic, being a product of secondary metabolism, probably would be produced most abundantly in the nutrient-rich take-all lesions, after a phase of rapid bacterial growth in the lesion. The antibiotic would further retard growth of the fungus in the lesion and possibly inhibit hyphae growing systemically in the stele.

BIOLOGICAL CONTROL OF PYTHIUM ROOT ROT WITH FLUORESCENT PSEUDOMONADS

Candidate fluorescent pseudomonads were isolated from roots of wheat and screened for suppressiveness of Pythium root rot in an *in vivo* assay (Weller and Graham, 1984) similar to the assay developed for selection of take-all suppressive bacteria (Weller et. al., 1985). Of 84 strains tested as seed treatments, 27% improved the height of seedlings grown in natural soil infested with *Pythium* as compared to untreated checks. That *Pythium* was the main pathogen controlled was verified by the fact that either fumigation (methyl bromide) or heating (60 C/30 min) the soil or metalaxyl, applied as a seed or soil treatment duplicated the growth response produced by the pseudomonads. Further, the bacterial treatment eliminated the stunting and twisting of leaves in the seedling phase characteristic of Pythium root rot. Finally, adding either of two *Pythium* species (*P. ultimum* var. *sporangiiferum* and *P. irregulare*) back to the soil at 500 propagules/g restored the symptoms of limited seedling growth.

Based on the results of the *in vivo* assay, strains of *P. fluorescens* biovar 1 (Q72a-80 and Q30Z80), biovar 2 (R8z-80) and *P. putida* (R5y-80) were selected for testing in the field in soil with an estimated *Pythium* population of 500 propagules/g. All four strains improved the growth of winter wheat by increasing stand, plant height, number of tillers or yield as compared to an untreated check; strain Qa72 was the best, resulting in a 26% greater yield (Weller and Graham, 1984). None of the strains tested in the *in vivo* assay were pre-screened for ability to inhibit *Pythium in vitro*; however, all four strains that were effective in the field were later shown to be inhibitory to *Pythium aristosporum* on KMB and all but R5y-80 were inhibitory to *G. graminis* var. *tritici*, *Rhizoctonia solani* (pathogenic to wheat), and *Fusarium culmorum* on KMB and PDA. As with selection of take-all suppressive pseudomonads, pre-screening for antibiotic and siderophore production probably would be useful in selecting effective strains.

In a parallel study, Becker and Cook (1984) demonstrated that 29% of 350 fluorescent pseudomonad strains from wheat roots improved the growth of wheat seedlings. *Pseudomonas* strain B324 or metalaxyl applied to wheat seeds or EDDHA added to the soil suppressed Pythium root rot and improved the growth of seedlings as compared to the untreated check; the growth response produced *in vivo* by strain B324 was eliminated by the addition od FeEDTA to the soil. Further, five mutants of strain B324 that lost siderophore production also lost some *in vivo* suppressive activity as compared to the parent. Siderophore production appears to be one mechanism by which fluorescent pseudomonads suppress Pythium root rot.

CONCLUSION

The future for the use of fluorescent pseudomonads for control of take-all and Pythium root rot is promising. Given that current chemical methods to control these diseases are inadequate and the trend is toward intensive wheat production and reduced tillage, biological control is a viable alternative. Problems still exist with maintaining strain stability, methods of application, storage of inoculum, and with consistency of activity from season to season and field to field. Nevertheless, as more is learned about the bacterial determinants of suppressiveness, genetic expression of the determinants, soil physical/chemical factors that affect root colonization, and the influence of host genotype on activity, more consistent results can be expected.

Current strains may be only prototypes of those that will be available in the future through better selection or genetic engineering of strains. Further, the use of mixtures of strains with complementary mechanisms of suppression should enhance biological control as long as strains are not deleterious to each other. The ideal for the future is a strain or combination of strains that can suppress several wheat root diseases and operate in the several diverse agroecosystems in which wheat is grown in the Pacific Northwest.

REFERENCES

Bateman, G.L., 1981. Effects of soil application of benomyl against take-all (*Gaeumannomyces graminis*) and footrot disease of wheat. Z. Pflanzenkr. Pflanzenschutz, 88:249-255.

Becker, J.O., and Cook, R.J., 1984. *Pythium* control by siderophore-producing bacteria on roots of wheat. Phytopathology, 74:806 (abstract).

Bedňařová-Civínová, M., Petříková, V., Staněk, M. and Vančura, V., 1981. Vliv bakterizace semen na *Gaeumannomyces graminis*, růst a výnos pšenice. Sb. UNTIZ-Ochr. Rostl., 17:89-96.

Bockus, W.W., 1983. Effects of fall infection by *Gaeumannomyces graminis* var. *tritici* and triadimenol seed treatment on severity of take-all in winter wheat. Phytopathology, 73:540-543.

Brown, M.E., 1981. Microbiology of roots infected with the take-all fungus (*Gaeumannomyces graminis* var. *tritici*) in phased sequence of winter wheat. Soil Biol. Biochem., 13:285-291.

Chamswaring, C., and Cook, R.J., 1985. Identification and comparative pathogenicity of *Pythium* species from wheat roots and wheat-field soils in the Pacific Northwest. Phytopathology, 75. (In press).

Colyer, P.D., and Mount, M.S., 1984. Bacterization of potatoes with *Pseudomonas putida* and its influence on postharvest soft rot diseases. Plant Dis., 68:703-706.

Cook, R.J. and Baker, K.F., 1983. The Nature and Practice of Biological Control of Plant Pathogens. American Phytopathology Society, St. Paul, MN. 539 pp.

Cook, R.J., and Haglund, W.A., 1982. Pythium root rot: A barier to yield of Pacific Northwest wheat. Washington State University Agricultural Research Center Research Bulletin No. XB0913. 20 pp.

Cook, R.J., and Rovira, A.D., 1976. The role of bacteria in the biological control of *Gaeumannomyces graminis* by suppressive soils. Soil Biol. Biochem., 8:269-273.

Cook, R.J., Sitton, J.W., and Waldher, J.T., 1980. Evidence for *Pythium* as a pathogen of direct drilled wheat in the Pacific Northwest. Plant Dis., 64:102-103.

Cook, R.J. and Zhang, B.X., 1985. Degrees of sensitivity to metalaxyl within the *Pythium* spp. pathogenic to wheat in the Pacific Northwest. Plant Dis., 67. (In press).

Howell, C.R. and Stipanovic, R.D., 1979. Control of *Rhizoctonia solani* on cotton seedlings with *Pseudomonas fluorescens* and with an antibiotic produced by the bacterium. Phytopathology, 69:480-482.

Howell, C.R., and Stipanovic, R.D., 1980. Suppression of *Pythium ultimum*-induced damping-off cotton seedlings by *Pseudomonas fluorescens* and its antibiotic, pyoluteorin. Phytopathology, 70:712-715.

Howie, W.J., 1985. Factors affecting colonization of wheat roots and suppression of take-all by pseudomonads antagonistic to *Gaeumonnomyces graminis* var. *tritici*. Ph.D. Thesis. Washington State University, Pullman. 82 pp.

Kleopper, J.W., Leong, J., Teintze, M., and Schroth, M.N., 1980a. *Pseudomonas* siderophores: a mechanism explaining disease-suppressive soils. Curr. Microbiol., 4:317-320.

Kloepper, J.W., Leong, J., Teintze, M., and Schroth, M.N., 1980b. Enhanced plant growth by siderophores produced by plant growth-promoting rhizobacteria. Nature, 286:885-886.

Kloepper, J.W., and Schroth, M.N. Relationship of *in vitro* antibiosis of plant growth-promoting rhizobacteria to plant growth and the displacement of root microflora. Phytopathology, 71:1020-1024.

Kloepper, J.W., Schroth, M.N., and Miller, T.D., 1980. Effects of rhizosphere colonization by plant growth-promoting rhizobacteria on potato plant development and yield. Phytopathology, 70:1078-1082.

Loper, J.E., Haack, C., and Schroth, M.N., 1985. Population dynamics of soil pseudomonads in the rhizosphere of potato (*Solanum tuberosum* L.). Appl. Environ. Microbiol., 49:416-422.

Loper, J.E., Suslow, T.V., and Schroth, M.N., 1984. Lognormal distribution of bacterial populations in the rhizosphere. Phytopathology, 74:1454-1460.

Moore, K.J., and Cook, R.J., 1984. Increased take-all of wheat with direct drilling in the Pacific Northwest. Phytopathology, 74:1044-1049.

Ridge, E.H., 1976. Studies on soil fumigation-II. Effects on bacteria. Soil Biol. Biochem. 8:249-253.

Rovira, A.D. and Wildermuth, G.B., 1981. The nature and mechanisms of suppression. In: "Biology and Control of Take-all," 385-415. (M.J.C. Asher and P. Shipton, Eds.). Academic Press, London, New York. 538 pp.

Scher, F.M., and Baker, R., 1982. Effect of *Pseudomonas putida* and a synthetic iron chelator on induction of soil suppressiveness to Fusarium wilt pathogens. Phytopathology, 72:1567-1573.

Schroth, M.N., and Hancock, J.G., 1981. Selected topics in biological control. Annu. Rev. Microbiol., 35:453-476.

Schroth, M.N., and Hancock, J.G., 1982. Disease-suppressive soil and root-colonizing bacteria. Science, 216:1376-1381.

Shipton, P.J., 1975. Take-all decline during cereal monoculture. In: "Biology and Control of Soil-Borne Plant Pathogens. " 137-144.(G.W. Bruehl, Ed.) American Phytopathological Society, St.Paul, MN, 216 pp.

Shipton, P.J., Cook, R.J., and Sitton, J.W., 1973. Occurrence and transfer of a biological factor in soil that suppresses take-all of wheat in eastern Washington. Phytopathology, 63:511-517.

Sivasithamparam, K., and Parker, C.A., 1978. Effect of certain isolates of bacteria and actinomycetes on *Gaeumannomyces graminis* var. *tritici* and take-all of wheat. Aust. J. Bot., 26:773-782.

Smiley, R.W., 1978. Colonization of wheat roots by *Gaeumannomyces graminis* inhibited by specific soil microorganisms and ammonium-nitrogen. Soil Biol. Biochem., 10:175-179.

Smiley, R.W., 1979. Wheat-rhizoplane pseudomonads as antagonists of *Gaeumannomyces graminis*. Soil Biol. Biochem., 11:371-376.

Sneh, B., Dupler, M., Elad, Y., and Baker, R., 1984. Chlamydospore germination of *Fusarium oxysporum* f. sp. *cucumerinum* as affected by fluorescent and lytic bacteria from Fusarium suppressive soil. Phytopathology, 74:1115-1124.

Suslow, T.W., 1982. Role of root colonizing bacteria in plant-growth. In: "Phytopathogenic Prokaryotes" 187-233. (M.S. Mount and G.H.Lacy, Eds.). Academic Press, London.

Suslow, T.V., and Schroth, M.N., 1982. Rhizobacteria of sugar beets: effects of seed application and root colonization on yield. Phytopathol,ogy, 72:199-206.

Vojinović, Ž.D., 1972a. Biological antagonism as the cause of decline of *Ophiobolus graminis* Sacc. in prolonged wheat monoculture. J. Sci. Àgric. Res. 25:31-41.

Vojinović, Ž.D., 1972b. Antagonists from soil and rhizosphere to phytopathogens. Institute Soil Science, Beograd, Yugoslavia. Final Technical Report. 130 p.p.

Vojinović, Ž.D., 1973. The influence of micro-organisms following *Ophiobolus graminis* Sacc. on its further pathogenicity. European Mediterranean Plant Protection Organization Bulletin, 9:91-101.

Vraný, J., Vančura, V., and Staňek, M., 1981. Control of microorganisms in the rhizosphere of wheat by inoculation of seeds with *Pseudomonas putida* and by foliar application of urea. Folia Microbiol., 26:45-51.

Weller, D.M., 1983. Colonization of wheat roots by a fluorescent pseudomonad suppressive to take-all. Phytopathology, 73:1548-1553.

Weller, D.M., 1984. Distribution of a take-all suppressive strain of *Pseudomonas fluorescens* on seminal roots of winter wheat. Appl. Environ. Microbiol., 48:897-899.

Weller, D.M., 1985. Application of fluorescent pseudomonads to control root diseases. In: "Ecology and Management of Soilborne Plant Pathogens." (C.A. Parker, A.D. Rovira, K.J. Moore, P.T.W. Wong and J.F. Kollmorgen, Eds.). APS Press, St. Paul, MN. (In press).

Weller, D.M., and Cook, R.J., 1981. Pseudomonads from take-all conducive and suppressive soils. Phytopathology, 71:264. (Abstract).

Weller, D.M., and Cook, R.J., 1983. Suppression of take-all of wheat by seed treatments with fluorescent pseudomonads. Phytopathology, 73:463-469.

Weller, D.M., and Graham, M.C., 1984. Application of fluorescent pseudomonads to improve the growth of wheat. Phytopathology, 74:806 (Abstract).

Weller, D.M., Howie, W.J., and Cook, R.J., 1985. Relationship of *in vitro* inhibition of *Gaeumannomyces graminis* var. *tritici* and *in vivo* suppression of take-all by fluorescent pseudomonads. Phytopathology, 75, (Abstract) (In press).

Weller, D.M., Zhang, B.X., and Cook, R.J., 1985. Application of a rapid screening test for selection of bacteria suppressive to take-all of wheat. Plant Dis., 69, (In press).

Wilkinson, H.T., Cook, R.J., and Alldredge, J.R., 1985. Relation of inoculum size and concentration to infection of wheat roots by *Gaeumannomyces graminis* var. *tritici*. Phyropathology, 75:98-103.

Wong, P.T.W., and Baker, R., 1984. Suppression of wheat take-all and Ophiobolus patch by fluorescent pseudomonads from a Fusarium-suppressive soil. Soil Biol. Biochem., 16:397-403.

Xu, G.-W., 1984. Selection of fluorescent *Pseudomonas* strains antagonistic to *Erwinia carotovora* and their effect on colonization and infection of *Solanum tuberosum* L. by *Erwinia carotovora*. M.S. Thesis. Washington State University, Pullman. 72 pp.

BIOLOGICAL CONTROL OF FUSARIUM WILTS BY *PSEUDOMONAS PUTIDA* AND ITS ENHANCEMENT BY EDDHA

F.M. Scher

Research Scientist, Allelix Inc.
6850 Goreway Drive
Mississauga, Ontario, Canada L4V 1P1

Pseudomonas putida, isolated from Salinas Valley Fusarium-suppressive soil has been shown to reduce Fusarium wilt incidence of flax, radish, cucumber and carnation. The mechanism of biological control was suggested, at least in part, to be as a result of competition for iron between the pathogen and siderophores produced by *P. putida*. Addition of iron to soil (as FeEDTA) nullified this biological control whereas addition of EDDHA (an Fe^{3+} chelate with high binding constant) suppressed Fusarium wilt and gave an additive control response when incorporated with *P. putida*. EDDHA was not fungicidal but did inhibit Fusarium germ tube elongation. *P. putida* populations in fallow and radish-rhizosphere soil were enhanced by EDDHA but *P. putida* did not utilize EDDHA as a nutritional source *in vitro*. The mechanism of *P. putida* enhancement by EDDHA was suggested to be selective management of iron availability in soil, and resultant depression of non-iron acquiring microbial competitors. EDDHA incorporation into soil may give selective advantage to siderophore-producing antagonists.

Despite long cultivation of a variety of Fusarium wilt-susceptible crops in the Salinas Valley of California, the disease does not occur there (Smith and Snyder, 1971). In greenhouse tests, Fusarium wilt of sweet potato was less severe in Salinas soil than in a similar soil taken from an area where wilt was known to occur; thus, the Salinas soil was determined to be Fusarium-suppressive.

Suppressiveness of the soil was alleviated by steam heating (54°C, 30 min) and was transferable to conducive soil, indicating that biological rather than strictly physical factors were responsible for the observed suppressiveness (Scher and Baker, 1980). *Pseudomonas* spp. retrieved from Fusarium fungal mats buried in soil conferred suppressiveness to conducive soils and were sensitive to steam heating. Scher and Baker (1980) suggested that *Pseudomonas* spp. were responsible for suppressiveness in the Salinas Valley soil.

When *P. putida* isolate A12, from Fusarium-suppressive soil, was added to conducive soil infested with *Fusarium oxysporum,* incidence of Fusarium wilt of flax, radish and cucumber was significantly reduced (Fig. 1). Similar results were reported for other *Pseudomonas* strains from the Salinas soil (Dupler and Baker, 1984, (Strain NIR): Sneh et al., 1984).

Scher and Baker (1980) indicated that there was no antibiosis *in vitro*

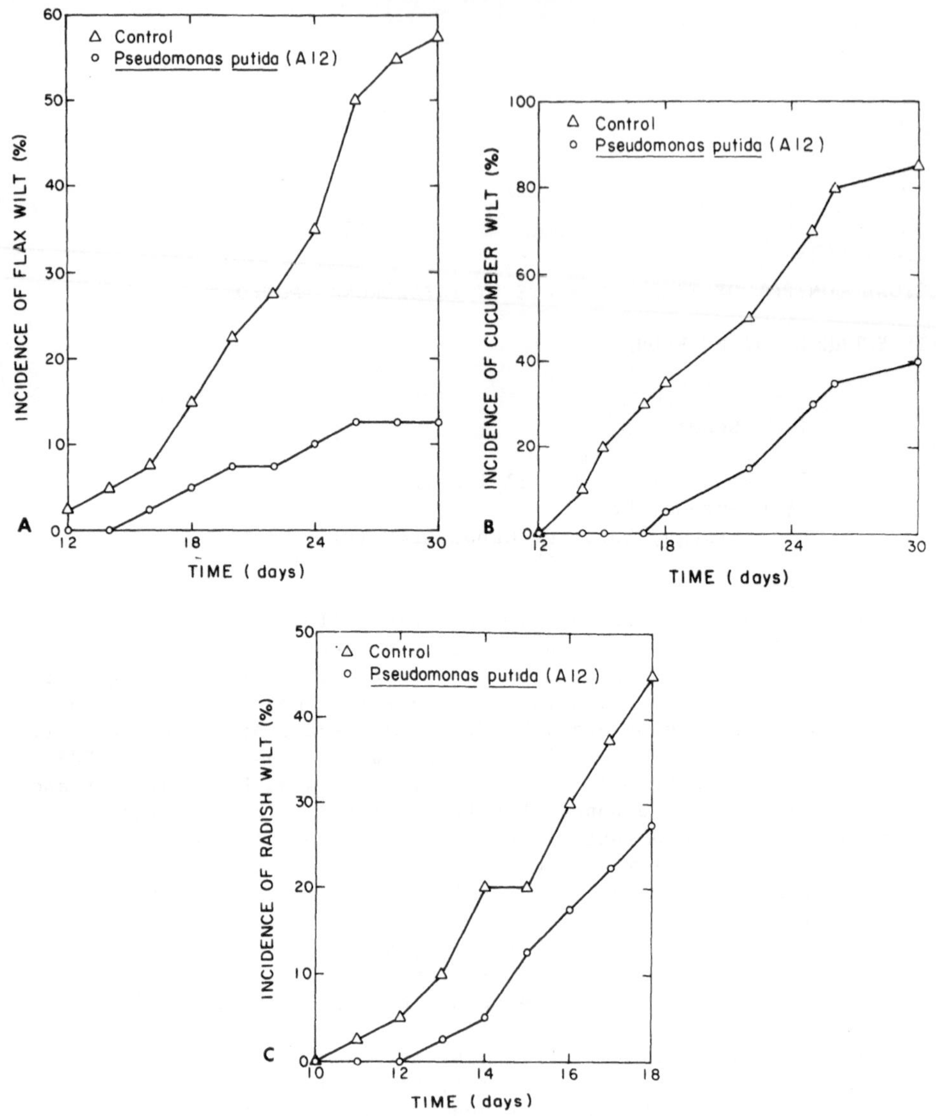

Fig. 1. Effect of addition of *Pseudomonas putida* (A12) at LOG 7 cells per gram of soil on Fusarium wilt incidence in A, *Fusarium oxysporum* f.sp. *lini*-, B. F.sp. *cucumerinum*-. and C, F.sp. *conglutinans*-infested conducive soil. No *P. putida* was added to the control. Wilt incidence in the *P. putida* treatments at the end of each experiment was significantly different (P=0.05) from the control for all three host systems. Data represent mean disease incidence of 40 (A and C) or 20(B) plants. (Reprinted with permission from Phytopathology, 72:1567-1573, Scher and Baker, 1982).

between *Pseudomonas* and *Fusarium*, and Sneh et al., (1984) reported that *P. putida* strains A12 and NIR did not cause lysis of the pathogen. Thus, competition as a biocontrol mechanism was generally believed to be responsible for suppressiveness.

The hypothesis that Fusarium wilt biocontrol by *Pseudomonas* spp. was by competition for iron (Fe^{3+}) was introduced by Kloepper et al., in 1980.

They utilized *Pseudomonas* sp. strain B10, originally isolated from take-all suppressive soil. Dipping flax transplants into a suspension of B10 significantly increased their survival in Fusarium-infested soil, unless Fe^{3+} in the form of Fe^{3+} ethylenediaminetetracetic acid (Fe^{3+} EDTA), was added to the soil. Thus, Fe^{3+} eliminated biological control by B10.

Fluorescent pseudomonads produce high-affinity Fe^{3+} chelators, termed siderophores, under low Fe^{3+} conditions (Neilands, 1973). Siderophores enhance microbial acquisition of iron in iron-deficient environments and have been found in soil (Powell et al., 1980). Addition of B10 or its siderophore, pseudobactin, caused conducive soils to become suppressive to Fusarium wilt (Kloepper et al., 1980), whereas addition of Fe^{3+}-pseudobactin had no such effect. Kloepper et al., suggested that pseudobactin acted by effectively chelating the already limiting supply of Fe^{3+} in soil, making it unavailable to the *Fusarium*. In an independent study, over 100 fluorescent pseudomonads tested inhibited growth of *Geotrichum candidum* on Fe^{3+}-limited agar, but not on agar supplemented with iron (Misaghi et al., 1981).

Teintze et al., (1981) characterized pseudobactin as a mixed catechol/hydroxamate siderophore consisting of a hydroxamate group, an o-dihydroxy group and an α-hydroxy acid. The ferric form had an absorption peak at 400nm. *P. putida* A12, from Salinas Valley soil, was examined by Scher and Baker (1982) and found to produce a similar siderophore which had an absorption peak at 410nm (ferric form). In addition, they reported the production of a siderophore, having a wide peak at 450nm (ferric form), by *F. oxysporum*. The latter is more typical of hydroxamate-type fungal siderophores (Emery, 1965) which generally have lower affinities for Fe^{3+} than do bacterial siderophores.

Two iron chelators were employed by Scher and Baker for comparative purposes in measuring Fe^{3+} affinity. One of these, Fe^{3+} EDTA is rather a poor iron chelator (stability constant of $\log_{10}K=25$). It binds Fe^{3+} well only at pH values below 7: above pH 7, its affinity for Ca^{2+} and Zn^{2+} exceeds that for Fe^{3+}. Conversely, ethylenediaminedi-O-hydroxyphenylacetic acid (EDDHA or EDDA) has a high affinity for Fe^{3+} (stability constant of $\log_{10}K=33.9$) over the pH range of 4-10 (Lindsay, 1979).

Pseudomonas siderophores are repressed by micromolar amounts of Fe^{3+} in agar medium; however, when EDDHA (2mg/ml) was added to agar containing 10^{-4}M Fe^{3+} siderophore production by A12 did occur (Scher and Baker, 1982). This indicated that EDDHA depleted the available Fe^{3+} in the agar to a low enough to induce siderophore production, that EDDHA was not toxic to A12 and that A12 was capable of removing Fe^{3+} from EDDHA.

Scher (1982) demonstrated that A12's siderophore, present in culture supernatant, bound Fe^{3+} previously held by EDDHA (Fig. 2) At time zero Fig. 2a) Fe^{3+} EDDHA exhibited its characteristic broad peak at 450-525nm. Instantaneously, upon addition of A12 supernatant, some Fe^{3+} was transferred to the A12 siderophore, which began to form its characteristic peak at 410nm. After 24 hours an equilibrium was reached whereby much, but not all, of the Fe^{3+} previously bound to EDDHA was bound to A12 siderophore. (Note that the amount of Fe^{3+} present was not sufficient to saturate either the siderophore or EDDHA). Thus, it is suggested that A12 siderophore has an affinity for Fe^{3+} higher than that of EDDHA, and a resulting stability constant log K>33.9.

From a study of these stability constants there emerges a hierarchy of Fe^{3+} binding compounds which, hypothetically, can be used to explain and predict microbial Fe^{3+} acquisition: Pseudomonas siderophores>EDDHA>

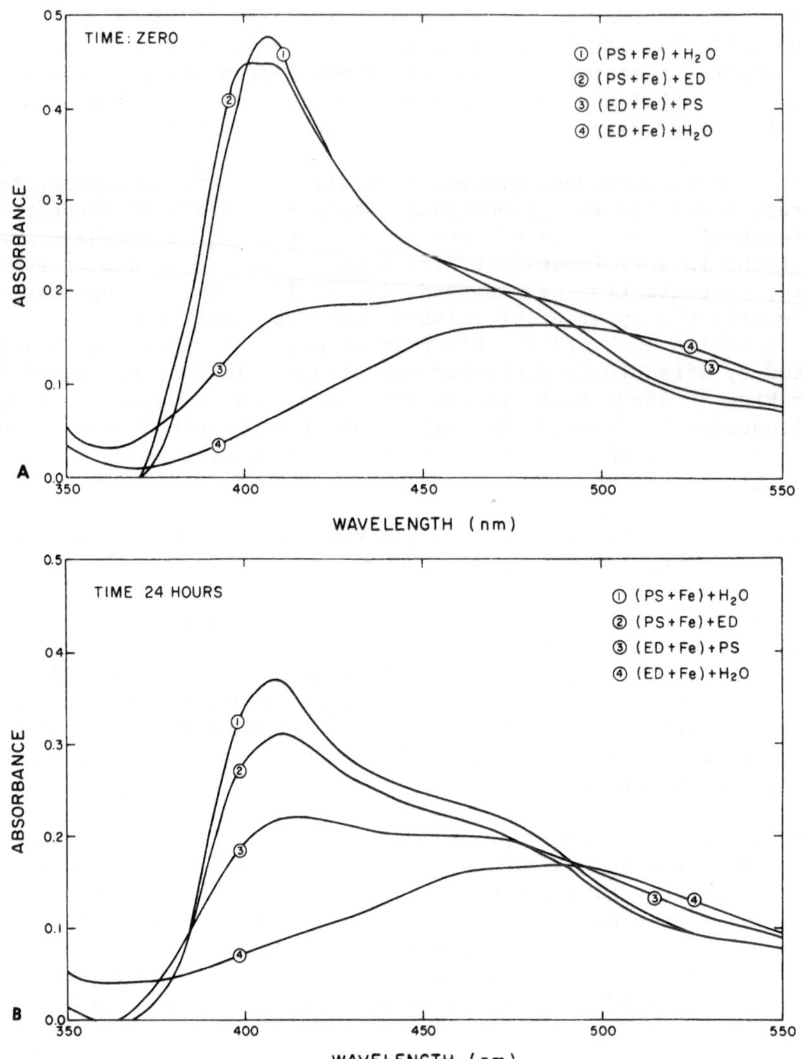

Fig. 2. Absorption spectra for *Pseudomonas putida* A12 culture filtrate and EDDHA upon addition of Fe^{3+} as $FeCl_3$. Brackets () in figure legends indicate that compounds inside brackets were mixed together prior to addition of compound outside brackets. Lines 1 and 4 are controls, whereas 2 and 3 show interactions between the Pseudomonas siderophore present in the supernatant (PS) and EDDHA (ED). A. Time:zero, B. Time:24h.

Fusarium siderophores>EDTA.

Such binding affinities predict that Fe^{3+} held by EDDHA would not be available to *Fusarium*. FeEDDHA (100μg/g soil) was introduced into conducive soil infested with *F. oxysporum* f. sp. *conglutinans*. A12 was added to half the treatments. Radish wilt incidence in the control was 45% (Fig. 3)

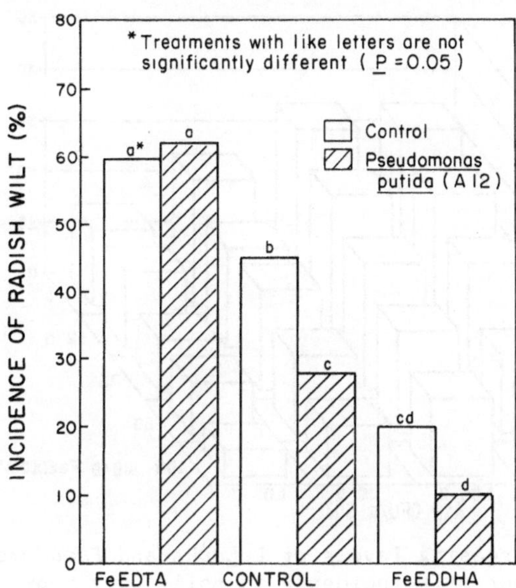

Fig. 3. Radish wilt incidence when FeEDDHA or FeEDTA were introduced into soil (100µg/g soil) with or without *Pseudomonas putida* A12 (LOG 7 cfu/g soil). Data represent mean disease incidence of 40 plants (Reprinted with permission from Phytopathology, 72:1567-1573, Scher and Baker, 1982).

and was reduced to 20% by FeEDDHA. Thus, FeEDDHA alone reduced Fusarium wilt incidence. A12 alone reduced disease incidence to 27.5%, whereas FeEDDHA plus *P. putida* reduced disease incidence to 10%, a significant reduction compared to A12 alone.

In contrast, the addition of FeEDTA to soil resulted in more disease than in the control. This is similar to the report by Kloepper et al., (1980) and may be explained in several ways. First, Fe^{3+} bound to EDTA was available to *Fusarium* via its own siderophore and thus increased its inoculum potential; second, Fe^{3+} was released upon addition of the chelate to soil and repressed antagonists' siderophore production; or, third, EDTA was toxic to antagonists such as *Pseudomonas* (previously reported by Wilkinson, 1967).

When FeEDDHA and A12 were added to soil at several levels, it was observed that disease control by *P. putida* and FeEDDHA was additivie (Fig. 4). Disease incidence of cucumber was reduced from 35% to 5% by addition of 100µg FeEDDHA and Log 7 cfu A12/g soil.

In a greenhouse experiment, *P. putida* NIR (from Salinas soil) reduced Fusarium wilt of carnations induced by *F. oxysporum* f. sp. *dianthi* (Fig. 5). The bacteria were applied in two ways, as a soil drench resulting in Log 7 cfu/g soil, or by dipping rooted cuttings into a Log 9 cfu/ml suspension prior to planting. Carnations were grown in above-ground benches in a soil-peat-perlite mix infested with the pathogen. Disease incidence after 10 months was 34% in the control. FeEDDHA (500 µg/g) caused a lag in disease development but was not protective after 8 mo. Maximum control was attained with *P. putida* as a root dip (12% disease incidence). FeEDDHA noticeably enhanced disease control by *P. putida* applied to soil, reducing disease from 22% to 12%, but not when added alone or accompanying a root

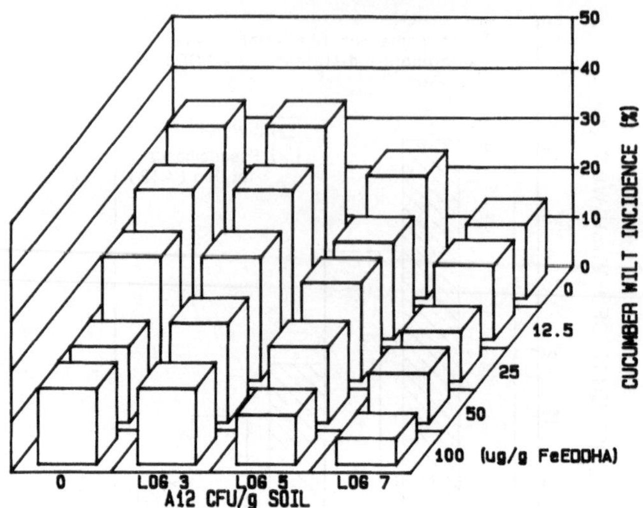

Fig. 4. Effect of increasing levels of FeEDDHA and *Pseudomonas putida* A12 of cucumber wilt incidence in soil. Data represent mean disease incidence of 20 plants. (Data used with permission from Phytopathology, 72:1567-1573, Scher and Baker, 1982)

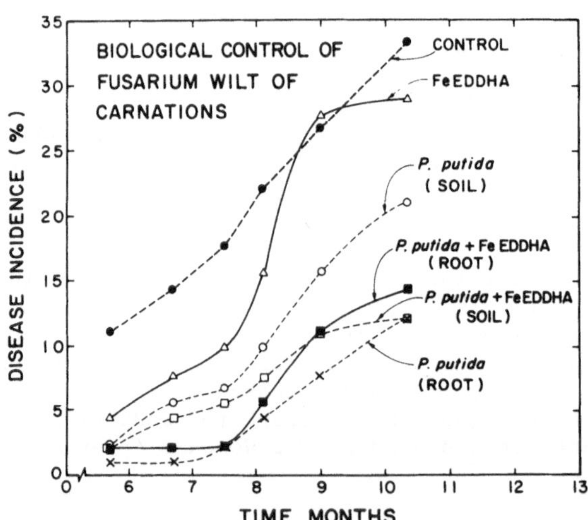

Fig. 5. Effect of FeEDDHA (500 μg/g) and *Pseudomonas putida* NIR on carnation wilt incidence in above-ground benches. NIR was either mixed into soil at approximately LOG 7 cfu/g soil (soil) or applied by dipping rooted cuttings into a LOG 9 cfu/g suspension prior to transplanting (root).

dip. FeEDDHA was applied before transplanting and was observed to leach out of the above-ground benches.

Scher et al., (1984 determined that the mechanism of Fusarium wilt control by *P. putida* or FeEDDHA was not fungicidal since neither affected the population density of *Fusarium* in soil. Fungistatic effects of EDDHA on *Fusarium*, which would not result in lowered population densities, were not ruled out. Scher and Baker (1982) reported that EDDHA inhibited Fusarium germ tube elongation and that its effect was reversed by FE^{3+}. In a later study they found that crudely purified A12 siderophore had the same effect on germ tube elongation (unpublished). Sneh et al., (1984) suggested that there was a direct correlation between relative siderophore production by fluorescent pseudomonads and their inhibition of Fusarium chlamydospore germination.

In addition, Sneh et al., (1984), reported that addition of *P. putida* strains A12 or NIR to soil reduced chlamydospore germination. Adding Fe^{2+} or other cations partially counteracted the inhibition. They reported that FeEDDHA inhibited chlamydospore germination in rhizosphere, but not non-rhizosphere soil. This supported the results of Scher and Baker (1983) that EDDHA did not affect chlamydospore germination in bulk soil, and suggests that the beneficial activity of FeEDDHA occurred in the rhizosphere.

In addition to direct competition for Fe^{3+} between *Fusarium* and EDDHA, a second mechanism of EDDHA-induced disease reduction was proposed by Scher et al., (1984). Addition of FeEDDHA to raw soil significantly increased rhizosphere and fallow soil population densities of the biocontrol agent *P. putida* NIR in soil (Fig. 6), however, *P. putida* could not utilize FeEDDHA nutritionally. Enhancement of *P. putida* population densities by EDDHA was suggested to be a result of its unique capacity to acquire Fe^{3+} from FeEDDHA. Upon addition of FeEDDHA to soil, it is likely that only microorganisms, like *P. putida* NIR and A12, that produce siderophores with a higher stability constant than EDDHA, can utilize the chelated Fe^{3+}. Thus, such microbes would be favored at the expense of other microbes without comparable iron-aquiring systems (such as *Fusarium*). It is

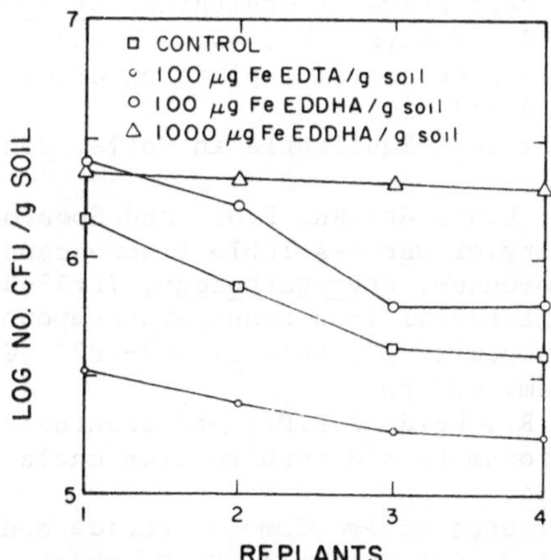

Fig. 6. Population densities of *P. putida* NIR in the rhizospheres of radishes undergoing successive weekly replanting. *P. putida* (LOG 4 cfu/g soil) and FeEDDHA (100 or 1000 µg/g soil) or FeEDTA (100 µg/g soil) were initially added to soil. Data points represent means from four pots, 10 plants each. Population densities in the FeEDTA and 1000 µg FeEDDHA/g soil treatments were significantly different than in the controls at all replanting times. Population densities in the 100 µg FeEDDHA/g soil treatment were significantly different from the control at replanting times 1 and 2. (Reprinted with permission from Can. J. Microbiol., 30:1271-1275, Scher, Dupler and Baker, 1984).

suggested that FeEDDHA may control Fusarium wilt by direct competition for Fe^{3+} with the pathogen and by enhancing activity of antagonists which produce siderophores capable of binding Fe^{3+} held to EDDHA.

Enhancement of *P. putida* biocontrol by EDDHA is a result of numerous dynamic processes, not all of which are clearly understood, which occur when both are present in the rhizosphere. Such enhancement represents a unique third step in studies of suppressive soils. First, suppressive soils were identified and verified in greenhouse studies, then microorganisms suspected to be responsible for suppressiveness were isolated and applied to conducive soils as biological control agents. Such agents have typically had limited success in reducing disease incidence to commercially acceptable levels. Thus, it should be our goal to take a third step and identify compatible agents, such as EDDHA, which will enhance biocontrol agents and improve their effectiveness. By combining biocontrol and enhancing agents, we may be able to manage rhizosphere ecology to the advantage of the plant and more successfully integrate biological control into commercial disease-control programs.

ACKNOWLEDGEMENTS

The author wishes to thank Ms. Marcelia Dupler and Dr. Ralph(Tex) Baker for their part in these investigations, and Dr. C. Simonson for review of the manuscript.

REFERENCES

Dupler, M., and Baker, R., 1984. Survival of *Pseudomonas putida*, a biological control agent, in soil. Phytopathology, 74:195-200.

Emery, T., 1965. Isolation, characterization, and properties of fusarinine a hydroxamic acid derivative of ornithine. Biochemistry, 4:1410-1417.

Kloepper, J.W., Leong, J., Teintze, M., and Schroth, M.N., 1980. Pseudomonas siderophores: a mechanism explaining disease-suppressive soils. Curr. Microbiol., 4:317-320.

Lindsay, W.L., 1979. Chemical Equilibria in Soils. John Wiley & Sons, New York. 449 pp.

Misaghi, I.J., Stowell, L.S., Grogan, R.G., and Spearman, J.C., 1982. Fungistatic activity of water-soluble fluorescent pigments of fluorescent pseudomonads. Phytopathology, 72:33-36.

Neilands, J.B., 1973. Microbial iron transport compounds (siderochromes). In: "Inorganic Biochemistry". Vol. I. 167-202. (G.L. Eichhorn, Ed.) Elsevier, Amsterdam, 607 pp.

Powell, P.E., Cline, G.R., Reid, C.P.P., and Szaniszlo, P.J., 1980. Occurrence of hydroxamate siderophire iron chelators in soils. Nature, 287:833-834.

Scher, F.M., 1982. Influence of *Pseudomonas putida* and synthetic iron chelates on Fusarium wilt diseases. Ph.D. thesis. Colorado State University. 96 p.

Scher, F.M., and Baker, R., 1980. Mechanism of biological control in a Fusarium-suppressive soil. Phytopathology, 70:412-417.

Scher, F.M., and Baker, R., 1982. Effect of *Pseudomonas putida* and a synthetic iron chelator on induction of soil suppressiveness to Fusarium wilt pathogens. Phytopathology, 72:1567-1573.

Scher, F.M., and Baker, R., 1983. Fluorescent microscopic technique for fungi in soil and its application to studies of a Fusarium-suppressive soil. Soil Biol. Biochem., 15:715-718.

Scher, F.M., Dupler, M., and Baker, R., 1984. Effect of synthetic iron chelates on population densities of *Fusarium oxysporum* and the biological control agent *Pseudomonas putida* in soil. Can. J. Microbiol. 30:1271-1275.

Smith, S.N., and Snyder, W.C., 1971. Relationship of inoculum density and soil types of severity of Fusarium wilt of sweet potato. Phytopathology, 61:1049-1051.

Sneh, B., Dupler, M., Elad, Y., and Baker, R., 1984. Chlamydospore germination of *Fusarium oxysporum* f. sp. *cucumerinum* as affected by fluorescent and lytic bacteria from a Fusarium-suppressive soil. Phytopathology, 74:1115-1124.

Teintze, M., Hossain, M.B., Baines, C.L., Leong, J., and van der Helm D., 1981. Structure of ferric pseudobactin, siderophore from a plant growth promoting Pseudomonas. Biochemistry, 20:6446-6457.

Wilkinson, S.G., 1967. The sensitivity of pseudomonads to ethylenediamine-tetra-acetic acid. J. Gen. Microbiol., 47:67-76.

PYOVERDINE-FACILITATED IRON UPTAKE AMONG FLUORESCENT PSEUDOMONADS

D. Hohnadel and J.M. Meyer

Laboratoire de Biochimie Microbienne
Institut Le Bel, Université Louis Pasteur
4, rue Blaise Pascal, 67000 - Strasbourg, France

INTRODUCTION

Within the past few years, interest in fluorescent pseudomonads has considerably increased, some of these bacteria found in the plant rhizosphere having been shown to participate actively in plant growth promotion or plant disease protection for a large variety of crops (Geels and Schippers, 1983; Kloepper et al., 1980; Scher and Baker, 1980; Scher and Baker, 1982). Iron nutrition is suspected to play an important role in both phenomena, and indeed, the knowledge of the iron uptake mechanisms occurring in the fluorescent pseudomonads is an important step in the understanding of the relationships existing between plants and their microflora.

PYOVERDINE: A SIDEROPHORE FOR FLUORESCENT PSEUDOMONADS

The existence of pyoverdine (named previously "bacterial fluoresceine" or "fluorescin" (King, et al., 1948; Lenhoff, 1963), the green-yellow fluorescent pigment produced in some growth conditions by the so-called "fluorescent pseudomonads", has been known for some time, certainly more than a hundred years ago, when Schroeter (1870) described pigmented bacteria, reminiscent of descriptions of pseudomonads. Detection of the yellow-green pigment by using the King's B medium (King et al., 1954) rapidly became one of the most important taxonomic criteria of recognition for these bacteria (Palleroni, 1984; Stainer et al., 1966). The colour and fluorescence in UV light of these compounds has facilitated research aimed at understanding their physiological significance and conditions of production. Among other nutritional growth factors suspected to influence the biosynthesis of pyoverdine are the nature and concentration of the carbon and nitrogen sources (Gouda and Chodat, 1963; Gouda and Greppin, 1965; Sullivan, 1905), the degree of aeration of the growth media (Lenhoff, 1963), the presence of Mg^{++} (Georgia and Poe, 1931) or Zn^{++} (Baghdiantz, 1952: Chakrabarty and Roy, 1964). But iron level in the growth medium has taken a major place consecutively to the works of King et al., (1948), Totter and Moseley, 1953, Lenhoff, 1963, Palumbo, 1972, Lluch et al., 1973.

The inverse relationship existing between the level of pyoverdine excreted by cells grown in a medium and iron content of this growth

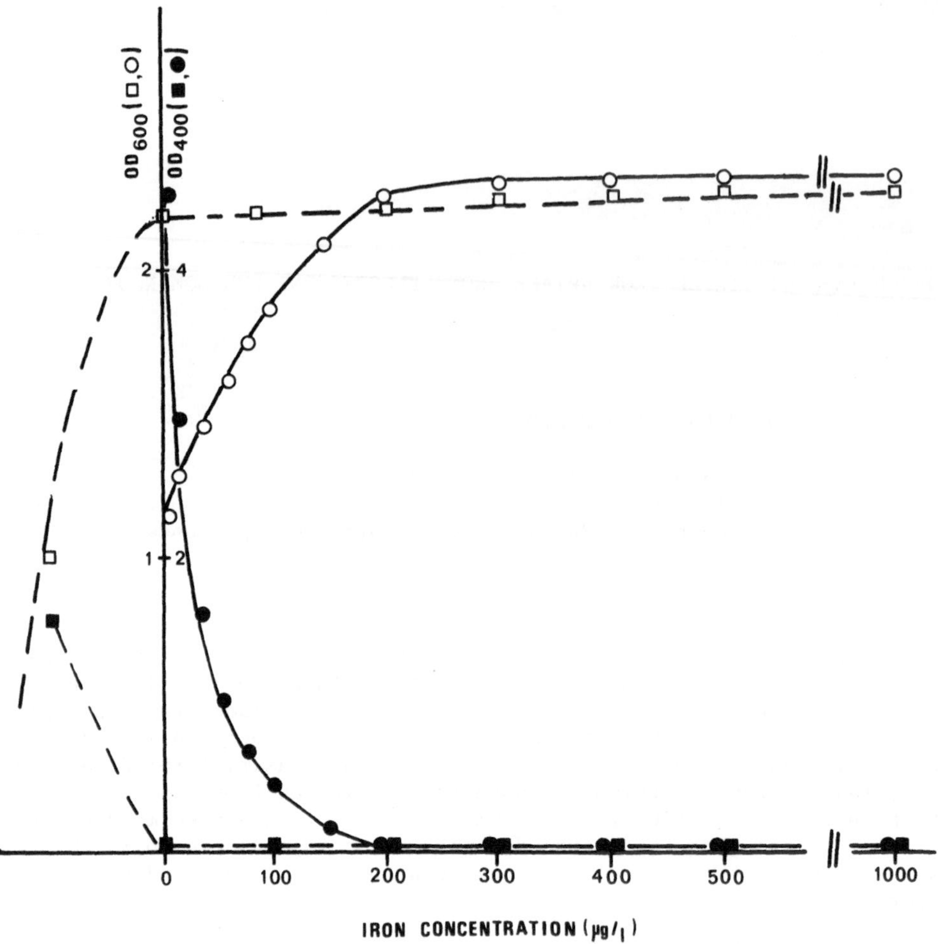

Fig. 1. Maximal growth and pyoverdine production of *Pseudomonas fluorescens* W as a function of added iron (III) concentration in succinate- (circles) and citrate- (squares) growth media. Growth (open symbols) was determined by measuring the OD_{600} at the end of the exponential phase, whereas pyoverdine (full symbols) was measured by the OD_{400} of the culture supernatant. Negative values on the abscissa correspond to the amount of pyoverdine and growth reached in citrate-medium pretreated with 8-hydroxyquinoline (Meyer and Abdallah, 1978).

medium appears in Fig. 1. When *P. fluorescens* (Lab strain W, held in the Czechslovack Collection of Mocroorganisms as CCM 2799) was grown in a succinate medium (PO_4HK_2, 6 g/l, PO_4H_2K,3 g/l, $SO_4(NH_4)_2$, 1 g/l, SO_4Mg,7 H_2O, 0.2 g/l, succinic acid, 4 g/l, Ph 7,0), an optimal production of pyoverdine was observed after growth, whereas supplementation of this medium with 200 μg (∿4 μM) of iron ($FeCl_3$) completely depressed pyoverdine biosynthesis. Thus, the lack of iron in the growth medium appeared to be an essential condition for pigmentation to occur. A conflicting conclusion was drawn following the finding that when citrate (or malate) was used as a carbon source instead of succinate, even in iron deficient media no pyoverdine was synthesised, supporting the chromogenic-antichromogenic substrate hypothesis (Gouda and Chodat, 1963). But, the accumulation of pyoverdine in citrate medium depleted for contaminating iron by 8-hydroxy-quinoline or bathophenantroline treatments (Fig. 1) demonstrated that pyoverdine biosynthesis is in fact directly related with the iron requirement of the bacterial cells. This view was confirmed by following the bacterial yield in these different media, a study which concluded that pyoverdine biosynthesis is specific for cells grown in iron limited conditions (Fig. 1). Citrate medium was shown to contain similar amounts of contaminating iron compared to the succinate medium. But, when measuring the intracellular iron concentration for cells grown in the presence of an excess of iron, it was effectively shown that the cell iron requirement depended on the carbon source and was much lower (140 µg/1g cell dry weight) for citrate- compared to succinate-grown cells (228 µg/1g cell dry weight (Meyer and Abdallah, 1978). Thus, not only the iron level in the growth medium, but also the choice of the carbon source which determines the iron requirement of the bacteria, appears to be of fundamental importance for production of pyoverdine by fluorescent pseudomonads. Similar results were obtained when studying other fluorescent pseudomonad spp., with however, some differences relating to iron concentration which inhibited completely pyoverdine biosynthesis in a succinate medium. For example, *Pseudomonas aeruginosa* (strain PA01, ATCC 15692) still produced some pigmentation at an iron concentration of 4 µM in a succinate growth medium and the inhibiting concentration was >20 µM. The iron requirement appears to vary considerably between strains. Consequently, growth parameters have to be carefully determined for each strain tested in order to determine the most favourable conditions for pyoverdine production.

The observation that the characteristic yellow-green colour of a *P. fluorescens* culture in a succinate medium changed very quickly to a brown red color with concomitant disappearance of the fluorescence when iron chloride was added, suggested a strong affinity of pyoverdine for iron. The complex was easily extractible by conventional methods using a chloroform-phenol solvent and after an ion exchange chromatography step followed by a decomplexation of pure pyoverdine was obtained (Meyer and Abdallah, 1978). With some minor variations, this method was applied to all the different strains tested in this Laboratory (over twelve) with however, one exception, *Pseudomonas putida* ATCC 12633. The fact that pyoverdine binds very tightly and specifically iron (III) ($K_A = 10^{32}$) was a strong argument that pyoverdine acts as a siderophore for the fluorescent pseudomonads according to the definition given by Neilands (1984). Moreover, the demonstration that pyoverdine facilitates iron uptake in *P. fluorescens* was in agreement with this conclusion (Meyer and Hornsperger, 1978).

The chemical structure of pyoverdine has been elucidated within the past few years. It is composed mainly of a quinoleinic chromophore and a peptidic chain moiety. The few examples so far described concerned with soil fluorescent pseudomonads, *Pseudomonas* B10, *Pseudomonas* 7RS1 (Teintze et al., 1981; Teintze and Leong, 1981; Yang and Leong, 1984) and *P. aeruginosa* ATCC 15692 (Wendenbaum et al., 1983) had previously demonstrated that pyoverdine (called pseudobactin in the case of *P.* B10 and 7RS1) is not a unique compound synthesized by all fluorescent pseudomonads. Even though the physicochemical properties of all the pyoverdines appear to be very closely related, if not identical, some differences can be noted especially in the peptide moiety. Moreover, the analysis performed on pyoverdines, isolated from a dozen different strains of fluorescent pseudomonads, some of them belonging to the same nomen-species, have revealed an amino acid composition of the peptide varying from one pyoverdine to an other. One exception has been found, namely the siderophores from *P. fluorescens* ATCC 13525 and *Pseudomonas chlororaphis* ATCC 9446 which appear identical at least at the level of the amino-acid composition (see the accompanying report in this volume by Abdallah et al., for a detailed description of the structural differences among pyoverdines). From these structure studies it can be concluded that the pyoverdine

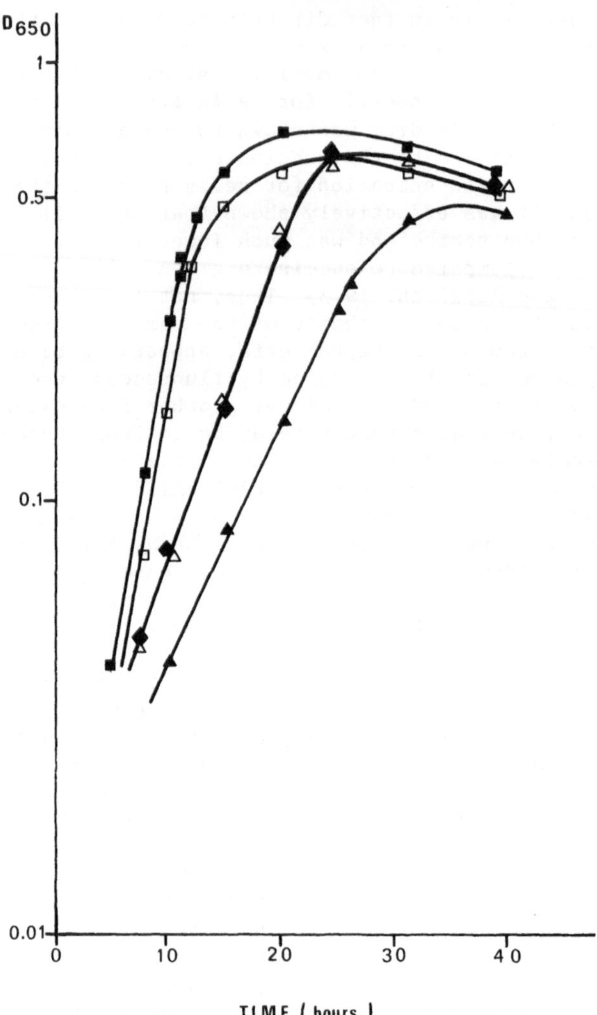

Fig. 2. Effects of pyoverdines on growth of *Pseudomonas fluorescens* ATCC 13525 in succinate-medium without addition (◆), or supplemented with pyoverdine (50 μM, final concentration) of *P. fluorescens* (ATCC 13525 (■), *P. aeruginosa* ATCC 15692 (□), *P. fluorescens* W (Δ) or *P. fluorescens* ATCC 17400 (▲).

system is usually strain specific, a conclusion which is supported by the genotype diversity observed for a large number of fluorescent pseudomonads during hybridization analysis with siderophore related gene probes (Lawson et al., 1985). The results described below, which concern cross-feeding studies as well as uptake or binding studies of iron complexed with pyoverdines of different origins, demonstrate this strain specificity at a physiological level.

GROWTH STIMULATION PROPERTIES OF PYOVERDINES

The addition of pure pyoverdine (50 μM) to a succinate growth medium has a strong stimulating effect on the growth of its producer strain by shortening the lag phase and increasing the growth rate. Fig. 2 illustrates the results obtained with *P. fluorescens* ATCC 13525 . This growth promoting

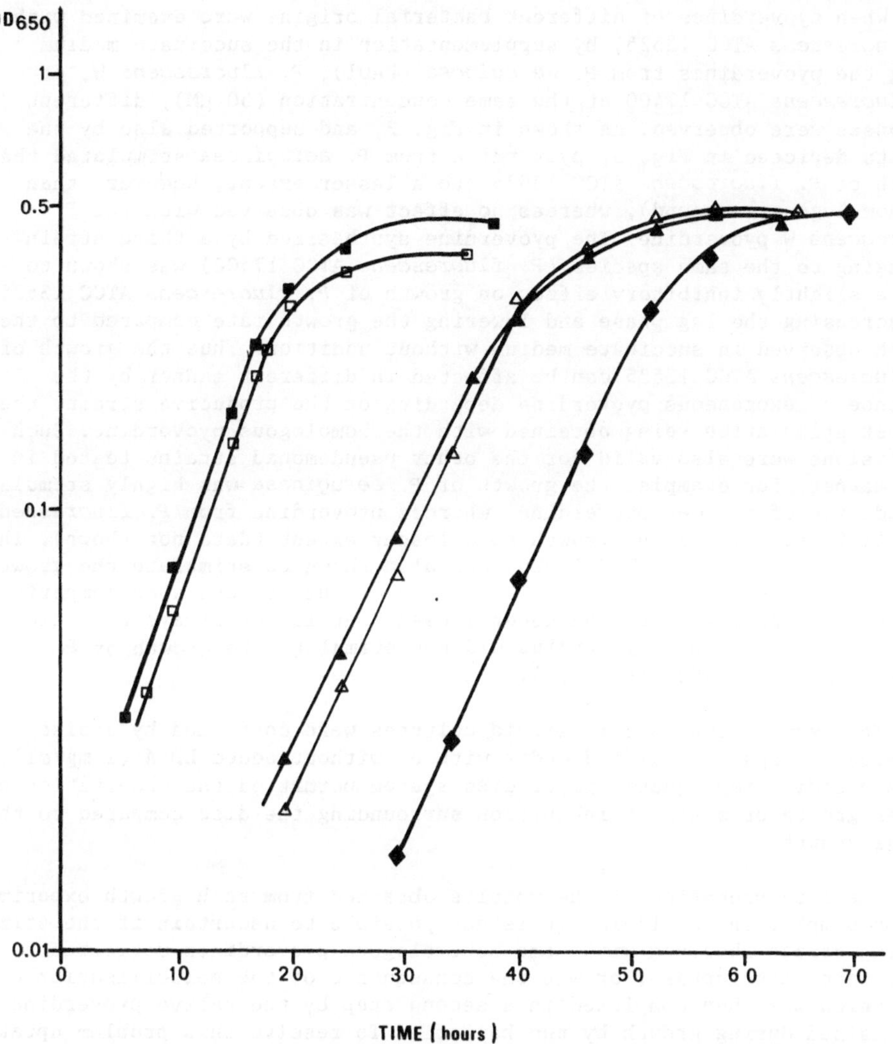

Fig. 3. Effects of pyoverdines on growth of *Pseudomonas fluorescens* ATCC 13525 in succinate-EDDA (1 mg/ml) medium. Symbols are the same as in Fig. 2.

effect of pyoverdine from *P. fluorescens* ATCC 13525 was not the consequence of some contaminating iron supplementation when adding the pyoverdine since it was shown that (i) the purified compound used was free of its precursor, the pyoverdine-iron complex, and (ii) the stimulation effect of pyoverdine was concentration dependent but reached rapidly a maximal value for about 50 μM of the added pyoverdine. It was also verified that pyoverdine by itself did not support any growth when added to a medium without succinate, ruling out an eventual growth stimulation effect in the succinate medium by the utilization of pyoverdine as a carbon source. A relationship between growth stimulation by pyoverdine and iron metabolism was suggested by growth experiments where EDDA (Ethylene-diaminedihydroxy-phenyl acetic acid) was added to the succinate medium at a final concentration of 1 mg/ml. This compound, known to chelate iron (III) has a strong effect on growth of *P. fluorescens* ATCC 13525 by prolonging the lag phase to over 20 hours, with, however, no long term effect on growth yield and growth rate. As shown in Fig. 3, the addition of pyoverdine to a succinate-EDDA medium suppressed completely the inhibitory effect of EDDA.

When pyoverdines of different bacterial origins were examined with *P. fluorescens* ATCC 13525, by supplementation in the succinate medium using the pyoverdines from *P. aeruginosa* (PAO1), *P. fluorescens* W, or *P. fluorescens* ATCC 17400 at the same concentration (50 µM), different responses were observed. As shown in Fig. 2, and supported also by the results depicted in Fig. 3, pyoverdine from *P. aeruginosa* stimulated the growth of *P. fluorescens* ATCC 13525 (to a lesser extent, however, than the homologous compound), whereas no effect was observed with the *P. fluorescens* W pyoverdine. The pyoverdine synthesized by a third strain belonging to the same species (*P. fluorescens* ATCC 17400) was shown to have a slightly inhibitory effect on growth of *P. fluorescens* ATCC 13525, by increasing the lag phase and lowering the growth rate compared to the growth observed in succinate medium without addition. Thus the growth of *P. fluorescens* ATCC 13525 can be affected in different manner by the presence of exogeneous pyoverdine depending on the producive strain, the highest stimulation being obtained with the homologous pyoverdine. Such conclusions were also valid for the other pseudomonad strains tested in a same manner. For example, the growth of *P. aeruginosa* was highly stimulated by addition of its own pyoverdine, whereas pyoverdine from *P. fluorescens* ATCC 13525 stimulated the growth to a lesser extent (data not shown). This strain, *P. fluorescens* ATCC 13525, was also shown to stimulate the growth of *P. fluorescens* W. Thus, the reciprocity of the effects when comparing two strains appears to be the general case, but is not always the rule since *P. fluorescens* W pyoverdine did not stimulate the growth of *P. fluorescens* ATCC 13525 (Fig. 2.).

The results obtained in liquid cultures were confirmed by a disc technique on agar solidified media with or without added EDDA (1 mg/ml). The pyoverdine impregnated paper disc system permitted the observation of faster growth or a slight inhibition surrounding the disc compared to the normal growth.

The interpretation of the results obtained from such growth experiments is not simple. In particular it is not possible to ascertain if the stimulating effect observed with some heterologous pyoverdines reflected a facilitated iron uptake, or was the consequence of the solubilization of iron which was then complexed in a second step by the native pyoverdine synthesized during growth by the bacteria. To resolve this problem uptake studies as well as binding studies were undertaken.

SPECIFICITY OF THE PYOVERDINE-FACILITATED IRON UPTAKE.

From uptake studies of labelled iron (^{59}Fe)-pyoverdine complex into *P. fluorescens* W cells previously grown in iron deficient conditions, it appears that iron translocation occurs by active transport since the incorporation was greatly decreased by uncoupler agents like dinitrophenol (DNP) or sodium azide plus iodoacetamide, or fully inhibited for cells incubated at 0°C or incubated in the absence of an energy source (Meyer and Hornsperger, 1978). Lower incorporation activities were observed when iron was complexed by other ligands like NTA, EDDA or citrate, demonstrating a preferential affinity of the cells for the pyoverdine-iron complex. Pyoverdines synthesized by other strains of fluorescent pseudomonads than the strain tested for iron uptake activities were also used as iron chelates. In growth studies, the results varied with the strain tested. For some strains like *Pseudomonas tolaasii* which is illustrated in Fig. 4, the uptake studies with different pyoverdines revealed a strict specificity for the native compound. Pyoverdins from *P. fluorescens* W and two related strains (ATCC 13525 and ATCC 17400), or from *P. aeruginosa* (PAO1), *P. chlororaphis* and *P. putida,* did not permit any iron incorporation in *P. tolaasii* cells when used as iron (III)-chelators. Similar conclusions

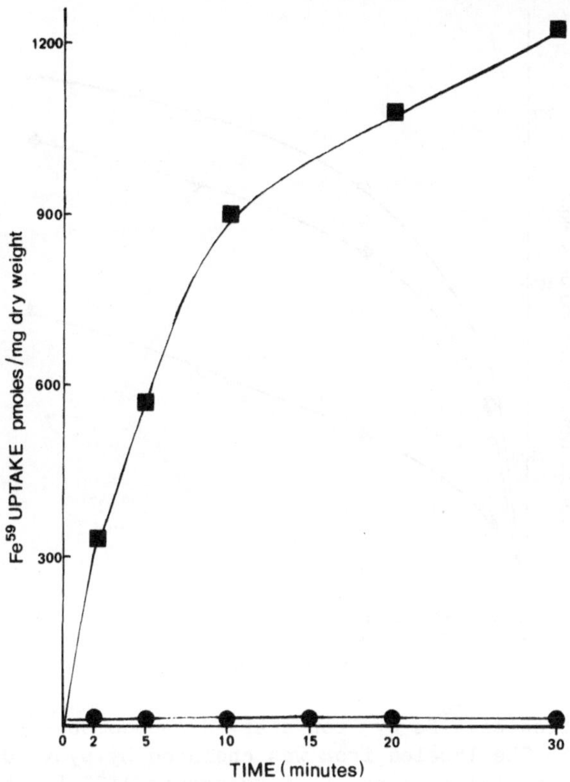

Fig. 4. Incorporation of ^{59}Fe by iron-starved *Pseudomonas tolaasii*. The radio-iron was chelated by pyoverdine of *P.tolaasii* (■) or by pyoverdines of different bacterial origins (●): *P. aeruginosa* ATCC 15692, *P. fluorescens* W, ATCC 13525 or ATCC 17400, *P. chlororaphis* ATCC 9446 or *P. putida* ATCC 12633.

were obtained from uptake studies performed with *P. fluorescens* ATCC 17400, *P. putida* ATCC 12633 or *P. fluorescens* W. On the contrary, as shown in Fig. 5, iron-deficient *P. fluorescens* ATCC 13525 cells were able to incorporate iron (III) when the metal was chelated with its own pyoverdine, but did in fact incorporate iron to a lesser extent when the ligands were pyoverdines of *P. chlororaphis* or *P. aeruginosa*. No incorporation occurred when the pyoverdines synthesized by *P. fluorescens* W, *P. fluorescens* ATCC 17400 or *P. putida* chelated the iron. The reverse relationships were verified with *P. aeruginosa* and *P. chlororaphis* strains. These strains were shown to incorporate iron when chelated with the three types of pyoverdines (*P. aeruginosa, P. chlororaphis* and *P. fluorescens* ATCC 13525 products). The greatest incorporation was found in each case with the homologous compound.

Binding experiments were performed in a same manner as the uptake studies, but with pure outer-membrane preparation of the different fluorescent pseudomonads cited above. The results lead to the same conclusions found with whole cells. For some strains (*P. fluorescens* W, *P. fluorescens* ATCC 17400, *P. putida* ATCC 12633, *P. tolaasii*), binding of the iron-pyoverdine complex on the outer membranes is strictly strain specific. For others (*P. aeruginosa* PA01, *P. chlororaphis* ATCC 9446, *P. fluorescens* ATCC 13525) the binding studies demonstrated similar relationships as demonstrated in the uptake studies of these three strains.

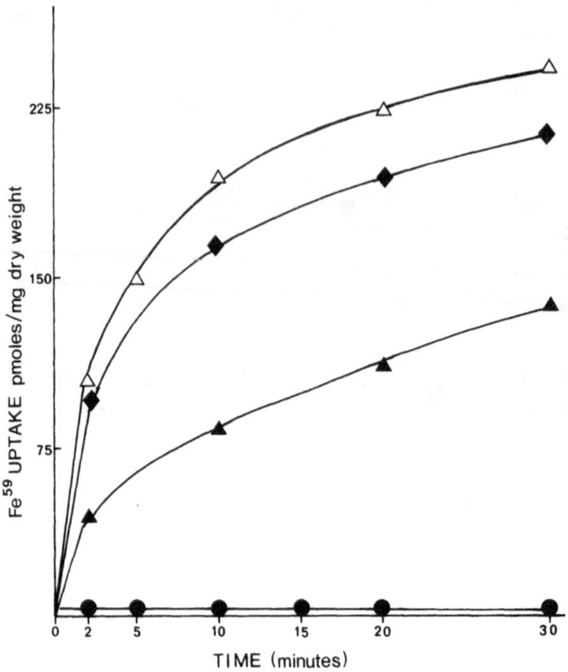

Fig. 5. Incorporation of ^{59}Fe by iron-starved *Pseudomonas fluorescens* ATCC 13525. The labeled iron was chelated by pyoverdines of various bacterial origins. *P. fluorescens* ATCC 13525 (Δ). *P. chlororaphis* ATCC 9446, (◆), *P. aeruginosa* ATCC 15692 (▲), or by pyoverdins from *P. fluorescens* W, *P. fluorescens* ATCC 17400 or *P. putida* ATCC 12633 (●).

CONCLUDING REMARKS

The siderophore-like nature of pyoverdine, the yellow-green fluorescent pigment characteristic of the fluorescent pseudomonads was defined first for a *P. fluorescens* strain (Meyer and Abdallah, 1978; Meyer and Hornsperger, 1978). Recently similar evidence was obtained for the pyoverdine synthesized by strain *P. aeruginosa* ATCC 15692 (Cox and Adams, 1985). This physiological role is now accepted for all the other strains which produce similar compounds and this view is supported by the results described above. For all the different strains tested, it was effectively established that pyoverdine is only synthesized by iron-starved cells, chelates iron (III) very tightly and participates actively in cellular iron uptake.

The heterologous uptake experiments described involving various pseudomonads and their respective pyoverdines, demonstrate that the pyoverdine-facilitated iron uptake is usually strain specific. Each strain synthesizes a chemically defined compound and has an affinity usually restricted to its own pyoverdine. Thus each strain has a specific pyoverdine facilitated uptake system, and is unable to utilise other pyoverdines, even when these are related by common chemical features, e.g., the presence of a common chromophore (Teintze et al., 1981; Wendenbaum et al., 1983). This specificity is also reflected in the differences between strains in electrophoretic mobility of the outer membrane proteins which appear during iron starvation and which may be the receptors for iron-pyoverdine complexes. (Meyer et al., 1979). The pyoverdine-facilitated iron uptake system of the pseudomonads appears to be a convenient model for the relationships between outer membrane receptors and their substrates at a molecular level.

In view of these results it would be of interest to know if there is a comparable specificity between soil fluorescent-pseudomonads and plants. If pyoverdine acts only as a strong iron scavenger, depriving saprophytic or pathogenic microflora in the rhizosphere of this essential nutrient, the beneficial effect should be non-specific.

It must be emphasized before concluding that some pseudomonads were seen to synthesize more than one siderophore when grown in iron-deficient conditions. In addition to pyoverdine, *P. fluorescens* ATCC 13525 excretes ferribactin (Maurer et al., 1968; Philson and Llinas, 1982a) which was shown to have some structural homology with pyoverdine, the chromophore moiety being replaced by amino-acids (Philson and Llinas (1982b). In contrast, the pyochelin of *P. aeruginosa* PA01 is structurally unrelated with the pyoverdine produced in the same conditions of growth by this strain (Cox et al., 1981). Recently pyochelin has been detected also in some other *Pseudomonas* strains (Sokol, 1984). The ability of the plant growth promoting fluorescent pseudomonads to synthesize siderophores other than pyoverdines, remains unknown.

ACKNOWLEDGEMENTS

We thank Dr. Bruce C Hemming for reviewing the manuscript and G. Seyer for appreciated technical assistance.

REFERENCES

Baghdiantz, A., 1952. Role of zinc in appearance of component II of the pigment of *Pseudomonas fluorescens* (Flugge-Migula). Archives Scientifiques de Genève, 5:47-48.

Chakrabarty, A.M., and Roy, S.C., 1964. Effect of trace elements on the production of pigments by a pseudomonad. Biochem. J., 93:228-231.

Cox, C.D., Rinehart, K.L., Moore, M.L., and Cook, J.C., 1981. Pyochelin: novel structure of an iron-chelating growth promoter for *Pseudomonas aeruginosa*. Proceedings of the National Academy of Sciences, U.S.A. 78:4256-4260.

Cox, C.D. and Adams, P., 1985. Siderophore activity of pyoverdin for *Pseudomonas aeruginosa*. Infec. Immun., 48:130-138.

Geels F.P., and Schippers, B., 1983. Reduction of yield depressions in high frequency potato cropping soil after seed tuber treatments with antagonistic fluorescent *Pseudomonas* spp. Phytopathol. Z. 108: 207-214.

Georgia, F.R., and Poe, C.F. 1931. Study of bacterial fluorescence in various media. I. Inorganic substances necessary for bacterial fluorescence. J. Bacteriol., 22:349-361.

Gouda, S., and Chodat, F., 1963. Glyoxalate et succinate, facteurs déterminant respectivement l'hypochromie et l'hyperchromie des cultures de *Pseudomonas fluorescens*. Pathologia Microbiologia, 26:655-664.

Gouda, S., and Greppin, H., 1965. Biosynthèse pigmentaire chez *Pseudomonas fluorescens* en fonction de la concentration du substrat hydrocarboné ou aminé. Archives Scientifiques de Genève. 18:716-721.

King, E.A., Ward, M.K., and Raney, D.E., 1954. Two simple media for the demonstration of pyocyanin and fluorescin. J. Lab. Clin. Med., 44:301-307.

King, J.V., Campbell, J.J.R., and Eagles, B.A., 1948. Mineral requirements for fluorescin production by *Pseudomonas*. Can. J. Res. 26C: 514-519.

Kloepper, J.W., Leong, J., Teintze, M., and Schroth, M.N., 1980. Enhanced plant growth by siderophores produced by plant growth promoting rhizobacteria. Nature, 286:885-886.

Lawson, E.C., Johnsson, C.B. and Hemming, B.C., 1985. Genotypic diversity of fluorescent pseudomonads as revealed by southern hybridization analysis with siderophore-related gene probes (this volume).

Leinhoff, H.M., 1963. An inverse relationship of the effects of oxygen and iron on the production of fluorescin and cytochrome C by *Pseudomonas fluorescens*. Nature, 199:601-602.

Lluch, C., Callao, V., and Olivares, J., 1973. Pigment production by *Pseudomonas reptilivora*. I. Effect of iron concentration in culture media. Archivs für Mikrobiologie, Deutschland. 93:239-243.

Maurer, B., Muller, A., Keller-Schierlein, W., and Zahner, H., 1968. Ferribactin, ein siderochrom aus *Pseudomonas fluorescens* Migula, Archivs für Mikrobiologie, 60:326-339.

Meyer, J.M., and Abdallah, M.A., 1978. The fluorescent pigment of *Pseudomonas fluorescens:* biosynthesis, purification and physicochemical properties. J. Gen. Microbiol., 107:319-328.

Meyer, J.M., and Hornsperger, J.M., 1978. Role of pyoverdine$_{Pf}$, the iron-binding fluorescent pigment of *Pseudomonas fluorescens*, in iron transport. J. Gen. Microbiol., 107:329-331.

Meyer, J.M., Mock, M., and Abdallah, M.A., 1979. Effects of iron on the protein composition of the outer membrane of fluorescent pseudomonads. FEMS Microbiol. Letts., 5:395-398.

Neilands, J.B., 1984. Siderophores of bacteria and fungi. Microbiol. Sci., 1:9-14.

Palleroni, N.J., 1984. Pseudomonadaceae. In: Bergey's Manual of Systematic Bacteriology, 1:141-199. (N.R. Krieg, Ed.). Williams and Wilkins, Baltimore, London.

Palumbo, S.A., 1972. Role of iron and sulfur in pigment and slime formation by *Pseudomonas aeruginosa*. J. Bacteriol., 111:430-436.

Philson, S.B., and Llinas, M., 1982a. Siderochromes from *Pseudomonas fluorescens*. I. Isolation and characterization. J. Biol. Chem., 257:8081-8085.

Philson, S.B., and Llinas, M., 1982b. Siderochromes from *Pseudomonas fluorescens*. II. Structural homology as revealed by NMR spectroscopy. J. Biol. Chem., 257:8086-8090.

Scher, F.M., and Baker, R., 1980. Mechanism of biological control in a Fusarium-suppressive soil. Phytopathology, 70:412-417.

Scher, F.M., and Baker, R., 1982. Effect of *Pseudomonas putida* and a synthetic iron chelator on induction of soil suppressiveness to Fusarium wilt pathogens. Phytopathology, 72:1567-1573.

Schroeter, S., 1870. Uber durch Bakterien gebildete Pigmente. Cohn's Beitrage für der Biologie der Pflanzen, 1:109-126.

Sokol, P.A., 1984. Production of the ferripyochelin outer membrane receptor by *Pseudomonas* species. FEMS Microbiol. Letts., 23:313-317.

Stanier, R.Y., Palleroni, N.J., and Doudoroff, M., 1966. The aerobic *Pseudomonas:* a taxonomic study. J. Gen. Microbiol., 43:159-271.

Sullivan, M.X., 1905. Synthetic culture media and the biochemistry of bacterial pigments. J. Med. Res., 14:109-160.

Teintze, M., Hossain, M.B., Barnes, C.L., Leong, J., and Van Der Helm, D., 1981. Structure of ferric pseudobactin, a siderophore from a plant growth promoting *Pseudomonas*. Biochemistry, 20:6446-6457.

Teintze, M., and Leong, J., 1981. Structure of pseudobactin A, a second siderophore from plant growth promoting *Pseudomonas* B10. Biochemistry, 20: 6457-6462.

Totter, J.R., and Moseley, F.T., 1953. Influence of the concentration of iron on the production of fluorescin by *Pseudomonas aeruginosa*. J. Bacteriol., 65:45-47.

Wendenbaum, S., Demange, P., Dell, A., Meyer, J.M., and Abdallah, M.A. 1983. The structure of pyoverdine Pa, the siderophore of *Pseudomonas aeruginosa*. Tetrahedron Letters, 24:4877-4880.

Yang, C.C., and Leong, J., 1984. Structure of pseudobactin 7RS1, a siderophore from a plant-deleterious *Pseudomonas*. Biochemistry, 23:3534-3540.

Author's Note: The material referred to here as EDDA is Ethylene-diamine dihydroxyphenyl acetic acid (EDDHA) and not Ethylene-diamine N-N' diacetic acid.

BACTERIAL SIDEROPHORES: STRUCTURE OF PYOVERDINS AND RELATED COMPOUNDS

P. Demange[1], S. Wendenbaum[1], A. Bateman[2], A. Dell[2], J.M. Meyer[3] and M.A. Abdallah[1*]

Laboratoire de Chimie Organique des Substances Naturelles[1]
Associé au CNRS, Département de Chimie
Université Louis Pasteur
1 rue Blaise Pascal, 67008-Strasbourg, France

Department of Biochemistry, Imperial College[2]
Imperial Institute Road, London, SW7 2AZ.

Laboratoire de Biochime Microbienne[3]
Institut Le Bel, Université Louis Pasteur
4 rue Blaise Pascal, 67000-Strasbourg, France

INTRODUCTION

The fluqrescent pseudomonads are bacteria which belong to the same intrageneric homology group, namely homology group No.1 (Palleroni et al., 1973). They are characterized by the biosynthesis, in iron-deficient conditions, of yellow-green, water-soluble fluorescent compounds, the pyoverdins, which are siderophores. Each strain of pseudomonad synthesizes mainly one pyoverdin, but closely related compounds can also be isolated from the culture media.

The structural determination of pyoverdins is a crucial stage in the study of the molecular mechanism of their action in iron transport and in the ensuing applications. The pyoverdins are chromopeptides possessing a peptide chain of 6 to 10 aminoacids bound to a chromophore derived from 2,3-diamino-6,7-dihydroxyquinoline (Philson and Llinas, 1982; Teintze et al., 1981, Wendenbaum et al., 1983). The structure pyoverdins has not been well researched due to the difficulties of obtaining crystals which can be used for an X-ray determination. Pseudobactin is the only example of a pyoverdin whose structure was solved by this method (Teintze et al., 1981).

In this paper we report the complete structure of the pyoverdin of *Pseudomonas aeruginosa* and other pyoverdins, and we describe how the techniques of FAB Mass Spectrometry and NMR Spectroscopy together comprise a powerful combination for the structural assignment of this type of peptide based siderophores.

We have used a similar approach in the characterization of fluorescent iron-chelating compounds isolated from the Azotobacteria. In iron-deficient conditions, *Azotobacter vinelandii* like the fluorescent pseudomonads excretes large amounts of a yellow-green, water-soluble compound, Azotobactin, which is the potential siderophore of this bacterium (Fekete et al., 1983; Page and Huyer, 1984). The structure of the pigment excreted by *Azotobacter vinelandii* strain O, which we call Azotobactin O, has been reported by Fukasawa et al., 1972. It is a chromopeptide (1) possessing a peptide chain bound to a fluorescent chromophore also derived from 2,3-diamino-6,7-dihydroxyquinoline. This compound presents some similarities with pyoverdins although it possesses only two chelating groups:- the catecholate from the chromophore and the hydroxyacid of β-hydroxyaspartic acid, instead of the three more common in such molecules. Moreover, no stereochemistry has been reported for Azotobactin O concerning either the chromophore or the aminoacids.

Chromophore-Aspartic acid-Homoserine-Serine-Homoserine-Citrulline-Serine-Glycine-Hydroxyaspartic acids

1

The structure elucidation of Azotobactin D, from strain D, is reported here, and we show that it differs significantly from the structure of Fukasawa et al., 1972.

PYOVERDIN Pa AND RELATED SIDEROPHORES OF *PSEUDOMONAS AERUGINOSA* (ATCC 15692)

Among the fluorescent pseudomonads, *P. aeruginosa* is of particular importance since it is pathogenic for weakened organisms (deeply burnt people for instance). In iron-deficient conditions, *P. aeruginosa* (ATCC 15692, PAO1) produces large amounts (*ca* 40 mg/l) of a mixture of one major fluorescent compound (Pyoverdin Pa) and two minor related pyoverdins (Pyoverdin Pa A and pyoverdin Pa B). These three pyoverdins were shown to transport iron into the cells of the bacteria.

Isolation and Purification of the Pyoverdin Pa Series

The procedure used was the same as previously described with slight modifications (Meyer and Abdallah, 1978) and was also used for the purification of the pyoverdins of other strains. The main steps are, after removal of the cells by centrifugation treatment of the supernatant with ethyl acetate (in order to remove a number of by-products), treatment of the supernatant with ferric chloride, concentration and extraction with 1:1 chloroform-phenol mixtures.

At this stage, the crude pyoverdin mixture is separated from most of the inorganic material. It was reextracted in aqueous phase before ion-exchange chromatography (CM-Sephadex pH 5.0). One major compound [pyoverdin Pa-Fe(III)] and two minor compounds [pyoverdin Pa A-Fe(III) and pyoverdin Pa B-Fe(III)] were isolated (Fig. 1).

For analytical purposes each compound was purified by reverse phase HPLC (ODS) using linear gradients of pyridine/acetic acid buffer - acetonitrile. The corresponding free ligands were also isolated after treatment of each iron complex with 8-hydroxyquinoline, ion-exchange column chromatography (CM-Sephadex, elution with a linear gradient 0.1 M to 1 M pH 5.0 pyridine/acetic acid buffer). For analytical purposes, each pyoverdin was repurified using reverse phase HPLC as above.

Physicochemical Properties of the Pyoverdins of *Pseudomonas aeruginosa*

Fast Atom Bombardment mass spectrometry gave a peak at M^+ = 1333 m.u.

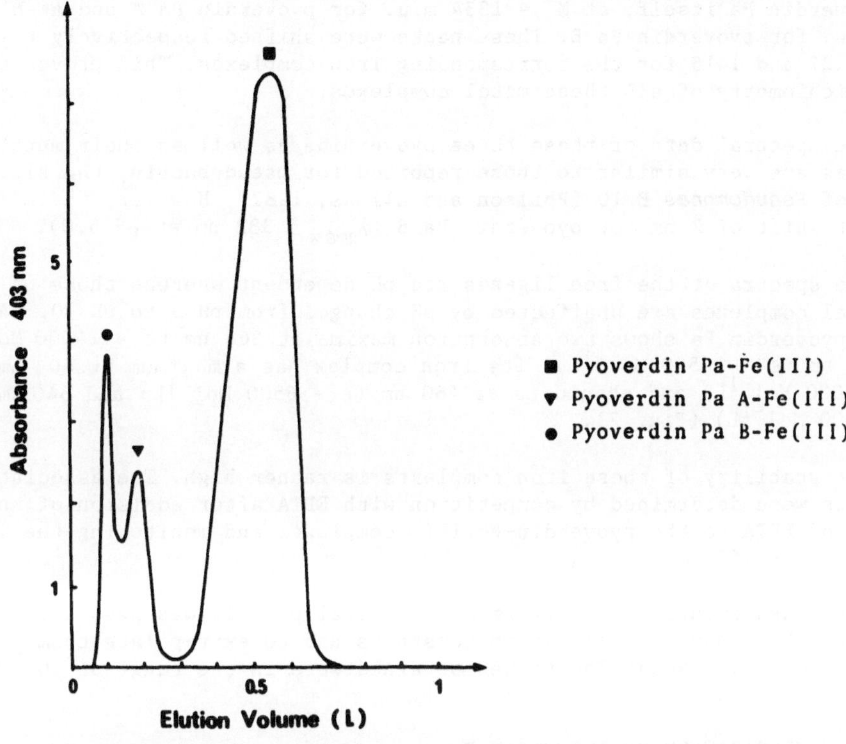

Fig. 1. Chromatography of pyoverdin Pa-Fe(III) on CM Sephadex C25, eluting with 0.1 M pyridine/acetic acid pH 5.0.

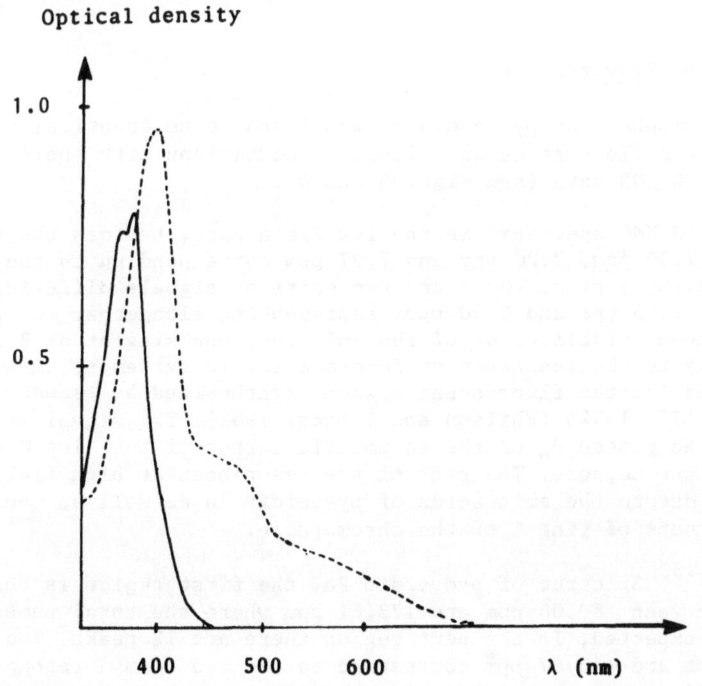

Fig. 2. Absorption spectra of pyoverdin Pa (——) and pyoverdin Pa - Fe(III) complex (- - -) in acetate buffer pH 5.0.

for pyoverdin Pa itself, at M^+ = 1334 m.u. for pyoverdin Pa A and at M^+ = 1362 m.u. for pyoverdin Pa B. These peaks were shifted respectively to 1386, 1387 and 1415 for the corresponding iron complexes. This proves the 1:1 stoichiometry of all these metal complexes.

The spectral data of these three pyoverdins as well as their metal complexes are very similar to those reported for pseudobactin, the siderophore of *Pseudomonas* B 10 (Philson and Llinas, 1982). However, there is a slight shift of 2 nm for pyoverdin Pa B (λ_{max} = 382 nm at pH 5.0).

The spectra of the free ligands are pH-dependent whereas those of the metal complexes are unaffected by pH changes from pH 3 to pH 10. At pH 5.0 pyoverdin Pa shows two absorption maxima at 364 nm (ε = 16000 $Mol^{-1}L$) and 380 nm (ε = 16500 $Mol^{-1}L$). Its iron complex has a maximum at 403 nm (ε = 19000 $Mol^{-1}L$) and shoulders at 460 nm (ε = 6500 $Mol^{-1}L$) and 540 nm (ε = 3500 $Mol^{-1}L$) (Fig. 2).

The stability of these iron complexes is rather high. The association constants were determined by competition with EDTA after addition of known amounts of EDTA to the pyoverdin-Fe(III) complexes and monitoring the absorption at 460 nm.

The measurements were performed at several pHs. It was possible to determine the apparent association constants and to extrapolate them (Anderegg et al., 1963). The values obtained were in the range of 10^{30}-10^{32}.

STRUCTURE ELUCIDATION OF PYOVERDIN Pa

Total acid hydrolysis of the siderophore indicated that it was constituted with one fluorescent chromophore which is modified during acid hydrolysis (see below), a peptide chain possessing serine (2), threonine (2), lysine (1), arginine (1) and N^{δ}-hydroxyornithine (2), and one mole of succinic acid.

NMR Spectra of Pyoverdin Pa

The chromophore of pyoverdin Pa was found to be identical to that of pseudobactin (Teintze et al., 1981) by comparison with their corresponding ^1H and ^{13}C NMR data (see Figs. 3 and 4).

In the ^1H NMR spectrum, in the low field part, besides the three singlets at 7.09 ppm, 7.20 ppm and 7.97 ppm corresponding to the protons on the quinolinium ring, there are two pairs of signals differing in intensity at 7.95 ppm and 8.30 ppm, representing altogether two protons. They give, upon acidification of the solution, one singlet at 8.22 ppm, corresponding to the resonance of formic acid, in agreement with what has been reported for the fluorescent pigment synthesized by *Pseudomonas fluorescens* ATCC 13525 (Philson and Llinas, 1982). The signal at 5.80 ppm represents the proton H_A of the asymmetric carbon of the Ring C of the fluorescent chromophore. The rest of the resonances at high field represent the signals due to the aminoacids of pyoverdin Pa as well as the two extra methylene groups of ring C of the chromophore.

In the ^{13}C Spectrum of pyoverdin Pa, the first region is that of the carbonyls between 180.08 ppm and 173.61 ppm where the total number of peaks is as expected. In the next region there are 12 peaks. Two of them at 162.61 ppm and 166.87 ppm correspond to the two formyl resonances, and one at 159.36 ppm to the carbon of the guanidine group of arginine. The nine signals left belong to the chromophore with three tertiary carbon atoms C-4, C-5 and C-8 at 142.08, 117.32 and 103.16 ppm, and six quaternary

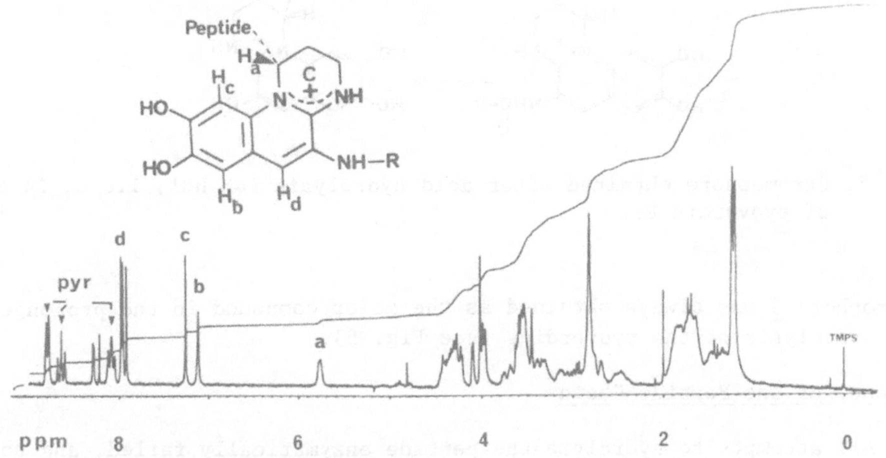

Fig. 3. ^{1}H NMR spectrum of pyoverdin Pa at 200 MHz in $^{2}H_2O$ with trimethylsilyl($^{2}H_6$)-propane sulphonate as an internal standard.

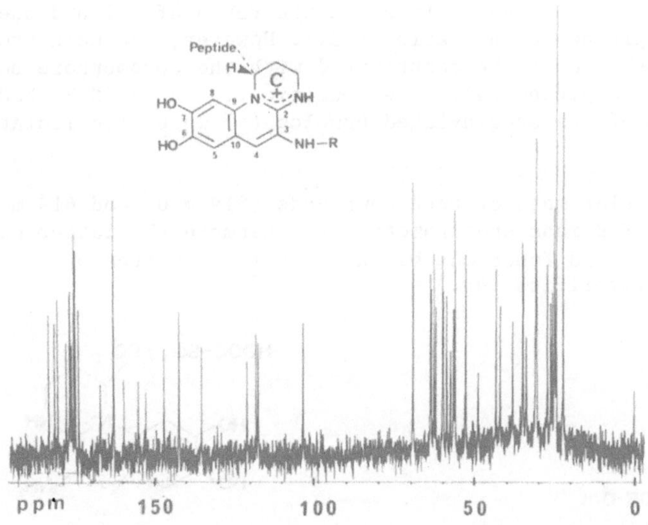

Fig. 4. ^{13}C NMR spectrum of pyoverdin Pa at 50 MHz in $^{2}H_2O$ using trimethylsilyl-($^{2}H_6$)-propane sulphonate as an internal standard.

carbons at 154.28, 152.48, 146.68, 134.77, 120.69 and 118.07 ppm, all at the same chemical shifts as those reported for the chromophore of pseudobactin (Teintze et al., 1981) or the siderochrome of *P. fluorescens* ATCC 13525 (Philson and Llinas, 1982). The rest of the signals from 69.27 ppm to 22.7 ppm correspond to the aminoacid resonances as well as to the three aliphatic carbon atoms of the ring of C of the chromophore.

The modification of the chromophore after acid hydrolysis (6M HCl, 110°C, 24 h) corresponds to the substitution of the amino group on the C-3 carbon atom of the quinoline ring by a hydroxyl group. This is very likely due to an attack by water on the carbon C-2 followed by hydrolysis of the enamine obtained and rearomatization. This new hydroxylated

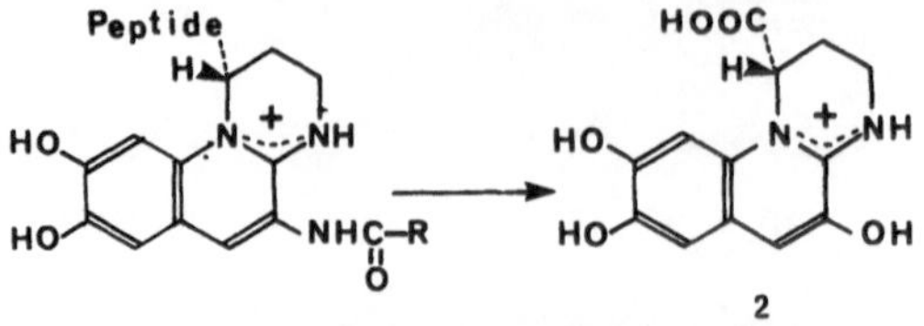

Fig. 5. Chromophore obtained after acid hydrolysis (6M HCl, 110°C, 24 h) of pyoverdin Pa.

chromophore 2 was always obtained as the major compound in the prolonged acid hydrolysis of the pyoverdins (see Fig. 5).

Sequence of the Peptide Chain

All attempts to hydrolyze the peptide enzymatically failed, and no free N- or C- terminus aminoacid could be detected. However, by mild acid hydrolysis (30 min, 100°C, 6M HCl) two dipeptides, Ser-OHorn and Lys-Ohorn, as well as a tetrapeptide Thr-Thr-Lys-OHorn could be isolated and purified by electrochromatography on cellulose sheets. The N-termini of the latter peptides were identified by dansylation.

Four fluorescent molecules could be isolated, one containing serine, two containing serine and arginine in the ratio of 1:1 and one containing serine and arginine in the ratio of 2:1. However, the main product of this hydrolysis was shown to be constituted with the chromophore bound to one serine and one arginine (3), after purification on a CM-Sephadex column. Small amounts of its succinylated homolog (4) were also isolated (see Fig. 6).

The molecular mass of both compounds (519 m.u. and 619 m.u.) were determined by FAB mass spectrometry. Furthermore the latter was shown to contain two acid functions because its mass shifted by 28 m.u. after methanol/HCl esterification.

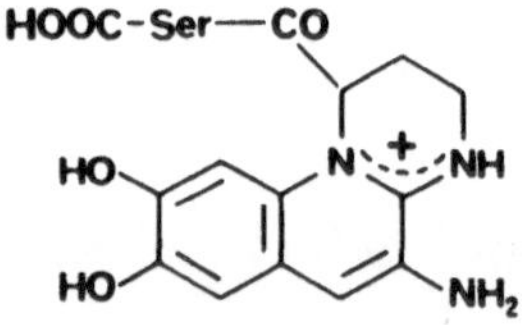

H_2N—Ser—δN-OH-Orn

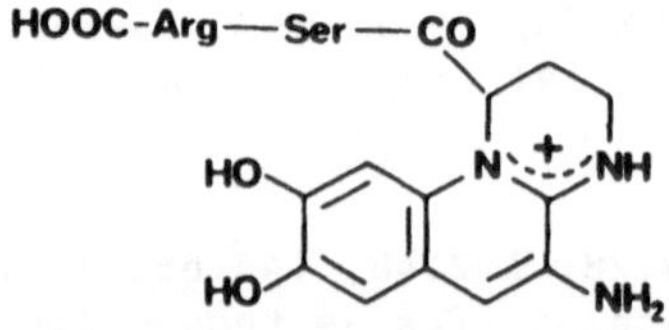

H_2N—Lys—δN-OH-Orn

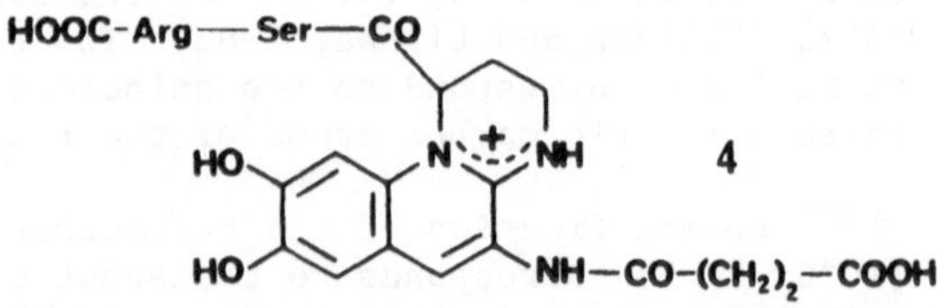

H_2N—Thr—(Thr, Lys, δN-OH-Orn)

Fig. 6. Compounds isolated after partial hydrolysis of pyoverdin Pa (6M HCl, 90°C, 30 min).

The time-course of mild acid hydrolysis of the siderophore was monitored by FAB mass spectrometry of aliquots of the hydrolysis mixture. A considerable proportion of the peptide sequence and the presence of the two formyl groups were established in this manner.

Very mild acid treatment (acidified methanol at pH 1, 20°C, 30 s) resulted in consecutive loss of the two formyl groups. More vigorous conditions (6M HCl, 90°C, time course up to 30 min), rapidly converted the succinamide moiety to succinic acid. This was followed by hydrolysis of the Arg-Ser bond and loss of succinic acid giving two major components having molecular masses at 519 m.u. and 678 m.u., consistent with the formerly mentioned compound (3), and the peptide (5). The latter was accompanied by a signal 18 m.u. higher, showing that partial opening of the N^{δ}-hydroxypiperidone ring had taken place. Further hydrolysis removed arginine from compound (3) and serine and N^{δ}-hydroxyornithine from the hexapeptide (5).

H_2N-Ser-N^{δ}-OHOrn-Thr-Thr-Lys-cyclic N^{δ}-OHOrn

5

Pyoverdin Pa, its metal complexes and the deformylated derivative exhibit a specific FAB cleavage across the saturated ring of the chromophore, giving a major fragment ion 302 m.u. below the molecular ion. The presence of this ion is therefore a diagnostic of the chromophore present in Pyoverdin Pa and related siderophores.

Both formyl groups are located on nitrogen atoms, one on the N^{δ} nitrogen atom of the hydroxyornithine in the middle of the peptide chain, and the other on the ε nitrogen group of lysine (and not on serine as previously suggested). This was established from examination of the ^{1}H and ^{13}C NMR spectra of the ^{15}N totally labelled pyoverdin Pa molecule (prepared from cultures in which the sole nitrogen source was $(^{15}NH_4)_2SO_4$) which showed the expected couplings between the formyl protons and carbons and ^{15}N.

Stereochemistry of the Different Chrial Groups of Pyoverdin Pa

The comparison of the circular dichroism (CD) spectra of pseudobactin and pyoverdin Pa which both exhibit the same positive Cotton effect in the 400 nm region, shows that their chiral center vicinal to the chromophores has the same absolute configuration which is (S).

The stereochemistry of the aminoacids was determined either after purification and measurement of their CD spectra, or by gas chromatography of their O-N-propyl-N-heptafluorobutyryl esters on a Chirasyl stationary phase capillary column (Chirasyl is a bound derivative of valine). Only the two serines were found to have an (R) configuration, all the other aminoacids having an (S) configuration, including the terminal N^{δ}-hydroxyornithine in its cyclic form (which was found to have the (R) configuration in Pseudobactin). Fig. 7 shows the complete structure of Pyoverdin Pa.

THE STRUCTURE OF PYOVERDIN Pa A AND PYOVERDIN Pa B

They both have the same peptide chain as pyoverdin Pa differing only in the chromophoric part of the molecules. Pyoverdin Pa A has identical electronic and NMR data as pyoverdin Pa. It differs only in one m.u. (M^+ = 1334 m.u.) from it. In fact it is the acid form of pyoverdin Pa

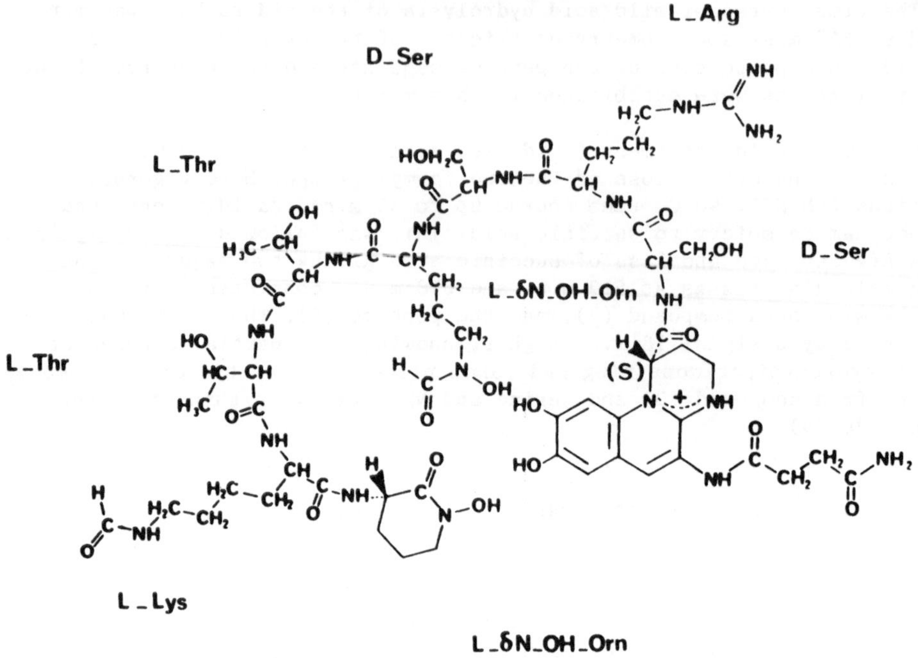

L_Arg

D_Ser

L_Thr

D_Ser

L_δN_OH_Orn

L_Thr

L_Lys

L_δN_OH_Orn

Fig. 7. The structure of pyoverdin Pa.

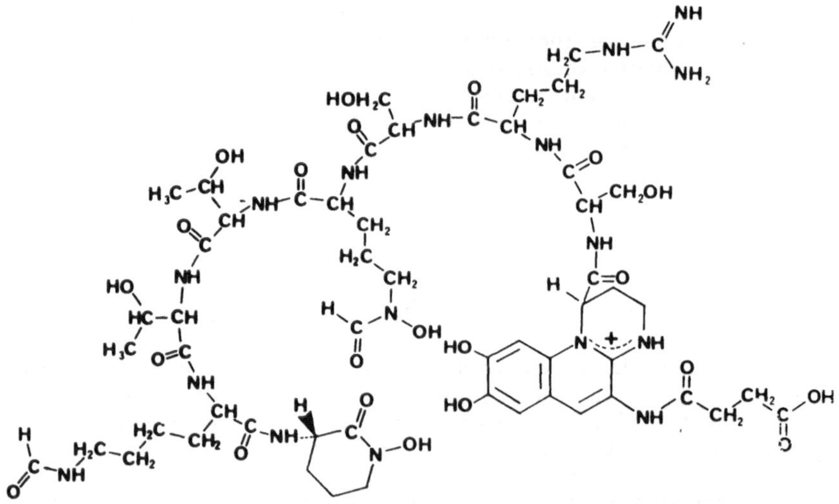

Figure 8. The structure of pyoverdin Pa A.

where the free amide of succinamide is fairly easily hydrolyzed to acid, very likely by an intramolecular mechanism (see Fig. 8).

This reaction is faster at higher pHs, and this can explain why in young cultures (where the Ph is still at 7.0) the amount of Pyoverdin Pa A is relatively low, whereas in older cultures (where the pH is nearly 9), pyoverdin Pa A is the major compound isolated. Therefore, pyoverdin Pa A is very likely a chemical by-product of pyoverdin Pa.

The other compound isolated, pyoverdin Pa B has also the same peptide chain and differs also from pyoverdin Pa in its chromophore. Its molecular mass, as measured by FAB mass spectrometry is 1362 m.u., and exhibits the characteristic chromophore loss at m/z = 1031. The molecular ion is 28 m.u. above that of pyoverdin Pa B, and this increment is lost in the formation of m/z = 1031. This 28 m.u. difference corresponds to an extra carbonyl group bound to the chromophore.

THE STRUCTURE OF AZOTOBACTIN D, SIDEROPHORE OF *AZOTOBACTER VINELANDII* STRAIN D

In iron deficient conditions, *A. vinelandii* like the fluorescent pseudomonads produces large amounts of a yellow-green water-soluble fluorescent compound, azotobactin, which presents striking similarities with the pyoverdins.

Isolation and Purification of Azotobactin

The purification procedure was different in this case since the Fe(III) complex of azotobactin was not readily extracted from the aqueous phase. Therefore the supernatant was treated with calculated amounts of barium chloride to remove most of the phosphates of the buffer and the solution was applied on a QAE-Sephadex column (chloride form). The fractions containing azotobactin were chromatographed first on a Biogel P2 column in order to remove most of the chlorides and then on a DEAE-Sephadex acetate type column eluted with pyridine/acetic acid buffer pH 5.0. In this process two compounds were isolated: azotobactin and azotobactin α. Both compounds exhibited very close physicochemical properties.

Physicochemical Properties of Azotobactin and Azotobactin α

Fast atom bombardment gave a peak at 1411 for azotobactin and 1393 for azotobactin α (18 m.u. below). The 1:1 stoichiometry of the corresponding iron complexes was also established by FAB mass spectrometry. The spectral data of azotobactin D are very similar to those reported for azotobactin O (Fukasawa et al., 1972). As for pyoverdin Pa, the spectra of the free ligands are pH dependent whereas the spectra of the iron complexes are unaffected by a change of pH. At pH 5.0, azotobactin D shows an absorption peak at 380 nm (ε = 23500 Mol^{-1} L) and a shoulder at 366 nm (ε = 19600 $Mol^{-1}L$). Its iron complex has a maximum at 412 nm (ε = 23000 $Mol^{-1}L$) and a shoulder at 450 nm (ε = 10000 $Mol^{-1}L$) and another shoulder at 550 nm (ε = 2000 $Mol^{-1}L$) (see Fig. 9).

STRUCTURE ELUCIDATION OF AZOTOBACTIN D

Total acid hydrolysis of azotobactin and azotobactin α indicated that both compounds were constituted with a fluorescent chromophore (unaltered during the hydrolysis procedure) and a peptide chain of 10 aminoacids composed of serine (2), homoserine (3), citrulline (1), Glycine (1), aspartic acid (1), N^{δ}-hydroxyornithine (1) and β-*threo*-hydroxyaspartic acid (1).

NMR Spectra of Azotobactin D

The ^{1}H NMR spectra show 3 singlets at 7.30 ppm, 7.4 ppm and 8.09 ppm corresponding to the three protons of the quinolinium ring. The signal at 6.03 ppm represents the H_A proton at the asymmetric carbon atom of the ring C bound to the quinoline moiety. The rest of the resonances at high field correspond to the signals due to the aminoacids of azotobactin (see Fig. 10).

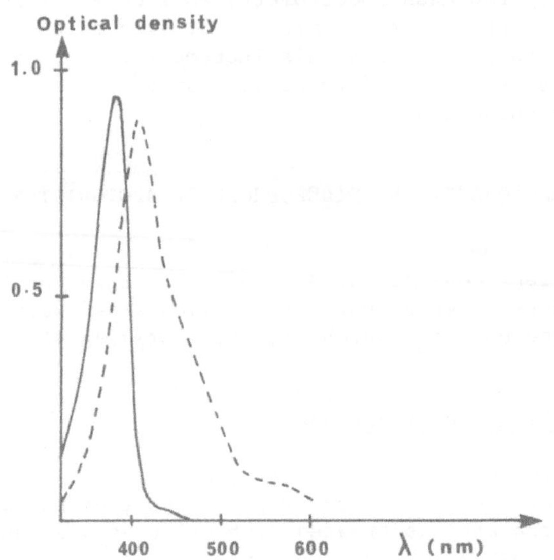

Fig. 9. Absorption spectra of azotobactin (——) and azotobactin - Fe(III) complex (- - -) in acetate buffer pH 5.0

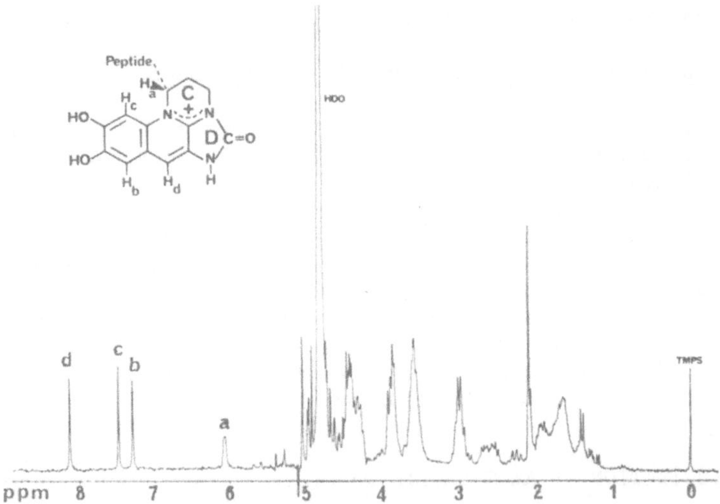

Fig.10. ^{1}H NMR spectrum of azotobactin D at 200 MHz in $^{2}H_2O$ using trimethylsilyl- ($^{2}H_6$)-propane sulphonate as an internal standard

In the ^{13}spectra of azotobactin, in the carbonyl region between 180.86 ppm and 164.34 ppm, the total number of peaks is as expected. In the next region there are 10 peaks. Nine of them belong to the quinoline ring with three tertiary carbon atoms (102.52, 115.83 and 123.32 ppm) and six quaternary carbon atoms (154.10, 148.58, 142.50, 131.16, 124.08, 122.52 ppm). The tenth signal corresponds to the carbonyl of ring D of the chromophore and occurs at 155.73 ppm (see Fig. 11). The rest of the signals from 74.80 ppm to 22.24 ppm correspond to the aminoacid resonances in addition to the three aliphatic carbon atoms of the ring C of the chromophore as well as an acetyl group resonance.

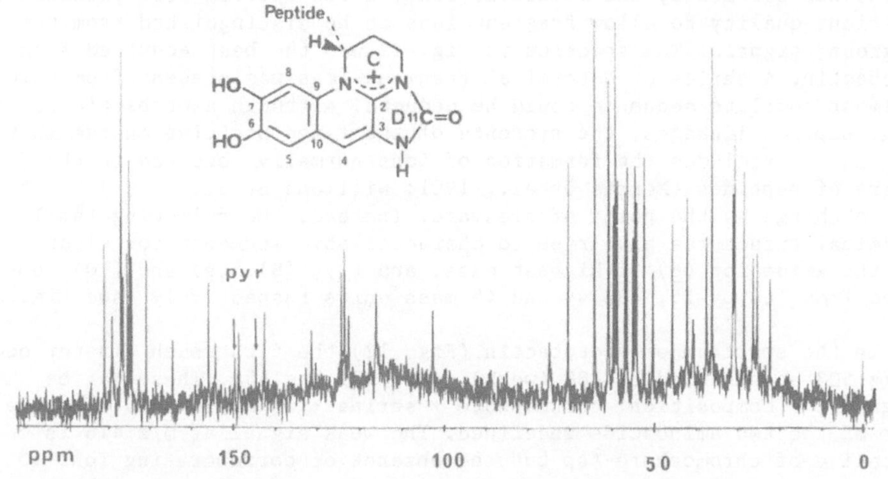

Fig. 11. ^{13}C NMR spectrum of azotabactin D at 50 MHz in $^{2}H_2O$ using trimethylsilyl-($^{2}H_6$)-propane sulphonate as an internal standard

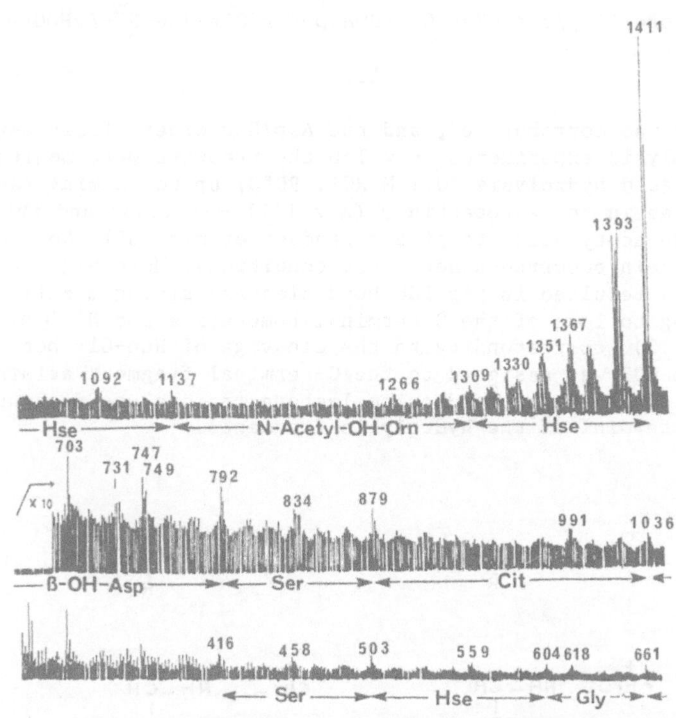

Fig. 12. FAB mass spectrum of azotobactin D

Determination of the Peptide Sequence

The peptide sequence of azotobactin was determined from the data afforded by FAB mass spectrometric analysis of the intact siderophore and products of partial hydrolysis. Although the majority of the ion current was carried by the molecular ions, a few spectra were obtained of sufficient quality to allow fragment ions to be distinguished from the background signals. The spectrum in Fig. 12 was the best acquired from azotobactin. A series of N-terminal fragment ions was present from which an almost complete sequence could be deduced. Although azotobactin contains normal peptide linkages, the presence of the fixed positive charge in the chromophore precludes the formation of ions normally observed in the FAB spectra of peptides (Morris et al., 1981; Williams et al., 1982) which carry a charge at the point of cleavage. Instead, the following family of N-terminal structures give rise to characteristic sequence ion clusters with the amide ion (6) at highest mass, and (7), (8), (9) and (10) separated from (6) by 15, 43, 44 and 45 mass units respectively (see Fig.13).

In the spectrum of azotobactin (Fig. 12) the first such cluster occurs at m/z 503 accompanied by 488 (weak), 460, 459 and 458. These can be assigned the composition: chromophore + serine + aspartic acid with the order of the two aminoacids undefined. The weak signal at m/z 416 is indicative of chromophore-Asp but the absence of corroborating ions 43, 44 and 45 below m/z 416 made this a very tentative assignment. The rest of the sequence can be deduced as shown in Table 1. Their companion fragment ions can be seen in Fig. 12.

In summary, the FAB fragmentation data suggested the following sequence for azotobactin D:

Chromophore-(Asp/ser)-Hse-Gly-HOAsp-Ser-Cit-Hse-N^{δ}-AcHOOrn-Hse

11.

This sequence was corroborated, and the Asp/Ser order firmly established by acid hydrolysis experiments in which the products were monitored by FAB-MS. Mild acid hydrolysis (0.1 M HCl, 90°C, up to 30 min) rapidly converted azotobactin to azotobactin α (m/z 1411 ⟶ 1393) and this was followed by de-acetylation to give a product at m/z 1351. No cleavage of the peptide chain occurred under these conditions. More vigorous conditions (6M HCl, 50°C) resulted in peptide bond cleavage giving m/z 1120 (12) (corresponding to loss of the C-terminal homoserine and N^{δ}-hydroxyornithine) and m/z 587 (13) (corresponding to the cleavage of Hse-Gly bond). A minor signal at m/z 807 was assigned to the C-terminal fragment arising from hydrolysis (14). A cyclic homoserine lactone residue was present at each of the new C-termini of the hydrolytic fragments.

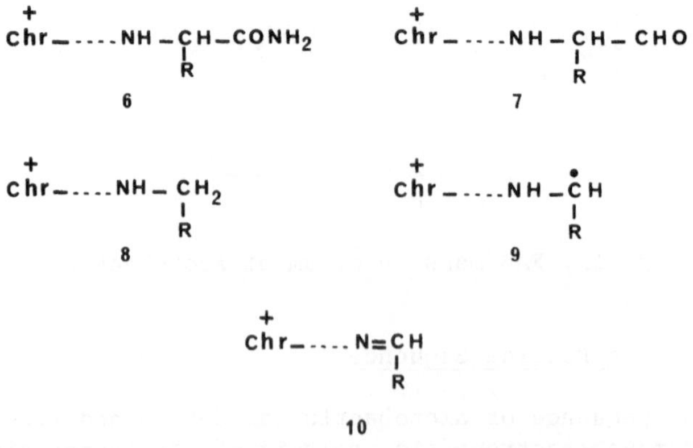

$\overset{+}{Chr}$—....NH—CH(R)—$CONH_2$ (6)

$\overset{+}{Chr}$—....NH—CH(R)—CHO (7)

$\overset{+}{Chr}$—....NH—CH_2(R) (8)

$\overset{+}{Chr}$—....NH—$\dot{C}$H(R) (9)

$\overset{+}{Chr}$—....N=CH(R) (10)

Fig. 13. Characteristic sequence ion clusters obtained in the FAB mass spectrum of azotobactin D.

Table 1. N-terminal amide ions of azotobactin and their interpretation

Ion	m/z
chr-(Asp + Ser)-$CONH_2$	503
chr-(Asp + Ser)-Hse-$CONH_2$	604
chr-(Asp + Ser)-Hse-Gly-$CONH_2$	661
chr-(Asp + Ser)-Hse-Gly-HOAsp-$CONH_2$	792
chr-(Asp + Ser)-Hse-Gly-HOAsp-Ser-$CONH_2$	879
chr-(Asp + Ser)-Hse-Gly-HOAsp-Ser-Cit-$CONH_2$	1036
chr-(Asp + Ser)-Hse-gly-HOAsp-Ser-Cit-Hse-$CONH_2$	1137
chr-(Asp + Ser)-Hse-Gly-HOAsp-Ser-Cit-Hse-N -AcHOOrn-$CONH_2$	1309

At higher temperatures (6M HCl, 90°C, 8 min) fragment (13) was the major product. More prolonged hydrolysis led to the formation of a species of molecular weight 417, precisely that expected for chromophore-Asp (15). This result provided firm evidence for the sequence 1 chromophore-Asp-Ser... rather than chromophore-Ser-Asp... in azotobactin. Compounds (13) and (15) were purified by HPLC, and upon aminoacid analysis, (13) gave one mole each of aspartic acid, serine and homoserine, while compound (15) gave only aspartic acid. The complete sequence of azotobactin D, deduced from the FAB results, is given below (16):

Chromophore-(Asp/ser)-Hse-Gly-HOAsp-Ser-Cit-Hse (lactone)
(12)
Chromophore-(Asp/ser)-Hse (lactone)
(13)
Gly-HOAsp-Ser-Cit-Hse-N^{δ}-AcHOOrn-Hse (lactone)
(14)
Chromophore-Asp
(15)
Chromophore-Asp-Ser-Hse-Gly-OHAsp-Ser-Cit-Hse-N^{δ}-AcHOOrn-Hse
(16)

Stereochemistry of the Chromophore and of the Aminoacids of Azotobactin D

The stereochemistry of the aminoacids was determined by derivation of acid hydrolyzate (HCl and HI) of azotobactin or its partial acid hydrolytic fragments, followed by gas chromatography on a capillary Chirasyl column. The derivatives were O-n-Propyl-N-heptafluorobutyryl esters or O-n-isoPropyl-N-trifluoroacetyl esters. As standards, (S)- and (R,S)-aminoacids constituting the peptide chain were separately treated the same way as the hydrolyzates of azotobactin D. It was found that all the homoserines were (S), citrulline and N^{δ}-hydroxyornithine were (R), aspartic acid and β-*threo*-hydroxyaspartic acid were (S) and both serines had opposite configurations. The same treatment performed on compound (13) established that the configuration of the serine on the chromophore side was (R).

The stereochemistry of the chromophore of azotobactin was established using CD technics by comparison with the chromophore of pyoverdin Pa which is the closest compound of known structure we could handle. Azotobactin D had a negative Cotton effect while pyoverdin Pa showed a positive one in the 300nm - 400nm range with a CD spectrum identical to that reported for

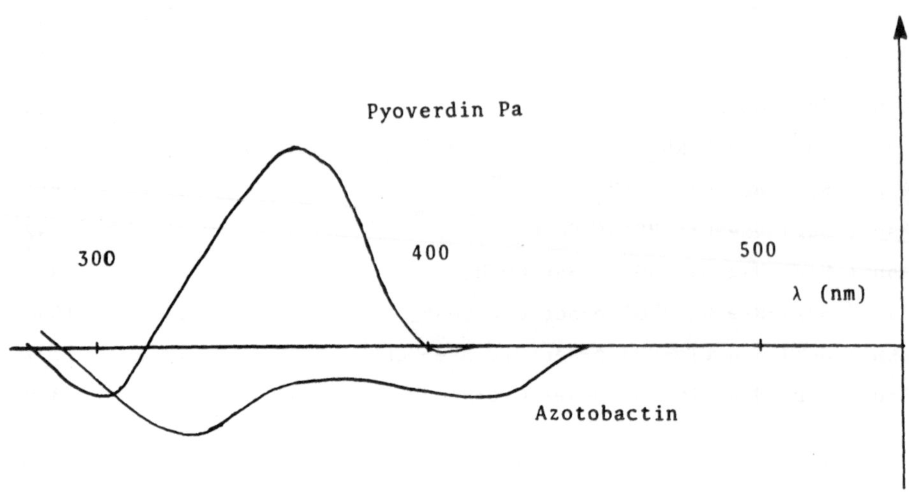

Fig. 14. CD spectra of pyoverdin Pa and azotobactin in acetate buffer pH 5.0.

pseudobactin (Teintze et al., 1981) (see Fig. 14). Total acid hydrolysis (6M HCl, 40 h, 110°C) of both siderophores cleaved the peptide chain leading to the modified hydroxylated chromophore (2) in the case of pyoverdin Pa and to the unchanged chromophore of azotobactin. Both compounds had still different CD spectra (data not shown). However, when the chromophore of azotobactin was subjected to prolonged acid hydrolysis with dilute acid (2M HCl, 150°C, 170 h), or when it was successively treated with ammonia (7M NH_4OH, 150°C, 3h) and then with hydrochloric acid (6M HCl, 150°c, 24 h) it yielded a compound which was shown to be identical to the trihydroxylated chromophore (2) deriving from pyoverdin Pa and having the same CD spectrum. This established that axotobactin D and pyoverdin Pa have chromophores with the same configuration (S). The complete structure of azotobactin D is reported in Fig. 15.

The Structure of Azotobactin α

The FAB spectrum of azotobactin α showed the same fragment ions as azotobactin up to and including the signal at m/z 1309. The 18 mass unit difference between the two siderophores must, therefore, reside at the C-terminal homoserine with conversion of homoserine to its lactone being the most likely explanation of the data. To confirm this hypothesis, the number of free acid groups in azotobactin α was established by FAB-MS monitoring of its esterification products.

Treatment of azotobactin α with an equimolar mixture of methanol and deuterated methanol containing anhydrous hydrochloric acid gave a 1:2:1 triplet at 1421, 1424 and 1427 consistent with the formation of a diester. Some methanolysis of the amide bonds occurred, despite the relatively mild esterification conditions affording the products whose molecular ions are present at m/z 1148, 1151, 1154 (1:2:1), 821, 824 (1:1) and 601, 604 (1:1). These correspond to the diester of peptide (12), the monoester of peptide (14) and the monoester of Peptide (13) respectively. The results clearly showed that azotobactin α contains only two free acid groups, both of which are present in its cleavage product (12) and neither of which, therefore, can be at the C-terminus. From the sequence, it can be seen that aspartic acid and hydroxyaspartic acid are the two residues carrying free carboxyl groups. Three free acid functions were expected for the

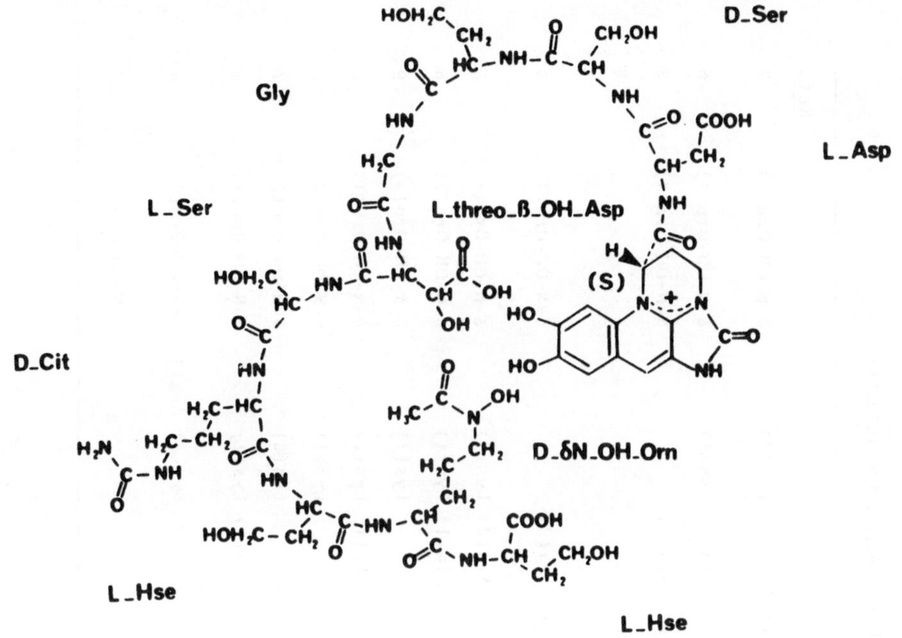

Fig. 15. The structure of axotobactin D.

siderophore containing a C-terminal carboxyl group. We conclude, therefore, that azotobactin α is lactonized at the C-terminus.

DISCUSSION

The techniques used to determine the structure of pyoverdin Pa and azotobactin were applied successfully to several pyoverdins from different strains of fluorescent pseudonomads, *Pseudomonas tolaasii, P. fluorescens* ATCC 13525 and ATCC 17400, *P. fluorescens, Pseudomonas* SB83, *Pseudomonas chlororaphis* ATCC 9446 etc. From this study it appeared clearly that all the pyoverdins so far investigated by us have the <u>same</u> chromophore to which is bound a <u>linear</u> peptide chain of 7 to 10 aminoacids (6 in the case of pseudobactin), containing in most cases serine and lysine and in every case <u>at least one</u> N^{δ}-hydroxyornithine. The other aminoacids are generally hydrophilic, acidic, neutral or basic and can have either (R) or (S) configuration: glycine, alanine, threonine, arginine, aspartic acid, glumatic acid in the case of most pyoverdins, and homoserine, and citrulline for Azotobactin (Table 2).

This chain is always blocked at its N-terminus by the chromophore and at its C-terminus by a cyclic N^{δ}-hydroxyornithine (for pyoverdins). In the middle of the chain there is always either a second N^{δ}- hydroxyornithine (acetylated or formylated on its δ nitrogen atom) or in some cases e.g. (*A. vinelandii, P. fluorescens*) β -*threo*-hydroxyaspartic acid which constitutes the third bidentate chelating group.

In the case of pyoverdins, an acyl group is always bound to the chromophore on the nitrogen at the 3 position. It is generally a succinamide group such as in pyoverdin Pa, but in one case of pyoverdin (*P. fluorescens* W) this acyl residue derived from malic acid. In the case of azotobactin, this acyl residue appears to be restricted to a carbonyl group linking the nitrogen atoms in position 2 and 3 of the chromophore and forming an extra imidazolone ring.

Table 2. Aminoacid composition of several pyoverdins and azotobactins

													Ref.
P. fluorescens ATCC 13525				Ser(2)				Gly(1)		Lys(2)		δ-N-OH-Orn(2)	7
P. B 10	β-*threo*-OHAsp(1)		*allo*-Thr(1)						Ala(2)	Lys(1)		δ-N-OH-Orn(1)	6
P. 7SR1	β-*threo*-OHAsp(1)		Thr(1)	Ser(3)				Gly(1)	Ala(1)			δ-N-OH-Orn(1)	8
A. vinelandii strain O	β-OHAsp(1)	Asx(1)		Ser(2)	Hse(2)		Cit(1)	Gly(1)					8
P. fluorescens W	β-*threo*-OHAsp(1)			Ser(1)				Gly(3)	Ala(2)			δ-N-OH-Orn(1)	8
P. fluorescens ATCC 17400	β-OHAsp(1)			Ser(1)		Glu(1)		Gly(2)	Ala(2)	Lys(1)		δ-N-OH-Orn(1)	9
P. SB83			Thr(1)	Ser(1)					Ala(1)	Lys(1)		δ-N-OH-Orn(2)	6
P. A6			Thr(2)	Ser(1)		Glu(1)		Gly(2)		Lys(1)		δ-N-OH-Orn(2)	9
P. L1	β-OHAsp(1)		Thr(2)	Ser(1)					Ala(1)	Lys(2)		δ-N-OH-Orn(1)	8
P. tolaasii			Thr(2)	Ser(5)						Lys(1)		δ-N-OH-Orn(2)	10
P. chlororaphis				Ser(2)				Gly(1)		Lys(2)		δ-N-OH-Orn(2)	7
P. putida	β-OHAsp(1)	Asx(1)	Thr(1)	Ser(1)		Glu(1)			Ala(1)	Lys(1)		δ-N-OH-Orn(1)	8
P. aeruginosa ATCC 15692			Thr(2)	Ser(2)						Lys(1)	Arg(1)	δ-N-OH-Orn(2)	8
A. vinelandii strain D	β-*threo*-OHAsp(1)	Asp(1)		Ser(2)	Hse(3)		Cit(1)	Gly(1)				δ-N-OH-Orn(1)	10

All these compounds constitute a new class of siderophores and the pyoverdins present these common features: a chromophore, 6 to 10 aminoacids and 3 bidentate chelating groups. These structural similarities between all these pyoverdins agree very well with their general behaviour and they act as siderophores for the bacteria which generate them, but they can also behave as growth factors or as antagonists towards other microorganisms.

REFERENCES

Anderegg, G., L'Epplantenier, F., and Schwarzenbach, G., 1963. Hydroxmatkomplexe. III. Eisen(III)-Austach zwischen Sideraminen und Komplexonen. Diskussion der Bildungskonstanten der Hydroxamatkomplexe. Helvetica Chimica Acta, 46:1409-1422.

Fekete, F.A., Spence, J.T., and Emery, T., 1983. Siderophores produced by nitrogen-fixing. *Azotobacter vinelandii* OP in iron-limited continuous cultures. Appl. Environ. Microbiol., 46:1297-1300.

Fukasawa, K., Goto, M., Sasaki, K., Hirata, Y., and Sato, S., 1972. Structure of the yellow-green fluorescent peptide produced by iron-deficient *Azotobacter vinelandii* strain O. Tetrahedron, 28:5359-5365.

Meyer, J-M., and Abdallah, M.A., 1978. The fluorescent pigment of *Pseudomonas fluorescens*. Biosynthesis, purification and physico-chemical properties. J. Gen. Microbiol., 107:319-328.

Morris, H.R., Panico, M., Barber, M., Bordoli, R.S., Sedgwick, R.D. and Tyler, A.N., 1981. Fast atom bombardment: a new mass spectrometric method for peptide sequence analysis. Biochem. Biophys. Res. Comm., 101:623-631.

Page, W.J. and Huyer, M., 1984. Derepression of the *Azotobacter vinelandii* siderophore system using iron-containing minerals to limit iron-repletion. J. Bacteriol., 158:496-502.

Palleroni, N.J., Kunisawa, R., Contopoulo, R. and Doudoroff, M., 1973. Nucleic acid homologies in the genus *Pseudomonas*. Int. J. Syst., Bacteriol., 23:333-339.

Philson, S.B. and LLinas, M., 1982. Siderochromes of *Pseudomonas fluorescens* II. Structural homology as revealed by NMR spectroscopy. J. Biol. Chem., 257:8086-8090.

Teintze, M., Hossain, M.B., Barnes, C.L., Leong, J., and van der Helm, D., 1981. Structure of ferric pseudobactin, a siderophore from a plant growth promoting *Pseudomonas*. Biochemistry, 257:8086-8090.

Wendenbaum, S., Demange, P., Dell, A., Meyer, J-M., and Abdallah, M.A., 1983. The structure of pyoverdin Pa, the siderophore of *Pseudomonas aeruginosa*. Tetrahedron Letters, 24:4877-4800.

Williams, D.H., Bradley, C.V., Santikarn, S., and Bojesen, G., 1982. Fast atom bombardment mass spectrometry. A new technique for the determination of molecular weights and aminoacid sequences of peptides. Biochem. J., 201:105-117.

METHODS OF STUDYING PLANT GROWTH STIMULATING PSEUDOMONADS: PROBLEMS AND PROGRESS

B. Schippers[1], F.P. Geels[1,2], P.A.H.M. Bakker[1], A.W. Bakker[2], P.J. Weisbeek[3] and B. Lugtenberg[4]

Willie Commelin Scholten Phytopathological Laboratory[1]
(Department of Plant Pathology, University of Utrecht
University of Amsterdam), Javalaan 20, 3742 CP Baarn
The Netherlands

present address: Mushroom Experimental Station[2]
Horst (L), The Netherlands

Department of Molecular Cell Biology[3]
University of Utrecht, Transitorium 3, Padualaan 8
3584 CH Utrecht, The Netherlands

Department of Molecular Botany, University of Leiden[4]
Nonnensteeg 3, 2311 VJ Leiden, The Netherlands

CROPPING FREQUENCY: A KEY FACTOR

The stimulation of plant growth by fluorescent pseudomonads has recently been reviewed (Schippers et al., 1985; Schippers et al., in press). Field- and pot experiments in the Netherlands strongly suggest that potato growth stimulation by seed tuber bacterization with pseudomonads selected for efficient siderophore mediated Fe^{3+}-uptake, is most noticable in soils where frequent potato cropping causes a significant reduction in potato yield (Geels and Schippers, 1983b; Geels and Schippers, 1983c; Schippers et al., 1985). Such yield reductions were more severe with increasing potato cropping frequency in long term rotational experiments (Hoekstra, 1981; Lamers, 1981; Schippers et al., 1985). They are largely caused by as yet unknown harmful rhizosphere microorganisms (HMO) rather than by well-known soil-borne pathogens (Scholte, et al.,1985; Schippers et al., 1985). The yield reductions were reproduced in pot experiments which could be almost eliminated if potato tubers were treated with selected pseudomonads before planting (Geels and Schippers, 1983a; Geels and Schippers, 1983b). In pot experiments significant stimulation of potato growth by bacterization of tubers was only observed in soil from fields frequently cropped with potato and not in soil from fields with no recent history of potato cropping. Similarly, in field experiments yield increases by tuber bacterization were only obtained in soils frequently cropped with potato. In some experiments yield increases were detected only during the first three months following planting and differences between treated tubers and controls were not significant at harvest-time. However, the rifampicin resistant *Pseudomonas* isolates used could be reisolated from roots of treated plants until harvest although their numbers gradually declined during the season (Geels et al., submitted for publication). Failure to increase yield in the field is one of the major problems. This is possibly due to inadequate colonization by PGSP of root parts that are responsible for tuber development later in the season. Failure of HMO to develop their harmful activity could be another factor, because yield reductions in short rotations were not apparent to the same extent every season and in every field plot.

A bioassay was developed to study both the microbial origin of yield reduction in short potato rotations and the mechanisms of growth stimulation by PGSP. The root-observation box developed by Van Vuurde and Schippers, (1980) was used to study root development in soils from different rotations. Within two weeks, root development of rooted potato stem cuttings was impaired in field soil cropped continuously (1:1) of every third year (1:3) with potato, compared to that in 1:6 soil from the same experimental fields. Availability of N, P and K and the physical structure were similar for all soils. Roots inhibited in growth showed no disease symptoms such as lesions or discolorations, etc. (Bakker et al., submitted for publication). Bacterization before planting of rooted stem cuttings with *Pseudomonas* isolate WCS 358 increased subsequent root growth in 1:1 and 1:3 soil by more than 100%. The root growth, however, did not always reach the level obtained with non-bacterized roots in 1:6 soil (Fig. 1).

A transposon Tn5 mutant of *P.* WCS 358, which had lots its fluorescence and its ability to produce siderophores (Tn5 Sid did not stimulate root

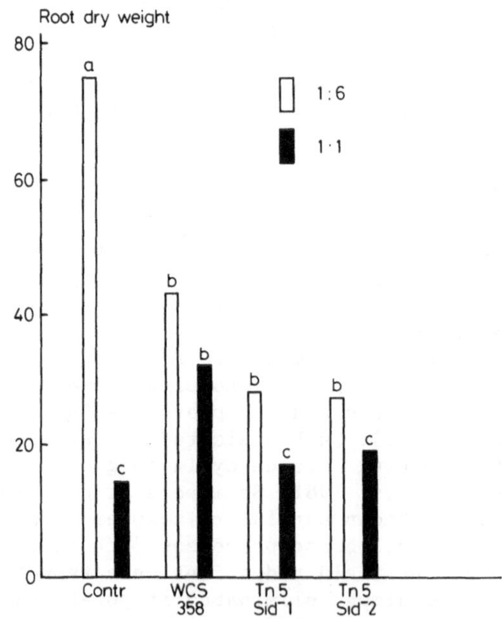

Fig. 1. Root dry weight in mg after 1 week of growth of rooted potato stem cuttings in soil from a field continuously cropped to potato (1:1) compared to that in 1:6 potato cropping frequency soil. The suppression of root development in 1:1 soil is partly compensated by treatment before planting of rooted sprouts with *P. putida* WCS 358, but not with its transposonmutants that lost their ability to produce siderophores (Tn5 Sid^-). Both WCS 358 and Tn5 Sid^- mutants inhibit root development in 1:6 soils similarly. This inhibition thus is not caused by siderophores of WCS 358 (P.A.H.M. Bakker).

growth (Fig. 1). Wild type *P.* WCS 358 and its Tn5 Sid$^-$ mutant were reisolated from potato roots in equal numbers (Bakker et al., submitted for publication). Siderophore production apparently plays an important role in root growth stimulation in short potato rotation soil. Based on this assumption, King's Medium B enriched with 8-hydroxy quinoline (8 oHQ) was used for further isolation and selection of plant growth stimulating pseudomonads (Geels et al., 1985). The selective action of 8 oHQ is based on iron stress induced by strong chelation of available iron.

Both isolate WCS 358 and its Tn5 Sid$^-$ mutant inhibited root development in 1:6 soil (Fig. 1). This may explain why bacterization with WCS 358 in 1:1 soil does not increase root development to the level obtained with non-bacterized roots in 1:6 soil. Isolate WCS 358 and its Tn5 Sid$^-$ mutant inhibit root growth in 1:6 soil almost equally. Therefore, siderophore production seems not to be responsible for impaired root growth in 1:6 soil (Bakker et al., submitted for publication).

Whether high cropping frequency is also a prerequisite for growth stimulation of other crops by fluorescent pseudomonads needs to be examined. In our experiments, yield increases in radish by seed bacterization with selected fluorescent pseudomonads was most noticeable in soil frequently cropped to radish (Geels, et al., 1985; Schippers et al., 1985).

HARMFUL MICRO-ORGANISMS

Inhibition of root development in short potato rotation soil in the absence of macro- and microscopically detectable symptoms could be due to an inhibition of the energy metabolism of potato root cells by certain microbial metabolites. Cyanide, a secondary metabolite of certain rhizosphere micro-organisms, could be the responsible metabolite. It is produced from e.g. glycine, which has often been found in root exudates. Cyanide inhibits cytochrome oxydase respiration in potato roots (A.W. Bakker, unpublished) resulting in a decrease in the efficiency of ATP production.

In *in vitro* experiments, concentrations as low as 5 μM HCN decreased the potato root O_2-consumption by 40% when the alternative electron transport pathway was inhibited by salicylhydroxamic acid (Bakker, submitted for publication). About 50% of the fluorescent pseudomonads isolated from potato rhizosphere both in 1:1 and 1:6 potato soil produced cyanide on King's Medium B supplemented with glycine (Bakker, submitted for publication).

Inhibition of the energy metabolism of root cells impairs energy demanding uptake of ions such as phosphate. Phosphorus concentrations in potato plants from soil frequently cropped to potato were significantly lower than in plants from a 1:6 potato cropping frequency (Hoekstra, 1981; J. Vos, personal communication).

Production of HCN by pseudomonads studied so far depends on the availability of iron (Castric, 1975; A.W. Bakker, unpublished results). Stimulation of potato root growth by PGSP thus could be a result of suppression of HCN production by HMO caused by siderophore competition for iron (Bakker, submitted for publication). If this hypothesis is valid, increasing yield reductions with increasing potato cropping frequency could either be explained by increasing numbers of HCN producing micro-organisms or by increasing HCN production without increase in numbers of responsible micro-organisms. The latter is most likely since no clear difference in numbers of HCN-producing pseudomonads was found in the potato rhizosphere of 1:1 and 1:6 soils. Most probably, an as yet unknown compound which stimulates HCN production by HMO in the potato rhizosphere accumulates in soil with increasing frequency of potato cropping (Bakker, submitted for publication).

The present hypothesis on the mechanism of plant growth promotion by fluorescent pseudomonads states that siderophores, induced by limiting Fe^{3+} concentrations in soil, bind Fe^{3+} thereby making this essential element unavailable for many other rhizosphere micro-organisms, including HMO (Schippers et al., in press). This hypothesis implicates suppression of all rhizosphere micro-organisms producing less siderophores or less efficient ones, including beneficial populations. Such a non-selective impact on the rhizosphere microflora is possible. However, a more general disturbance of the microbial equilibrium in the rhizosphere is difficult to reconcile with the beneficial effect of treatment with fluorescent pseudomonads. A more comprehensible hypothesis is that the production of phytotoxic metabolites by specific populations of rhizosphere micro-organisms is suppressed by the plant growth stimulating pseudomonads.

MOLECULAR MICROBIOLOGICAL APPROACHES

To increase the efficiency of applied fluorescent pseudomonads used as biological control agents, information is needed on (1) mechanisms of root colonization by pseudomonads and their host specificity and dependence on environmental factors, (2) genetics of the siderophore system and of factors controlling root colonization. With this information we may improve growth stimulation by pseudomonads and/or select, wild type isolates that are most suitable for certain applications. With this aim, a cooperative research project was initiated in 1984 between the 'Willie Commelin Scholten' Phytopathological Laboratory at Baarn (Department of Plant Pathology, Universities of Utrecht and Amsterdam), the Department of Molecular Cell Biology (University of Utrecht) and the Department of Molecular Botany (University of Leiden).

By using transposon Tn5 as the mutagenic agent and complementary studies with cosmid genebanks, the presence of at least 13 genes directly involved in siderophore production of isolate WCS 358 was indicated (Marugg et al., 1985; Weisbeek et al., this volume). Each of the plant growth promoting isolates *Pseudomonas putida* WCS 358 and *Pseudomonas fluorescens* WCS 374 produce a siderophore that, based on fluorescens and absorption spectra at different pH's, have a similar structure to pseudobactin (Geels et al., 1985; Marugg et al., 1985).

A highly specific antiserum was obtained for isolates WCS 374. Using immuno fluorescence microscopy (IMF) cells of this isolate were conclusively identified to occur on the root surface one week after bacterization in non-sterile soil. However, compared to data obtained with rifampicin resistant mutant of this isolate, the number of cells per cm root length determined with IMF was 100 to 1000 x less (unpublished data). Immunofluorescence technique opens possibilities for qualitative studies of root colonization but to date is unreliable in allowing quantification of specific isolates on root surfaces. Kanamycin/streptomycin resistanc Tn5 mutants of isolate WCS 358 selected for non-impaired *in vitro* growth are used in addition to rifampicin resistant mutants to study root colonization.

Pseudomonas-plant interactions are studied to identify mechanisms of root colonization. Physical adhesion of *Pseudomonas* cells to sterile potato roots has been observed and is used for the isolation of adhesive-defective mutants. A preliminary search for fimbriae on our *Pseudomonas* isolates was negative. Flagella-less mutants of isolate WCS 374 have been isolated and are being studied *in vitro* and *in vivo* for their role in root colonization (L. de Weger, unpublished data).

Outer membrane protein patterns of *Pseudomonas* isolates were shown to be a powerful tool for isolate identification. Cell envelope proteins regulated by ferric ions appeared to be of special value in this respect (Weger et al., 1985).

In conclusion, the comparitive analysis of soils with different cropping frequencies and the use of molecular biological methods significantly contribute to progress in understanding plant growth promotion by fluorescent pseudomonads. Many problems, however, must be solved before reliable applications of growth stimulating pseudomonads can be expected.

ACKNOWLEDGEMENT

Part of these investigations were supported by the Netherlands Technology Foundation (STW).

REFERENCES

Bakker, A.W., and Schippers, B. Yield reductions in short potato rotations: A possible role of cyanide produced by rhizossphere pseudomonads. (submitted for publication).

Bakker, P.A.H.M., Bakker, A.W., Marugg, J.D., Weisbeek, P.J. and Schippers, B. A bioassay to study the mechanisms of potato growth promotion by *Pseudomonas* spp. in short rotations. (submitted for publication).

Castric, P.A., 1975. Hydrogen cyanide, a secondary metabolite of *Pseudomonas aeruginosa*. J. Microbiol., 21:613-618.

Geels, F.P., Lamers, J.L., Hoekstra, O., and Schippers, B. Potato plant response to seed tuber bacterization in the field in various rotations. Netherlands Journal of Plant Pathology, (submitted for publication).

Geels, F.P. and Schippers, B., 1983a. Selection of antagonistic fluorescent *Pseudomonas* spp. and their root colonization and persistence following treatment of seed potatoes. Phytopathol. Z. 108: 193-214.

Geels, F.P. and Schippers, B., 1983b. Reduction of yield depressions in high frequency potato cropping soil after seed tuber treatments with antagonistic fluorescent *Pseudomonas* spp. Phytopathol. Z. 108: 207-214.

Geels, F.P. and Schippers, B., 1983c. Reduction of yield depressions in high frequency potato cropping by fluorescent *Pseudomonas* spp. Les Collogues de l'INRA, 18: 231-238.

Geels, F.P., Schmidt, E.D.L., and Schippers, B., 1985. The use of 8-hydroxyquinoline for the isolation and pre-qualification of plant-growth-stimulating rhizosphere pseudomonads. Biol. Fert. Soils, 1. (in press).

Hoekstra, O., 1981. 15 jaar 'De Schreef'. Resultaten van 15 jaar vruchtwisselingsonderzoek op het bouwplannenproefveld 'De Schreef' Publikatie PAGV, 11:1-93.

Lamers, J.G., 1981. Continueteelt in nauwe rotaties van aardappelen en suikerbieten. Publikatie PAGV., 12:1-65

Marugg, J.D., van Spanje, M., Hoekstra, W.P.M., Schippers, B., and Weisbeek, P.J., 1985. Isolation and analysis of genes involved in siderophore-biosynthesis in the plant growth-stimulating *Pseudomonas putida* strain WCS 358. J. Bacteriol., (submitted for publication).

Scholte, K., Veenbaas-Rijks, J.W. and Labruyère, R.E., 1985. Potato growing in short rotations and the effect of Streptomyces spp., *Colletrotrichum coccoides, Fusarium tabacinum* and *Verticillium dahliae* on plant growth and tuber yield. Potato res., 28, (in press).

Schippers, B., Geels, F.P., Hoekstra, O., Lamers, J.G., Maenhout, C.A.A., and Scholte, K., 1985. Yield depressions in narrow rotations caused by unknown microbial factors and their suppression by selected pseudomonads. In: "Ecology and management of soil-borne plant pathogens", 127-130. (C.A. Parker, A.D. Rovira, K.J. Moore, P.T.W. Wong, eds.). The American Phyotological Society, St. Paul, Minnesota.

Schippers, B., Lugtenberg, B. and Weisbeek, P.J. Plant growth control by fluorescent pseudomonads. In: "Non-conventional approaches to plant disease control". (I. Chet, ed.), Wiley and Sons, New York. (in press).

Schroth, M.N., and Hancock, J.G., 1982. Disease-suppressive soil and root-colonizing bacteria. Science, 216:1376-1381.

Suslow, T.V., 1982. Role of root-colonizing bacteria in plant growth. In: "Phytopathogenic prokaryotes". Vol. I. pp. 187-223. (M.S. Mount and G.H. Lacey, eds. Academic Press, New York-London-Paris-San Diego-San Francisco-San Paulo-Sydney, Tokyo-Toronto.

Van Vuurde, J.W.L., and Schippers, B., 1980. Bacterial colonization of seminal wheat roots. Soil Biol., Biochem., 12:559-565.

Weger de, L.A., van Boxtel, R., van der Burg, B., Gruters, R., Geels, F.P., Schippers, B., and Lugtenberg, B., 1985. Outer membrane proteins of plant growth-stimulating root-colonizing *Pseudomonas* spp. (submitted for publication).

EMERGENCE-PROMOTING RHIZOBACTERIA:

DESCRIPTION AND IMPLICATIONS FOR AGRICULTURE

J.W. Kloepper, F.M. Scher, M. Laliberté, and B. Tipping

Allelix Inc.
6850 Goreway Drive
Mississauga, Ontario, L4V 1P1, Canada

INTRODUCTION

Specific strains of root-colonizing bacteria, termed plant growth-promoting rhizobacteria (PGPR), have recently been used as experimental inoculants to increase yield of sugar beet (Suslow and Schroth, 1982), radish (Kloepper and Schroth, 1978) and potato (Burr et al., 1978; Howie and Echandi, 1983; Kloepper et al., 1980). In each case, evidence of enhanced plant growth was observed sometime during the early-growth season prior to harvest. Increased early-season development of potato was manifested by increased stolon lengths on PGPR-treated plants (Kloepper et al., 1980; Kloepper and Schroth, 1981A; Kloepper and Schroth, 1981B). The earliest indication of PGPR-enhanced growth of sugar beet was a significant increase in seedling weight (Suslow and Schroth, 1982); seedling emergence was not affected by PGPR treatments.

During a field screening program for new PGPR strains for soybean, we observed that some bacteria induced increases in seedling emergence of 100% greater than controls (Kloepper and Scher, unpublished). This emergence promotion was repeated in the field, using the same bacterial strains, only when soil temperatures were below 20°C.

The objective of the work described herein was to expand upon our initial observations and determine if a specific group of PGPR could be identified which would increase emergence of seedlings in cold field soils. We termed such bacteria "EPR" (Emergence-promoting rhizobacteria). Our strategy was three fold: to develop an emergence assay in which we could screen phychrotrophic rhizobacteria for EPR activity; to determine the consistency of emergence enhancement by select EPR in the assay; and finally to field test strains which were consistent in the assay. We chose soybean and canola (rapeseed) as model crops.

METHODS

Isolation of Bacterial Strains

Two isolation procedures were used during the course of this study. For procedure one, canola and soybean seeds were surface-sterilized by rinsing for 5 minutes in 95% ethanol, rinsed in sterile water, soaked

5 min. in 1.5% sodium hypochlorite, and rinsed in sterile water again. Seeds were planted in various soil samples at 10 to 14°C. Roots from developing seedlings were removed, washed in sterile water to remove loosely adhering soil particles and ground in 5 ml sterile 0.1 M $MgSO_4$. Serial ten-fold dilutions were plated onto Pseudomonas Agar F (PAF)(Difco Labs, Detroit, MI, USA 48232) and plates were incubated 2 weeks at 14°C. Colonies were purified on PAF at 20°C.

For isolation procedure two, roots of plants collected in the eastern Northwest Territories, Canada, were washed to remove soil particles and placed directly onto asparagine soft agar (ASA). ASA contained 1g L-asparagine, 2g Bacto Agar and 1000 ml distilled water and was previously used to assess bacterial chemotaxis as an indicator of root-colonization capacity (Scher, unpublished). Bacteria which grew out from root segments on ASA were then purified on PAf plates at 20°C.

Strains isolated using procedures one and two were restreaked on PAF plates, and examined for rapid growth at 4, 10 and 14°C. Strains which developed an observable lawn in 24 h at 14°C, 48 h at 10°C and 4-5 d at 4°C were further tested for growth on exudate agar at 20°C. Exudate agar was prepared by mixing 10% soybean or 20% canola seed exudates with 2% washed purified agar (Difco). Exudates were prepared as described previously (Scher et al., 1985).

Identification and Storage of Bacteria

Purified bacterial strains were stored in glycerol at -80°C prior to being tested in the assays. Strains which induced emergence increases were rechecked for purity on PAF, and 10 copies of each strain were returned to -80°C storage. A new vial of bacteria was used for each emergence assay. Identification was done only for strains which repeated emergence-promoting activity. All strains were Gram-negative and were further tested for reaction profiles on API 20E test strips (Analytab Products, Ayerst Laboratories, Inc. Plainview, N.Y., U.S.A.). Additional tests included growth on MacConkey medium, type of metabolism in OF glucose medium, production of fluorescent pigment, gelatine hydrolysis, nitrate reduction, starch hydrolysis, oxidase reaction, production of DNase, and lipase production (Tween 80 hydrolysis). Methods for all of the above biochemical tests were those recommended by the American Society for Microbiology. The identifications of emergence-promoting strains are listed in Table 1.

Canola Emergence Assay

The following assay was developed to assess emergence of canola (*Brassica campestris* cv 'Tobin'). Field soil was collected from the Allelix Field Research Centre near Caledon, Ontario and consisted of a clay loam with 2% organic matter, pH 7.0, total exchange capacity (M.E.) 14, and with the following nutrient levels in ppm: nitrate nitrogen 4, phosphorous 1, potassium 2, calcium 70, magnesium 16, sodium 0.5, boron 0.4, iron 550, manganese 130, copper 2, and zinc 7. Soil was thoroughly mixed in a 1:1 ratio with perlite and the resulting mix was used throughout the study.

Test bacteria were grown on PAF plates at 10°C for 3 days, scraped off plates, and mixed in 0.1 M $MgSO_4$. Canola seeds were agitated in the bacterial suspensions for 2 h at 10°C prior to planting 20 seeds in each of 8 replicate 15 cm pots. Seeds were planted 2 cm deep and pots were watered immediately and placed at 9°C. Each experiment consisted of 6 to 8 bacterial treatments with one control. The control consisted of canola seeds soaked in 0.1 M $MgSO_4$ which had been poured over an uninoculated PAF plate. Pots were examined daily and the number of emerged seedlings was recorded.

Table 1. Identification of emergence promoting bacteria

Soybeans	Strain designation	Identification
	1-104	*Pseudomonas putida*
	1-226	*Pseudomonas fluorescens*
	1-206	*Pseudomonas fluorescens*
	2-16	*Serratia liquefaciens*
	2-18	*Serratia liquefaciens*
	2-20	*Serratia liquefaciens*
	2-22	*Pseudomonas putida biovar B*
	2-67	*Serratia liquefaciens*
	2-114	*Enterobacter aerogenes*
	17-114	*Pseudomonas putida*
	17-29	*Pseudomonas fluorescens*
	17-76	*Pseudomonas putida*
	17-34	*Pseudomonas fluorescens*
	G25-25	*Pseudomonas fluorescens*
	G25-26	*Pseudomonas putida*
	G25-44	*Pseudomonas putida*
	G20-20	*Pseudomonas fluorescens*
	G23-34	*Pseudomonas putida*
	G24-16	*Pseudomonas putida*
	G24-14	*Pseudomonas putida*
	G24-3	*Pseudomonas putida*
	G20-18	*Pseudomonas fluorescens*
	1-102	*Serratia liquefaciens*
Canola	**Strain designation**	**Identification**
	G1-1	*Beijerinckia* spp.
	G1-3	*Pseudomonas fluorescens*
	G1-4	*Beijerinckia* spp.
	52-30	*Pseudomonas putida*

A total of 50 bacterial strains isolated using procedure one (see previous section) and 60 strains isolated using procedure two were tested for emergence promotion relative to controls. Strains which demonstrated significant (P=0.05) emergence promotion were retested 3 times to determine the consistency of emergence promotion.

Soybean Emergence Assay

The initial assay which was used for selection of soybean EPR strains was used from June 1983 through February 1984. Candidate EPR strains were grown for 48 h on PAF plates at 14°C and scraped into 50 ml 0.1 M $MgSO_4$. One hundred fifty soybean seeds (cv. 'Maple Presto' or 'Maple Arrow') were added to each 50 ml suspension and were shaken at 100 RPM at 10°C for 3 h. Typical experiments consisted of 6 bacterial treatments with one $MgSO_4$ control, each with 9 to 10 replications. Each replication consisted of 12 seeds planted in a 12-well plastic seeding tray (Plant Products Ltd., Bramalea, Ontario) with overall dimensions of 18 cm wide x 27 cm long x 6 cm deep and with dimensions of individual wells of 6 cm long x 5 cm wide x 6 cm deep. Seeds were planted 3 cm deep in "conditioned field soil".

"Conditioned field soil" was prepared by mixing soil from the Allelix Field Research Centre (described in the previous section) in a 1:5 ratio with Promix C (Plant Products Ltd., Bramalea, Ontario). Soybean was seeded into flats containing the soil mixture and grown to the second true leaf stage when the plants were discarded. The same soil (termed "conditioned soil") was reblended and used in the soybean emergence assay.

After planting, each replicate seeding tray was watered and placed at 14°C. The number of emerged seedlings was recorded after 14 d, and data were analysed using a one-way analysis of variance to detect significant differences between treatment means. A total of 277 strains which were isolated using isolation procedure one, described above and 84 strains which were isolated using procedure two were tested for emergence-promoting activity in this assay. Strains which induced a significant increase in emergence in the first trial were retested twice using the same assay procedures.

A second assay was used from February 1984, through February 1985, for strains which demonstrated repeated EPR activity in the initial assay. Soybean seeds were shaken in bacterial suspensions or in 0.1 M $MgSO_4$ as described above and planted in a 1:1 mix of Allelix field soil:perlite. Five cm of the mix was placed in the bottom of 25 cm plastic azalea pots (Kord Plastics Ltd., Toronto); 20 seeds of the same treatment were placed on the soil: soil perlite mix was added to give a planting depth of 5 cm. Pots were immediately watered and placed at 12 to 14°C. Each pot of 20 seeds constituted a single replication, and 8 replications were used per treatment. Typical experiments consisted of 5 to 7 bacterial treatments with one 0.1 M $MgSO_4$ control. Emergence was recorded daily and strains were deemed "EPR" when they induced a 50% increase in emergence of controls for 3 consecutive days in 2 of 3 repeating experiments.

Preliminary Mode-of-Action Studies

Eleven soybean EPR strains were selected to test for possible siderophore action, which has been suggested to account for some plant growth-promoting rhizobacteria (PGPR) strains' mode-of-action. Emergence assays were set up as described above except that 16 replications were used instead of 8. Eight replications of each treatment and control were watered with 10^{-3} M $FeCl_3$, and 8 replications were watered with water immediately after seeding. Soil pH was recorded before and after addition of $FeCl_3$. Each experiment was repeated once.

RESULTS

Isolation of Bacterial Strains

Using isolation procedure one, 630 strains were obtained from soybean and 450 from canola. Of these strains, 277 soybean and 50 canola strains were found to grow on PAF in 4-5 days at 4°C and on exudate agars in 24 h at 20°C.

With isolation procedure two, 940 strains were obtained by direct isolation of chemotactic zones from roots on asparagine soft agar. Approximately 250 of these grew on PAF in 4-5 days at 4°C.

Canola Emergence Assay

Of 50 strains obtained using isolation procedure one, 3 induced

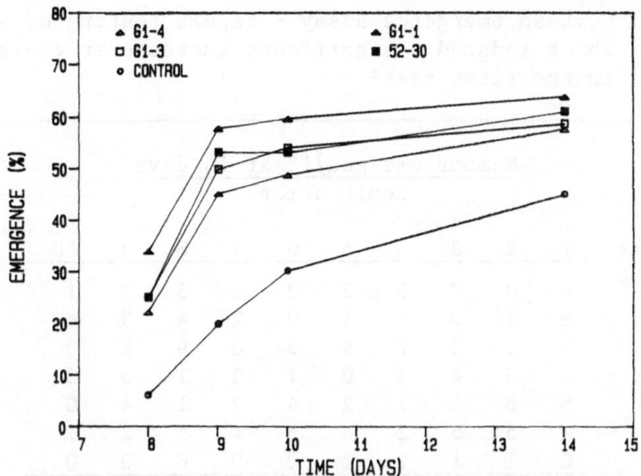

Fig. 1. Canola emergence promotion assay showing increases in emergence by 4 EPR. Assay was conducted in a field soil/perlite mix at 9°C with seeds shown to 2 cm depth. The percentage emergence values shown are the mean of 8 replications, each sown with 20 seeds. Similar results were obtained with the same 4 bacterial strains in 3 of 5 repeat experiments.

increases in emergence of 50% or greater than controls in first tests. None of these 3 increased emergence in 3 repeat experiments. Sixty strains obtained using isolation procedure two were tested in the canola assay and 10 increased emergence 40% or more in first trials. Four strains consistently increased emergence in 3 of 5 repeat trials (Fig.1). The percentage of emerged seedlings with bacterial treatments was 4 to 7 times greater than the percentage of emerged controls 8 days after seeding, 2 to 3 times greater at 9 days and 40 to 50% greater at 14 days.

Table 2. Soybean emergence assay initial selection untested strains*

	Number emerged/12 at 14 days										
	Replication										
Treatment	1	2	3	4	5	6	7	8	9	10	$\bar{x}$
1	2	4	2	3	2	5	3	3	2	4	3.0**
2	1	3	1	2	0	2	1	3	1	0	1.4
3	1	2	1	0	1	1	2	1	0	1	1.0
4	2	2	3	2	3	2	3	5	2	3	2.7**
5	1	2	2	1	0	2	1	1	1	0	1.1
6	0	1	1	2	3	1	1	2	0	1	1.2
Control	2	2	1	1	1	0	3	1	2	1	1,4

LSD 0.01 = 1.1 F = 8.1

* Assay was conducted at 14°C. The data shown are from one typical experiment. A total of 361 strains were tested over a 9 month period. See text for details.

Table 3. Soybean emergence assay - repeat testing of strains which induced a significant increase in emergence in the first test*

	Number emerged/12 at 14 days										
	Replication										
Treatment	1	2	3	4	5	6	7	8	9	10	$\bar{x}$
2-16	1	0	7	3	3	3	1	3	2	1	2.4*
2-17	6	1	0	1	1	0	2	4	3	6	2.4*
2-18	3	1	5	2	9	3	6	9	5	8	5.1**
2-19	1	3	4	1	0	1	3	1	3	8	2.5*
2-20	5	6	1	1	2	4	2	1	4	0	2.6*
2-21	2	5	5	2	4	5	4	5	2	7	4.1**
Control	0	1	1	1	2	0	0	0	0	0	0.5

LSD 0.05 = 1.8 F = 4.94
0.01 = 2.5

* Assay was the same used for data shown in Table 2. A total of 62 strains were retested. Data shown here are from one typical experiment.

Soybean Emergence Assay

Over a 9 month period, 277 strains isolated using procedure one and 84 strains isolated using procedure two were tested in the initial soybean emergence assay using "conditioned field soil". Sixty-two strains induced significant increases in emergence at 14 days compared to controls. Raw data from one typical experiment in which 2 EPR strains were selected, are shown in Table 2. All 62 strains were retested twice with the same assay,

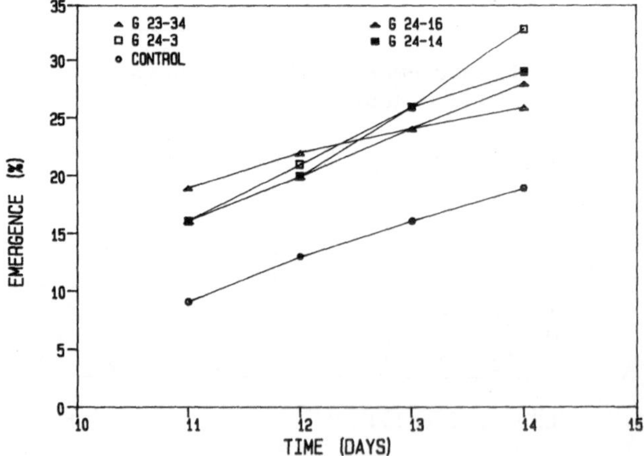

Fig. 2. Soybean emergence promotion showing increases in emergence by 4 EPR strains. The assay is the second soybean assay described in the text and was conducted in a field soil/perlite mix at 14°C. The percentage emergence values shown are the mean of 8 replications, each sown with 20 seeds.

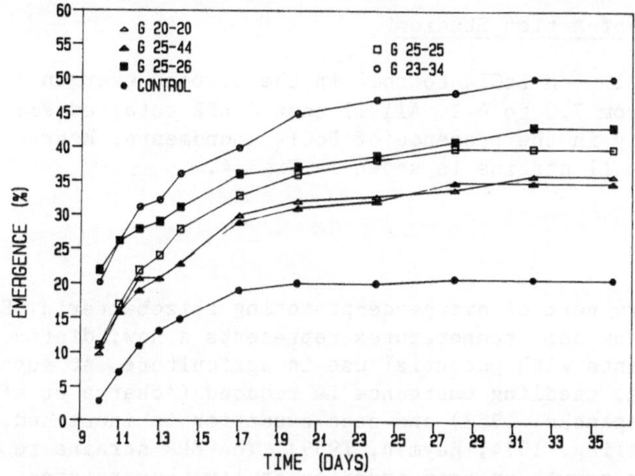

Fig. 3. Effect of 5 soybean EPR on early emergence rate and final percentage emergence. The assay is the same used in Fig. 2.

and 30 strains repeated emergence promotion in both repeat experiments (Table 3).

A second assay in field soil/perlite with 20 seeds per replication was used over a 12 month period to confirm the phenomenon of emergence promotion and to determine the effect of EPR on emergence on multiple days. Strains which induced a 50% increase in the percentage emergence of controls on each of 3 consecutive days were deemed to be EPR (Fig. 2). Twenty-three of the 30 strains which were originally selected for emergence-promoting activity repeated emergence promotion in at least 2 of 3 repeat experiments. Some of these strains increased both the rate of emergence and the final percentage emergence (Fig. 3) under the experimental conditions.

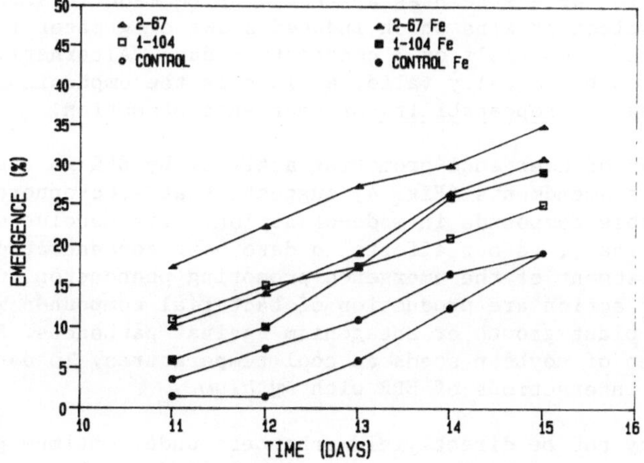

Fig. 4. Promotion of soybean emergence by 2 EPR strains with and without 10^{-3} M $FeCl_3$ amendments.

Preliminary Mode-of-Action Studies

Addition of 10^{-3} M $FeCl_3$ to soil in the soybean emergence assay changed the pH from 7.0 to 6.9. All 11 tested EPR retained emergence-promoting capacity in the presence of $FeCl_3$ amendments. Representative data for 2 of the 11 strains is shown in Fig. 4.

DISCUSSION

The discovery here of emergence-promoting rhizobacteria (EPR) which are operable at low soil temperatures represents a new, distinct class of microbial inoculants with potential use in agriculture. At suboptimal soil temperatures, seedling emergence is reduced (Acharya et al., 1983; Szyrmer and Szczepanska, 1982) and seed exudation is increased, (Schroth et al., 1966; Keeling, 1974; Hayman, 1969). The EPR strains reported herein were selected for growth on seed exudates at low temperatures, and hence may serve to reduce the total carbohydrates in the spermosphere which are available for growth of seedling pathogens.

The second procedure (described in the Methods Section) for isolation of candidate EPR strains, in which root segments were placed directly onto asparagine soft agar, allowed the direct isolation of motile strains which were chemotactically attracted to one of the major amino acids in seed exudates. (Scher et al., 1985). This procedure yielded a higher percentage of strains which were ultimately deemed to be EPR strains based on repeatable emergence promotion: for canola, 4 of 60 strains (7%) from procedure two, versus 0 of 50 from procedure one; for soybean, 14 of 277 strains (5%) from procedure one versus 9 of 84 (11%) from procedure two. It was surprising that EPR strains included 3 bacterial taxa not previously reported to be plant growth-promoting rhizobacteria (Table 1) i.e. *Enterobacter aerogenes, Serratia liquefaciens* and *Beijerinckia* spp.

The work reported here was concentrated on soybeans. However, the fact that a few canola EPR were also found suggests that EPR, like PGPR, can likely be found for any crop. The preferred method for detecting emergence promotion is the second soybean assay in which a field soil perlite mixture was used and emergence was recorded daily. We also found that it is important to conduct each assay 3 times and select strains which increase emergence in 2 of 3 tests.

The designation of a strain as an "EPR" can be made in several ways. We preferred to select strains which induced a 50% or greater increase in emergence relative to controls on 3 consecutive days. Alternative selection parameters are probably equally valid, as long as the emphasis is placed upon demonstrating the <u>repeatability</u> of emergence promotion.

The retention of emergence-promoting activity by EPR in the presence of ferric chloride amendments (Fig. 4) suggests that siderophores are not the primary operable compounds in mode-of-action. This conclusion is definitely preliminary, as our efforts to date have concentrated on the laborious establishment of the emergence-promoting phenomenon. Other possible modes of action are production of bacterial compounds which directly promote plant growth or antagonism against pathogens. *Pythium* is the major pathogen of soybean seeds at cool temperatures. To date we have not investigated interactions of EPR with *Pythium*.

While EPR may not be direct yield-enhancers under optimum plant growth conditions, as are PGPR, they will likely indirectly stimulate yields under adverse growing conditions. The development of plant cultivars with increased emergence rates at low soil temperatures has been identified as a high priority for canola (Acharya et al., 1983) and soybean (Szyrmer and Szczepanska, 1982) breeders in order to increase yields. In addition, EPR should prove useful as one component in strategies to increase stands, since increased emergence rates at cold temperatures result in increased final stands (Hatfield and Egli, 1974) and sometimes in yield (Khan et al., 1983).

Results of field tests with EPR are required in order to determine the ultimate usefulness of EPR in agriculture. However, three possible use areas may be good targets for EPR. Firstly, EPR would be beneficial at the extreme northern perimeter of a crop zone. For example, canola on the Canadian Prairies is concentrated in areas with 90 or more frost-free days. If an EPR-induced acceleration in seedling emergence translates into crop maturity of even 5 days sooner, the total hectarage available for canola growing would be increased substantially. Secondly, EPR are reasonable candidates for including in an integrated control strategy for some disease situations in which the host is most susceptible at the early season stages. For example, tests at the Agriculture Canada Research Centre at Harrow, Ontario, indicate that a reduced incidence of disease caused by *Phytophthora megasperma* var. *glycinea* occurs following treatments which accelerate seedling emergence and growth (T. Anderson, personal communication). EPR also have potential usefulness in non-irrigated crop lands where post-seeding rainfall is often limited. Under these conditions, accelerated emergence would likely result in increased root mass prior to the drought stress period.

The EPR strains described here are currently being tested in field trials. Multiple planting dates have been used on 2 canola and 2 soybean cultivars at 2 locations. The results of these tests will help clarify the potential for EPR in agriculture.

ACKNOWLEDGEMENTS

This work was partially funded by The Program for Industry Laboratory Projects of the National Research Council of Canada.

The authors thank Charmaine Rodrick-Semple for technical assistance during the first year of the program, Drs. C. Simonson and R. Lifschitz for help on the collecting trip to the High Arctic, and I. Zaleska and C. Singleton for technical assistance with the strain collection.

We thank Drs. D. Hume and W. Beversdorf, Crop Science Department, University of Guelph, for useful consultations on strategies for experimentation and for ongoing co-operation with field tests.

Drs. Simonson, Lifschitz and Polonenko reviewed the manuscript and provided useful suggestions.

REFERENCES

Acharya, S.N., Dueck, J. and Downey, R.K., 1983. Selection and heritability studies on canola/rapeseed for low temperature germination. Can. J. Plant Sci., 63:377-384.

Burr, T.J., Schroth, M.N. and Suslow, T. 1978. Increased potato yields by treatment of seed pieces with specific strains of *Pseudomonas fluorescens* and *P. putida*. Phytopathology, 68:1377-1383.

Hatfield, J.L. and Egli, D.B., 1974. Effect of temperature on the rate of soybean hypocotyl elongation and field emergence. Crop Sci. 14:423-426.

Hayman, D.S., 1969. The influence of temperature on the exudation of nutrients from cotton seeds and on pre-emergence damping-off by *Rhizoctonia solani*. Can. J. Botany, 47:1663-1669.

Howie, W.J. and Echandi, E., 1983. Rhizobacteria: influence of cultivar and soil type on plant growth and yield of potato. Soil Biol.Biochem. 15:127-132.

Keeling, B.L., 1974. Soybean seed rot and the relation of seed exudate to host susceptibility. Phytopathology, 64.1445-1447.

Khan, A.A., Peck, N.H., Taylor, A.G. and Samimy, C., 1983. Osmoconditioning of beet seeds to improve emergence and yield in cold soil. Agronomy J., 75:788-794.

Kloepper, J.W. and Schroth, M.N., 1978. Plant growth promoting rhizobacteria on radishes. Proc. 4th Int. Conf. Plant path. Bact., Angers, P. 879-882.

Kloepper, J.W. and Schroth, M.N., 1981A. Development of a powder formulation of rhizobacteria for inoculation of potato seed pieces. Phytopathology, 71:590-592.

Kloepper, J.W. and Schroth, M.N., 1981B. Relationship of *in vitro* antibiosis of plant growth-promoting rhizobacteria to plant growth and the displacement of root microflora. Phytopathology, 71:1020-1024.

Kloepper, J.W., Schroth, M.N. and Miller, T.D., 1980. Effects of rhizosphere colonization by plant growth-promoting rhizobacteria on potato plant development and yield. Phytopathology, 70:1078-1082.

Scher, F.M., Kloepper, J.W. and Singleton, C.A., 1985. Chemotaxis of fluorescent *Pseudomonas* spp. to soybean seed exudates *in vitro* and in soil. Can. J. Microbiol., 31:570-574.

Schroth, M.N., Weinhold, A.R. and Hayman, D.A., 1966. The effect of temperature on quantitative differences in exudates from germinating seeds of bean, pea and cotton. Can. J. Bot., 44:1429-1432.

Suslow, T.V. and Schroth, M.N., 1982. Rhizobacteria on sugar beets: effects of seed application and root colonization on yield. Phytopathology, 72:199-206.

Szyrmer, J. and Szczepauska, K. 1982. Screening of soybean genotypes for cold-tolerance during germination. Z. Pflanzenzuchtg., 88:255-260.

NATURE OF INTRAGENERIC COMPETITION BETWEEN PATHOGENIC AND NON-PATHOGENIC FUSARIUM IN A WILT-SUPPRESSIVE SOIL

C. Alabouvette[1], Y. Couteaudier[1], and P. Lemanceau[2]

INRA, 17 Rue Sully, F 21034 - Dijon Cédex[1]

ENITAH, Rue Le Nôtre F49000 - Angers[2]

INTRODUCTION

Over recent years, research on disease suppressive soils has been widely developed. Various models have been examined dealing with the major soil-borne diseases: damping-off, root rot, vascular wilts. Great progress has been made in our knowledge of the antagonistic populations and the microbial interactions which are basically responsible for these phenomena of soil suppressiveness. It is still difficult to make general statements about the microbiological mechanisms involved in soil suppressiveness, but many studies underline the role of competition for nutrients between soil microorganisms. Scher and Baker (1982) showed that fluorescent *Pseudomonas* competing for iron are responsible for the wilt-suppressiveness of the soils from the Salinas Valley. On the other hand we demonstrated that competition for carbon is involved in the mechanisms of suppression of the soils from the Chateaurenard region (Alabouvette et al., 1985a). However, it is difficult to demonstrate experimentally that two microorganisms are in competition and to determine the substrate for which they compete in soil.

The purpose of this paper is to briefly review the results obtained by our laboratory in the study of Chateaurenard suppressive soils and to indicate the approach we are now following to obtain a better understanding of competition for nutrients between pathogenic and non-pathogenic *Fusarium* in soils.

SOIL RECEPTIVITY TO FUSARIUM WILTS

The concept of soil receptivity to diseases was coined as we started working on fusarium wilt-suppressive soils. We realized that the absence of disease could not always be accounted for by the absence of pathogens. E.g. *Fusarium oxysporum* f. sp. *melonis,* that causes fusarium wilt, could be found in the Chateaurenard soils without causing disease. Therefore, soil can limit in some degree the occurence of a disease. This can be easily demonstrated by introducing into various soils increasing amounts of a given pathogen and comparing, under similar cropping conditions, expression of the disease on a population of susceptible host plants. At similar inoculum densities, severity of disease is found to vary significantly according to soils (Fig. 1), indicating the various levels of soil

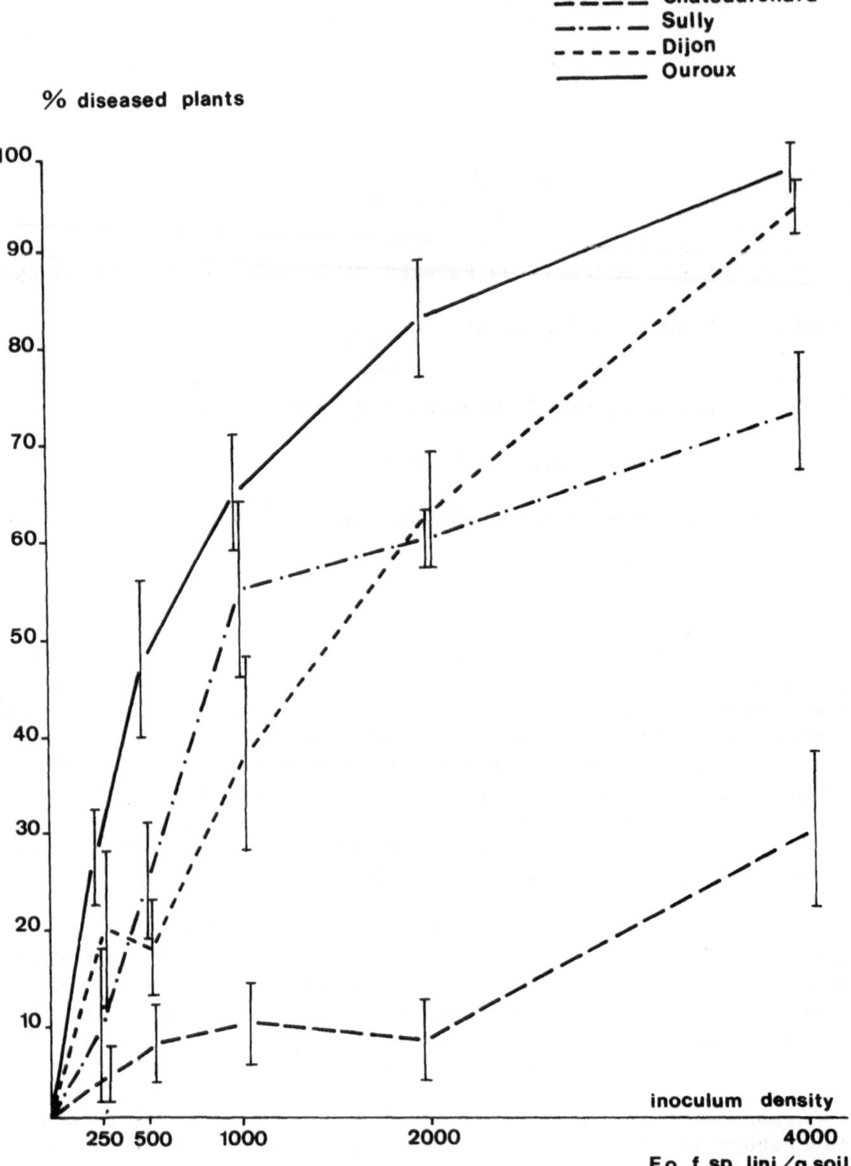

Fig. 1. Receptivity of 4 soils to fusarium wilt of flax: percentage of wilted plants 4 weeks after soil infestation with *Fusarium oxysporum* f. sp. *lini* at increasing concentrations. Mean of 12 replicates of 5 plants with its standard deviation.

receptivity to fusarium wilt. It is thus possible to identify disease-suppressive soils in which expression of disease is limited by natural phenomena. It is also possible to use this method to study the role of different populations of microorganisms and of different nutrients in the expression of the suppressiveness.

CHARACTERISTICS OF THE WILT SUPPRESSIVENESS OF THE SOILS FROM CHATEAURENARD

From a theoretical point of view, soil can inhibit disease expression owing to its physico-chemical or biological properties. Evaluating the effects of various biocidal treatments, we established that suppressiveness is based on microbiological interactions (Louvet et al., 1976). This does not mean that the soil physico-chemical properties are not involved in expression of wilt-suppressiveness. The Chateaurenard soil is a clay-loam with a high pH (>7); but the role of the clay and pH has not yet been demonstrated.

In contrast, it is easy to demonstrate that soil microorganisms are involved in disease suppression: wilt-suppressiveness can be transmitted by simply mixing a small proportion of suppressive soil with a previously heat-treated conducive soil (Louvet et al., 1976). Only microorganisms are likely to multiply and to confer a high level of suppressiveness to a conducive soil.

After demonstrating indirectly the role of the microflora in wilt-suppressiveness, we tried to identify precisely the microorganisms involved. Eliminating selectively certain components of the soil microflora by applying steam-air treatments, we found that heat-sensitive microorganisms, probably fungi play a role in the suppression mechanisms (Rouxel et al., 1977).

Isolation of the main species of fungi followed by their reintroduction into soil previously disinfected showed that the presence of non-pathogenic *F. oxysporum* and *Fusarium solani* is necessary to the expression of suppressiveness in the soils from Chateaurenard (Rouxel et al., 1979). These soils host a large population of indigenous *Fusarium* making up 25 to 40% of the fungal microflora. A relationship may be assumed to exist between the abundance of non-pathogenic *Fusarium* and wilt suppressiveness. Moreover, the suppressiveness inhibits the occurrence of all formae speciales of *F. oxysporum* but allows the expression of other soilborne pathogens (Alabouvette et al., 1980a). This specific action proves the existence of a particular mechanism which only affects the activity of *F. oxysporum* that causes wilt.

It may be concluded that suppressiveness of soils from Chateaurenard implies the activity of non-pathogenic *Fusarium*, inhibiting the expression of pathogenic *Fusarium*. As it was impossible to observe any relationship of antibiosis or hyperparisitism between pathogenic and non-pathogenic *Fusarium* we assumed that these two types of *Fusarium* are competing for nutrients in soil.

ROLE OF GENERAL SUPPRESSION AND SPECIFIC SUPPRESSION IN THE SUPPRESSIVENESS

Cook and Baker (1983) described two types of suppression mechanisms in soils called "specific" and "general" suppression. Specific suppression is based on the activity of a particular population of antagonistic microorganisms inhibiting the development of the pathogen. The role played by the non-pathogenic *Fusarium* in the suppressiveness of the soils from Chateaurenard corresponds with this definition and we can assimilate this mechanism to a specific suppression phenomenon. The general suppression refers to the suppressiveness based on the non-specific activity of the total soil microbial biomass acting by means of competition mechanisms. The general suppression is expressed with variable intensity in all soils and any specific mechanism is expressed against a back-ground of general suppression.

We demonstrated that the suppressiveness of the Chateaurenard soils is based both on specific and general suppression. We compared the kinetics of CO_2 release after glucose amendment in a suppressive and a conducive soil. The results showed that the levels of microbial biomass and activity are higher in the suppressive than in the conducive soil. Thus the general competition for nutrients is more intense in the suppressive soil and controls the intrageneric competition between pathogenic and non-pathogenic *Fusarium* (Alabouvette et al., 1985b).

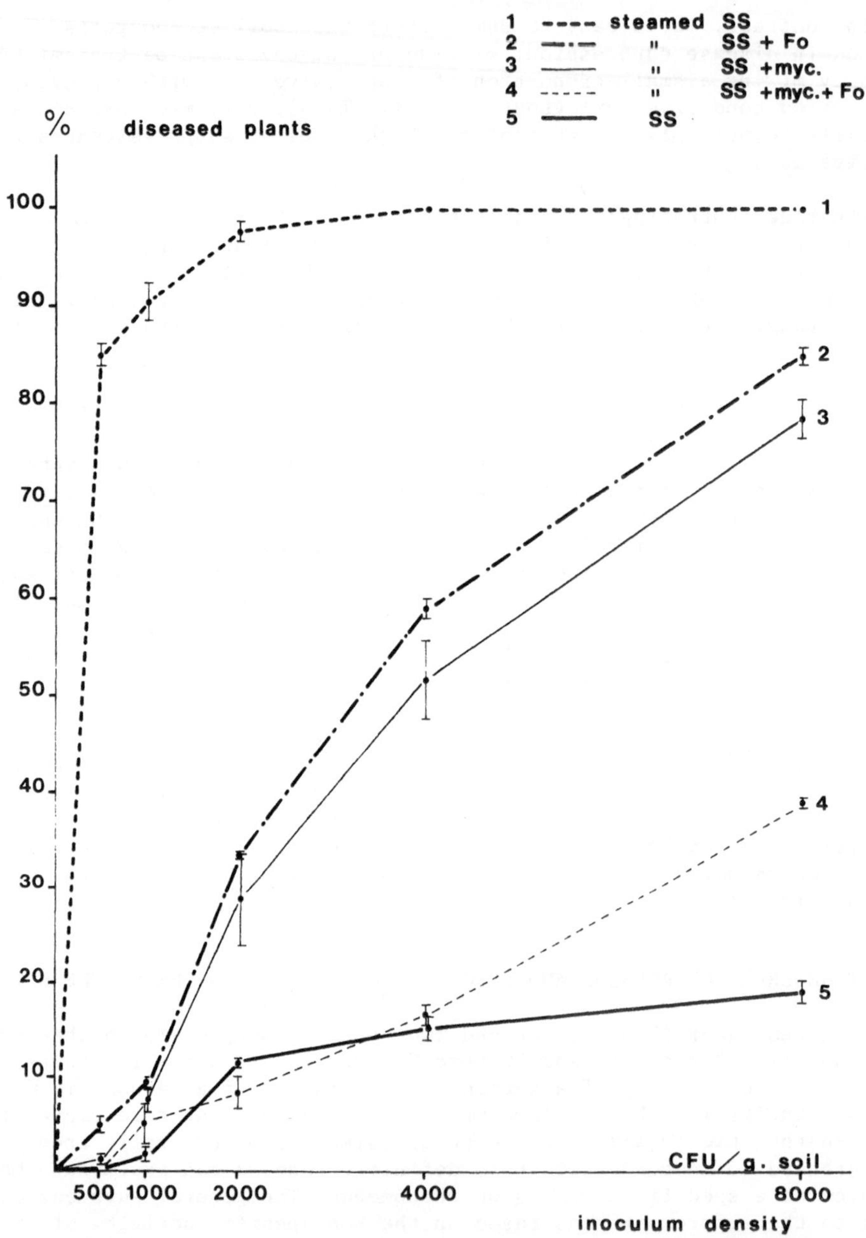

Fig. 2. Receptivity to fusarium wilt of flax of steamed suppressive soil (S.S.) recolonized by a strain of non-pathogenic *Fusarium oxysporum* (*F.o.*), a complex mycoflora without *Fusarium* (myc.) or with both populations together. Percentage of wilted plants 8 weeks after soil infestation with *F.o.* f. sp. *lini* at increasing concentrations. Mean of 6 replicates of 10 plants with its standard deviation.

To demonstrate the additive effects of specific and general suppression we compared the levels of wilt-suppressiveness induced by different populations of microorganisms introduced in the steamed suppressive soil. A strain of non-pathogenic *F. oxysporum* was added in the steamed soil, just after the steam-treatment or after the soil had been recolonized by a complex mycoflora without *Fusarium*. After 3 weeks of incubation the pathogenic *F. oxysporum* f. sp. *lini* was introduced at increasing concentrations and the level of soil receptivity estimated after 8 weeks of flax cropping. The results are shown in Fig. 2. Population of non-pathogenic *Fusarium* alone and the complex mycoflora without *Fusarium* both reduce conduciveness of the steamed soil. Addition of non-pathogenic *Fusarium* in soil recolonized by the complex mycoflora confers a high level of suppressiveness close to the level observed in the suppressive soil itself. A new mathematical model (Corman et al., unpublished) allowed the levels of soil receptivity to fusarium-wilts to be compared statistically. It showed that the different levels of suppressiveness observed in this experiment are significantly different and proved that the effects of general and specific suppression are additive.

Thus, the fusarium wilt-suppressiveness of soils from Chateaurenard relies on the complementary association of a general mechanism of nutrient competition between the whole soil micoflora and the entire *Fusarium* population with a specific mechanism of intrageneric competition between pathogenic and non-pathogenic *Fusarium*.

NATURE OF COMPETITION BETWEEN MICROORGANISMS IN THE SUPPRESSIVE SOIL

To determine the nature of competition between microorganisms and especially between pathogenic and non-pathogenic *Fusarium* in the suppressive soils from Chateaurenard, two types of experimental method were used: first the global effect of addition of a nutrient on the level of suppressiveness was studied and then the direct effects of amendment on specific populations in soil were analysed.

Competition for Carbon

We established that addition of glucose considerably increases the level of receptivity of soils to fusarium-wilts. (Fig. 3) and destroys the suppressiveness of the soils from Chateaurenard. The modification of the level of soil suppressiveness depends both on initial level of suppressiveness and concentration of glucose. It is necessary to add higher concentrations of glucose in the suppressive soil than in the conducive soil to increase their levels of receptivity to fusarium-wilts (Alabouvette et al., 1985c). This quantitative relationship probably indicates that competition phenomena are involved.

To better elucidate the role of competition for carbon in suppression we studied chlamydospore germination of various *Fusarium* strains in both suppressive and conducive soils. This demonstrated that fungistasis is more intense in suppressive than in conducive non-amended soil (Alabouvette et al., 1980b). Addition of glucose at increasing concentrations showed that a larger amount of energy (5 to 10 times more) is necessary in the suppressive soil to counteract fungistasis and induce the same percentage of germinated chlamydospores as in the conducive soil. This also showed that the saprophytic development of *Fusarium* following chlamydospore germination depends on the concentration of glucose added to soils (Alabouvette, 1985a).

The receptivity of these soils to fusarium-wilts appears then to be correlated with the intensity of competition for carbon, which affects

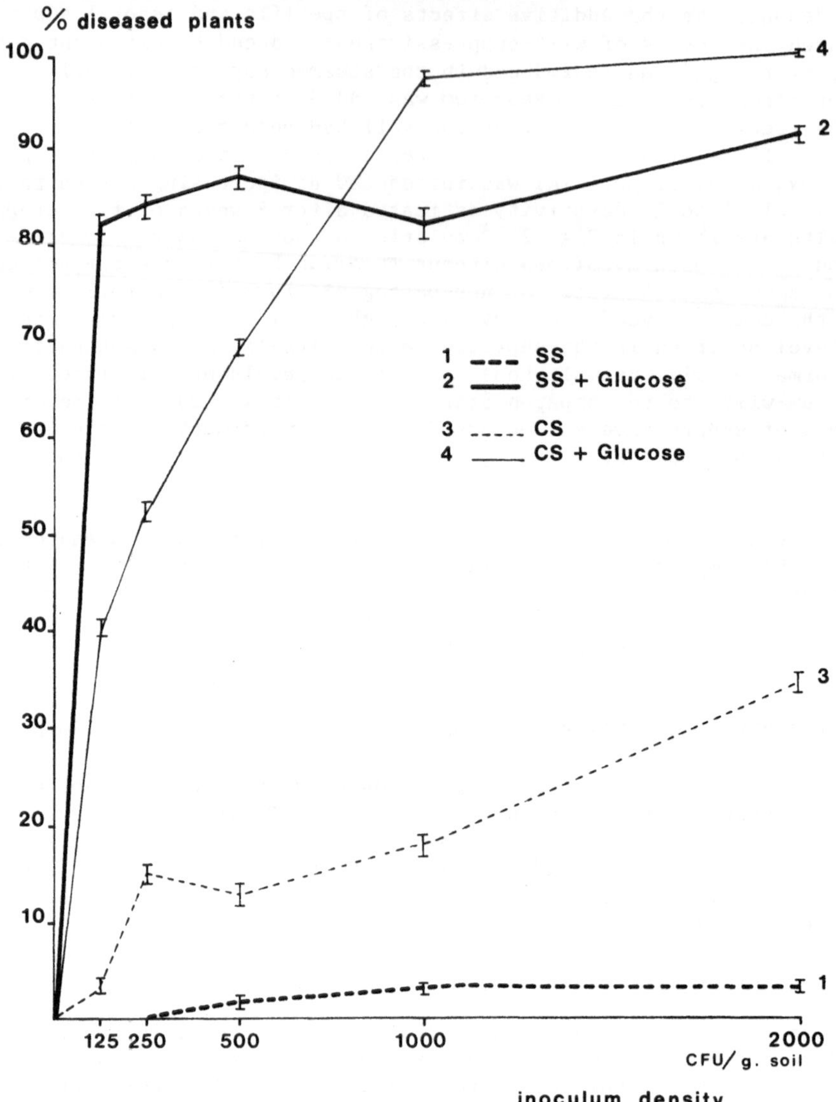

Fig. 3. Receptivity to fusarium wilt of flax of conducive soil (C.S.) and suppressive soil (S.S.) amended with glucose (10 mg/g soil). Percentage of wilted plants 8 weeks after soil infestation with *Fusarium oxysporum* f. sp. *lini* at increasing concentrations. Mean of 12 replicates of 5 plants with its standard deviation.

the saprophytic behaviour of *Fusarium* in soils. But competition for carbon is not a specific phenomenon and may affect other populations of soil microorganisms. It is probably in correlation with the general suppression phenomenon which depends on the total activity of the soil microbial biomass, but does not control the intrageneric competition between pathogenic and non-pathogenic *Fusarium*. In fact we were unable to show differences between strains of *Fusarium* for their saprophytic development in soil amended with glucose.

Pathogenic and non-pathogenic *Fusarium* spp. having very similar requirements probably compete for more uncommon and specific nutrients. The main question is how to determine the nature of the intrageneric competition, and the nutrients for which *Fusarium* spp. are competing in soil.

Competition for Iron

After the first papers dealing with competition for iron in

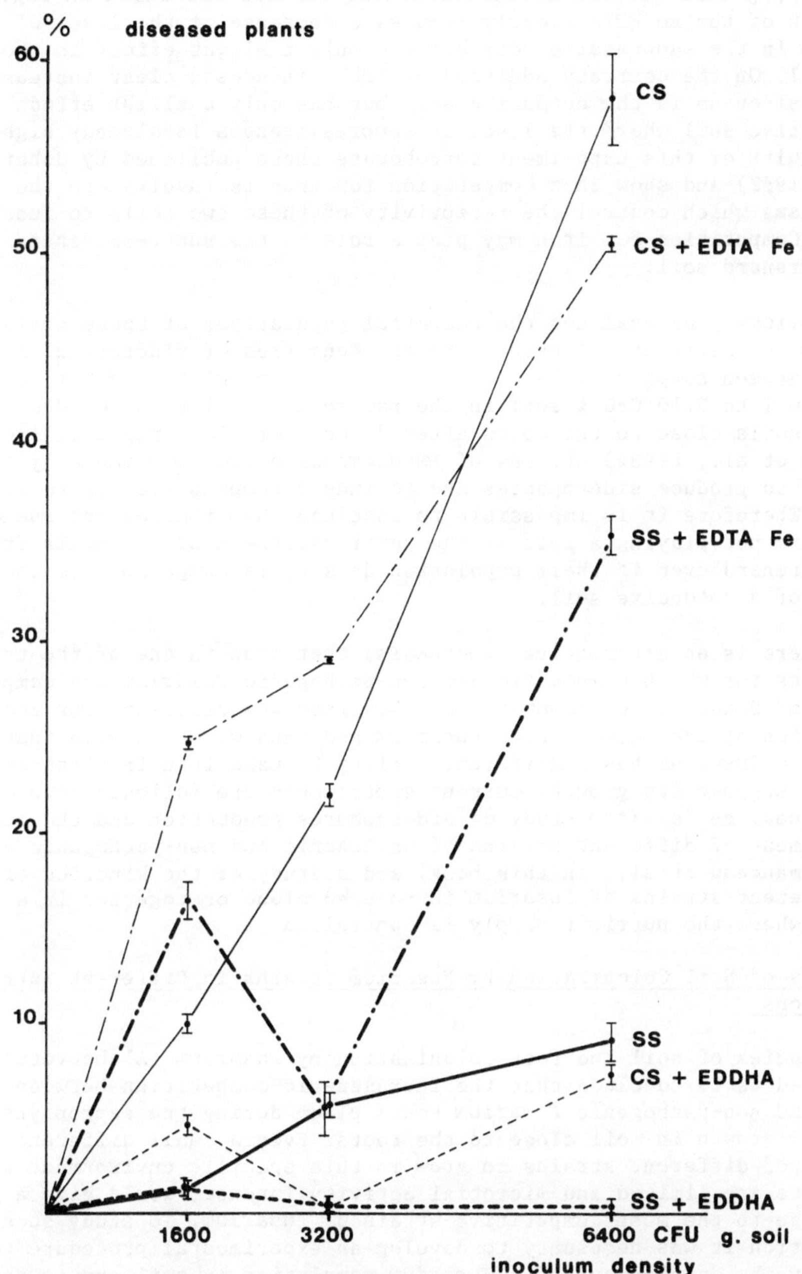

Fig. 4. Receptivity to fusarium wilt of flax of conducive soil (C.S.) and suppressive soil (S.S.) amended with EDDHA or Fe EDTA (300 µg/g soil). Percentage of wilted plants 8 weeks after soil infestation with *Fusarium oxysporum* f. sp. *lini* at increasing concentrations. Mean of 6 replicates of 10 plants with its standard deviation.

suppressive soils had been published (Kloepper et al., 1980; Scher and Baker, 1980) we tested this hypothesis with the suppressive soils from Chateaurenard. In a preliminary experiment FeEDTA and EDDHA was added (100 µg/g soil) at 3 times, planting and one and two weeks later, to the suppressive Chateaurenard soil and a conducive soil (both >pH7). The small differences in disease incidence observed were not significant. The experiment was repeated but the concentration of chelators was increased to 300 µg/g soil at each application. The results are shown in Fig. 4. Addition of the Fe EDTA clearly induces a decrease of the level of suppressiveness in the suppressive soil but has only a slight effect in the conducive soil. On the contrary addition of EDDHA induces a clear increase of suppressiveness in the conducive soil but has only a slight effect in the suppressive soil where the level of suppressiveness is already high. Thus the results of this experiment corroborate those published by Scher and Baker (1982) and show that competition for iron is involved in the mechanisms which control the receptivity of these two soils to fusarium wilts. Competition for iron may play a role in the suppressiveness of Chateaurenard soil.

Obviously we analyzed the bacterial populations of these soils and found no significant differences in the densities of fluorescent *Pseudomonas* between suppressive and conducive soils; population densities were low 1 to 3.10^3CFU/g soil in the raw soils and 1 to 3.10^5 CFU/g soil in the soils close to the roots after 3 weeks of flax cropping. According to Sneh et al., (1984) strains of *Pseudomonas* differ considerably in their ability to produce siderophores and to induce suppressiveness to fusarium-wilts. Therefore it is impossible to conclude that fluorescent *Pseudomonas* are not playing a role in the suppressiveness of the soils from Chateaurenard even if their population density is comparable to the population of a conducive soil.

There is an alternative hypothesis; that iron is one of the trace nutrients for which pathogenic and non-pathogenic *Fusarium* are competing. Scher and Baker (1980) demonstrated that iron was necessary for the elongation of the germ-tube of *Fusarium* and thus we may assume that every strain of *Fusarium* has a different ability to take iron from its environment to support its growth. Current experiments are following two different approaches: an *in vitro* study of siderophores production and the iron requirement of different strains of pathogenic and non-pathogenic *Fusarium* (cf. Lemanceau et al., in this book) and a study of the kinetics of growth of different strains of *Fusarium* introduced alone or together in a soil system where the nutrient supply is controlled.

Dynamics of Soil Colonization by Fusarium Strains in Different Nutritional Conditions.

Studies of soil and root colonization by *Fusarium* (Alabouvette et al., 1984) led us to conclude that the intrageneric competition between pathogenic and non-pathogenic *Fusarium* takes place during the saprophytic phase of growth in soil close to the roots. Even a small difference in the ability of different strains to grow in this specific environment where nutrients are limited and microbial activity intense, would give a great advantage to the most competitive strain of *Fusarium*. To study such competition it was necessary to develop an experimental procedure to quantify the development of a *Fusarium* population in soil and to analyze the interactions between two strains of *Fusarium* competing in different nutritional conditions.

We started with the simplest model, studying the kinetics of growth of a strain of *Fusarium* in a steamed soil. In these conditions, nutrients are not limiting when *Fusarium* is introduced at low population densities

(10^2 to 10^3 CFU g/soil); however, this study was necessary to establish the parameters of *Fusarium* growth. Frequent analysis based on soil-dilutions technics permitted a mathematical model to be elaborated (Son et al., unpublished) which allows the growth rate and the population density to be calculated at the plateau, when the population after 7 to 15 days has reached an equilibrium state. This well defined procedure enabled the ability of different strains to colonize a steamed soil to be compared. Fig. 5 illustrates the results obtained with a strain of pathogenic *F. oxysporum* and a non-pathogenic *F. solani* growing alone in soil.

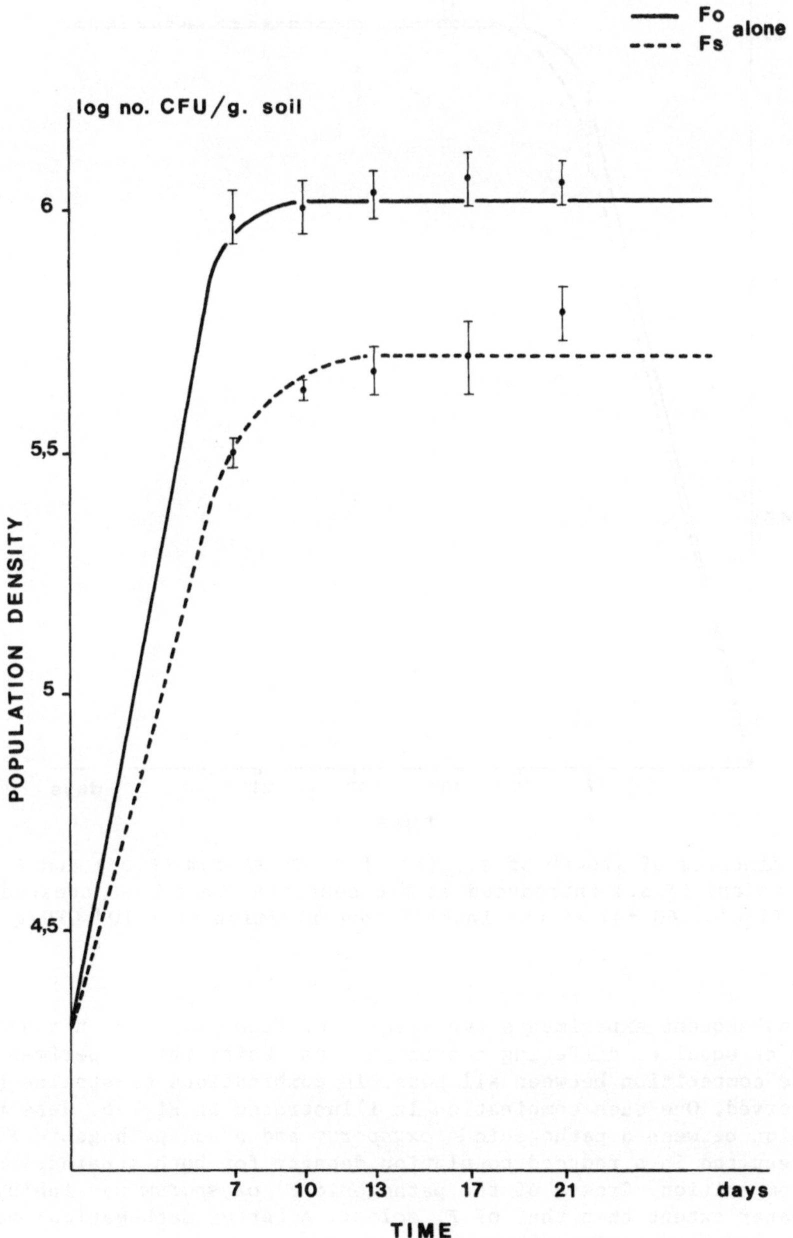

Fig. 5. Kinetics of growth of strains of *F. oxysporum* (*F.o.*) and *F. solani* (*F.s.*) introduced alone in a heat treated soil (120°C, 60 mn) at the initial concentration of 2.10^4 CFU/g soil.

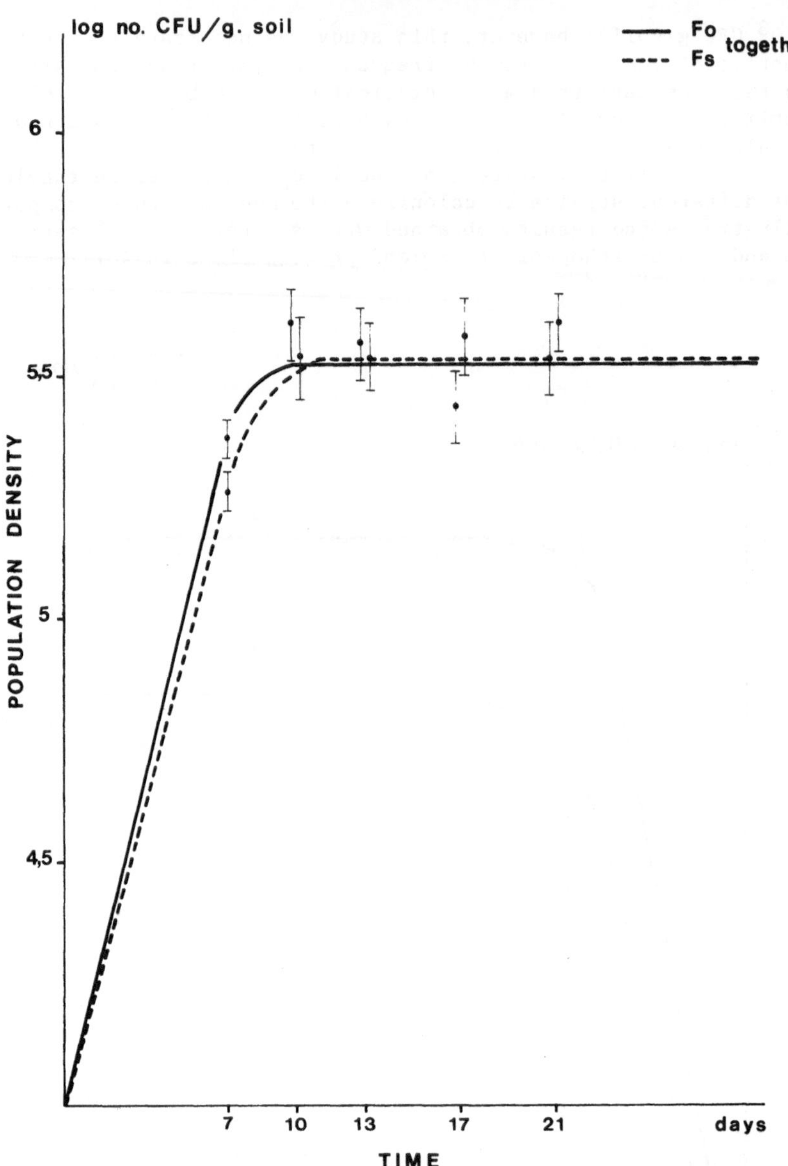

Fig. 6. Kinetics of growth of strains of *F. Oxysporum* (*F.o.*) and *F. solani* (*F.s.*) introduced at the same time in a heat treated soil (120°C, 60 mn) at the initial concentration of 1.10^4 CFU/g soil.

In subsequent experiments two strains of *Fusarium* were introduced together at equal or differing concentrations. Using this experimental procedure competition between all possible combinations of strains has been observed. One such combination is illustrated in Fig. 6. Here the interaction between a pathogenic *F. oxysporum* and a non-pathogenic *F. solani* resulted in a reduced population density for both strains, indicative of competition. Growth of the pathogenic *F. oxysporum* was inhibited to a greater extent than that of *F. solani*. A better mathematical model is still needed to analyze these interactions.

This procedure also permits the effect of specific nutrients on the interactions to be observed. Fig. 7 shows the behaviour of a strain of

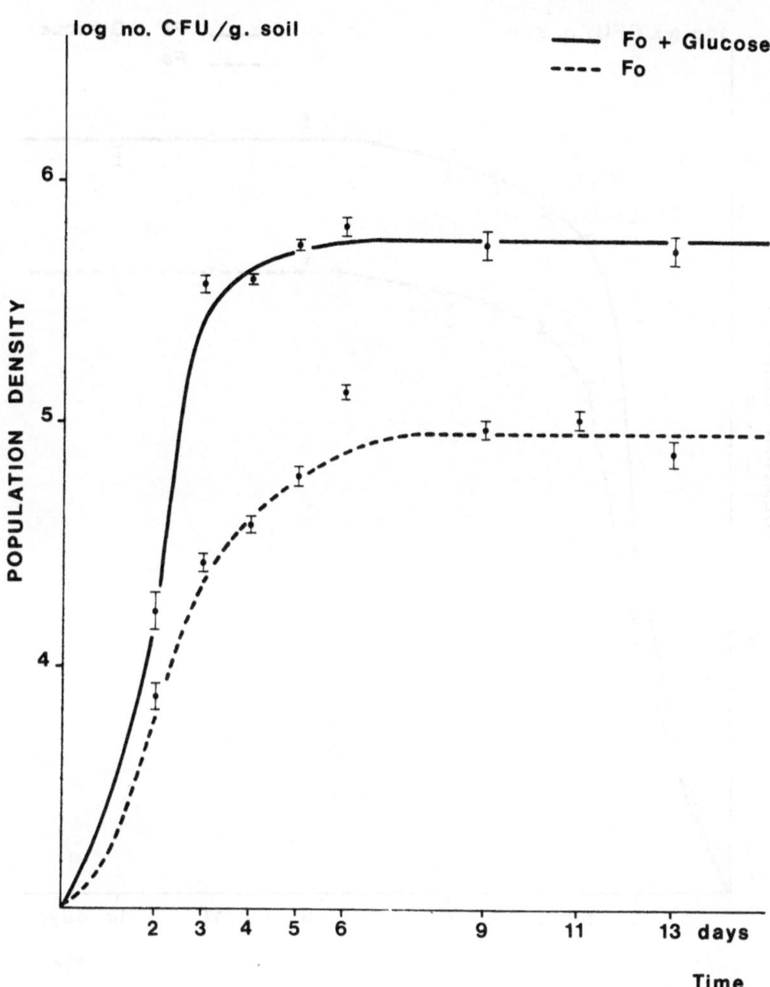

Fig. 7a. Kinetics of growth of a strain of *F. oxysporum* (*F.o.*) introduced at the initial concentration of 1.10^3 CFU/g soil in a heat treated soil (120°C, 60 mn) amended with glucose (10 mg/g soil).

F. oxysporum and *F. solani* growing in steamed suppressive soil amended with glucose (1 mg/g soil). This nutrient added at the same time as the *Fusarium* changes both the growth rate and the population density at the plateau for both *F. oxysporum* and *F. solani*. Even in steamed soil glucose amendment immediately stimulates the growth of these *Fusarium*. Finally, it is interesting to add the nutrient after the population has reached its plateau. Fig. 8 shows the effect of a glucose amendment (1 mg/g soil) on a population of a pathogenic strain of *F. oxysporum*, at the equilibrium state. It seems that this input of energy immediately produces a new development of the population which reaches a higher level. This type of experiment should help to determine which nutrients limit growth of *Fusarium* in soil and what concentration is needed to let the *Fusarium* grow. Experiments are now in progress with iron and nitrogen amendments. Using iron chelators at different concentrations it may be possible to compare the ability of different strains of *Fusarium* to compete for iron.

In the future we will use strains of *Fusarium* resistant to benomyl to study the competition in raw soil, in presence of wild populations of microorganisms, in a normal environment where the competition for nutrients

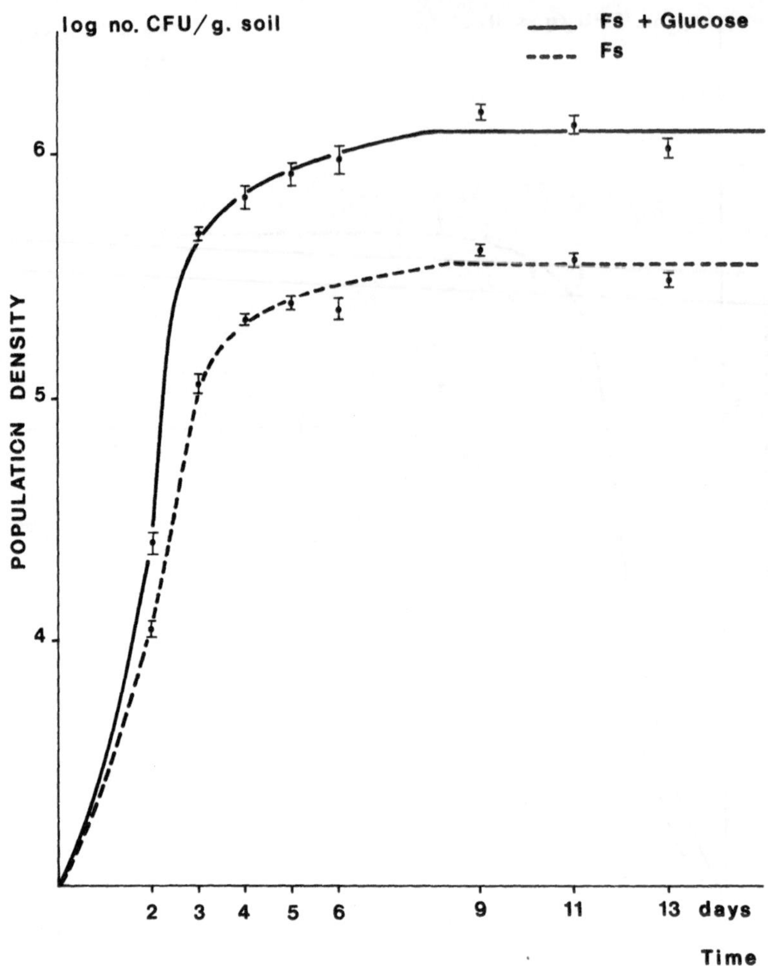

Fig. 7 b. Kinetics of growth of a strain of *F. solani* (*F.s.*) introduced at the initial concentration of 1.10^3 CFU/g soil in a heat treated soil (120°C, 60mn) amended with glucose (10 mg/g soil).

is really intense. We hope to be able to compare the competitive saprophytic ability of pathogenic and non-pathogenic strains of *Fusarium* in conditions where specific nutrients will be limiting.

CONCLUSION

This short review of our work devoted to fusarium wilt-suppressive soils from Chateaurenard shows all the complexity of studies dealing with analyses of mechanisms of suppression.

We first proved the role of intragenic competition between pathogenic and non-pathogenic *Fusarium* and then the role of the whole microflora acting as a nutrient sink. Studying relationships of competition between populations of microorganisms in soil, we obtained some experimental data which suggested that carbon is one of the nutrients for which these populations are competing. Then, using experimental procedures described by others, we obtained some evidence that competition for iron is also involved in these mechanisms of suppression.

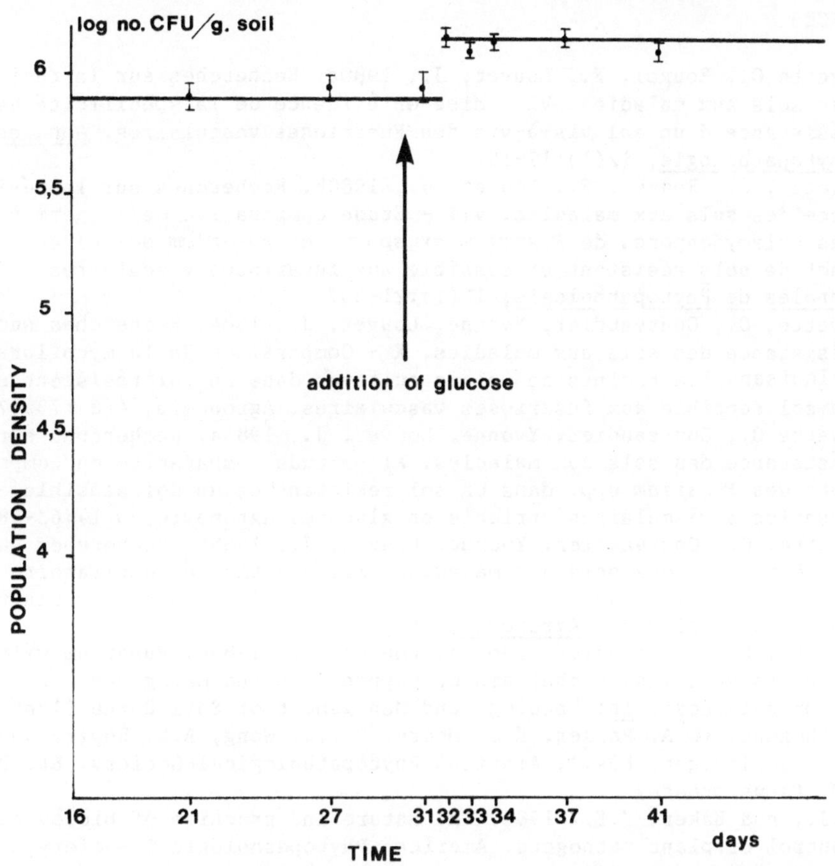

Fig. 8. Effect of glucose amendment of soil (10 mg/g soil) on the population density of a strain of *F. oxysporum* after it has reached its plateau.

All these results emphasize the role of competition for nutrients in the mechanisms of suppression of soils to fusarium wilts. Two basic problems arise; (i) how to demonstrate that two microorganisms are competing in soil and (ii) how to determine the nutrients for which they are competing. To answer these questions we proposed a new approach based on the study of kinetics of growth of fungal population in soil. The first data presented in this paper are encouraging but more research is needed before answering the basic questions and defining the laws of competition for nutrients between microorganisms in the soil ecosystem.

Finally, these studies showed that the various theories proposed to explain the phenomenon of soil suppressiveness to fusarium wilts are complementary rather than opposed. We strongly believe that such a powerful and stable suppressiveness as the suppressiveness of the soils from Chateaurenard is likely to result from the association of various microbial processes that act simultaneously, successivley or independently of each other, according to environmental conditions.

REFERENCES

Alabouvette C., Rouxel, F., Louvet, J., 1980a. Recherches sur la résistance des sols aux maladies. VI - Mise en évidence de la spécificité de la résistance d'un sol vis-à-vis des Fusarioses vasculaires. Annales de Phytopathologie, 12(1):11-19.

Alabouvette, C., Rouxel, F., Louvet, J., 1980b. Recherches sur la résistance des sols aux maladies. VII - Etude comparative de la germination des chlamydospores de *Fusarium oxysporum* et *Fusarium solani* au contact de sols résistant et sensible aux fusarioses vasculaires. Annales de Phytopathologie, 12(1):21-30.

Alabouvette, C., Couteaudier, Yvonne, Louvet, J., 1984. Recherches sur la résistance des sols aux maladies. X - Comparaison de la mycoflora colonisant les racines de melons cultivés dans un sol résistant ou un sol sensible aux fusarioses vasculaires. Agronomie, 4(8):735-740.

Alabouvette C., Couteaudier, Yvonne, Louvet, J., 1985a. Recherches sur la résistance des sols aux maladies. XI - Etude comparative du comportement des *Fusarium* spp. dans un sol résistant et un sol sensible aux fusarioses vasculaires enrichis en glucose. Agronomie, 5(1):63-68.

Alabouvette, C., Couteaudier, Yvonne, Louvet, J., 1985b. Recherches sur la résistance des sols aux maladies, XII - Activité respiratoire dans un sol résistant et un sol sensible aux fusarioses vasculaires enrichis en glucose. Agronomie, 5(1):69-72.

Alabouvette, C., Couteaudier, Yvonne, Louvet, J., 1985c. Fusarium-wilt suppressive soils: mechanisms of suppression and management of suppressiveness. In: "Ecology and Management of Soil Borne Plant Pathogens. (C.A. Parker, K.J. Moore, P.T.W. Wong, A.D. Rovira and J.F. Kollmorgen, Eds.), American Phytopathological Society, St. Paul MN, (sous presse).

Cook R.J., and Baker, K.F., 1983. The nature and practice of biological control of plant pathogens. American Phytopathological Society, St. Paul MN, USA, 539 pp.

Kloepper, J.W., Leong, J., Teintze, M., Schroth, M.N., 1980. *Pseudomonas* siderophores: a mechanism explaining disease-suppressive soils. Current Microbiologie, 4:317-320.

Louvet, J., Rouxel, F., Alabouvette, C., 1976. Recherches sur la résistance des sols aux maladies. I - Mise en évidence de la nature microbiologique de la résistance d'un sol au dévelopment de la Fusariose vasculaire du Melon. Annales de Phytopathologie, 8: 425-436.

Rouxel, F., Alabouvette, C., Louvet, J., 1977. Recherches sur la résistance des sols aux maladies. II - Incidence de traitements thermiques sur la résistance d'un sol à la Fusarioses vasculaire du Melon. Annales de Phytopathologie, 9:183-192.

Rouxel, F., Alabouvette, C., Louvet, J., 1979. Recherches sur la résistance des sols aux maladies. IV - Mise en évidence du rôle des *Fusarium* autochtones dans la résistance d'un sol à la Fusariose vasculaire du Melon. Annales de Phytopathologie, 11:199-207.

Scher, M.F., and Baker, R., 1980. Mechanisms of biological control in a *Fusarium* suppressive soil. Phytopathology, 70:412-417.

Scher, M.F., and Baker, R., 1982. Effects of *Pseudomonas putida* and synthetic iron chelator on induction of soil suppressiveness to Fusarium wilt pathogens. Phytopathology, 72:1567-1573.

Sneh, B., Dupler, M., Elad, Y., Baker, R., 1984. Chlamydospore germination of *Fusarium oxysporum* f. sp. *cucumerinum* as affected by fluorescent and lytic bacteria from a fusarium suppressive-soil. Phytopathology, 74:1115-1124.

ANTAGONISM AND SIDEROPHORE PRODUCTION BY BIOCONTROL AGENTS, PLANT GROWTH PROMOTING ORGANISMS AND THE GENERAL RHIZOSPHERE POPULATION

R. Campbell, A. Renwick and S.K.A.M. Coe

Department of Botany
University of Bristol
Bristol, BS8 1UG, England

INTRODUCTION

The biological control of plant pathogens and the phenomen of disease suppressive soils have been recently reviewed (Burr and Caesar, 1984; Cook and Baker, 1983, Schneider, 1982). Among the mechanisms which have been suggested for the microbial control of plant pathogens are antibiosis, enzymic lysis, competition for space and nutrients and the production of siderophores (Campbell and Faull, 1979: Hornby, 1979, Howell and Stipanovic, 1979, Kloepper et al., 1980b). Many potentially useful microorganisms have been isolated and described in the literature (Cook and Baker, 1983), though there are few commercially available. *Pseudomonas* species, especially fluorescent pseudomonads, have been used by many workers as they are antagonistic to many plant pathogens (Geels and Schippers, 1983a; Geels and Schippers, 1983b; Kloepper et al., 1980b; Weller and Cook, 1983), may cause increased plant growth (plant growth promoting rhizobacteria, PGPR: Kloepper et al., 1980a) and are successful at colonizing the roots of many plants (Geels and Schippers, 1983a). Some pseudomonads are antagonistic to the take-all fungus (*Gaeumannomyces graminis* var. *tritici*: Cook and Rovira, 1976; Weller and Cook, 1983) and are also present in soils suppressive to take-all.

Many biocontrol agents and PGPR's have been selected by *in vitro* tests (e.g. antagonism or antibiotic production on agar plates) and this does not give the best organisms for use under soil conditions in the presence of the growing host plant.

The role of iron in microbial metabolism was discussed by Neilands, (1981). In soil, iron is complexed as highly insoluble oxides and hydro-oxides, though the solubility increases at lower pH (Scher and Baker, 1982). However, many intensively cultivated arable soils, under cereals for example, are limed to maintain a high soil pH. Under such conditions iron may limit the growth of plants and microorganisms; many of the latter obtain iron by producing chelating compounds called siderophores (Neilands, 1981). Microorganisms producing siderophores with high stability constants could deprive potential plant pathogens of iron, resulting in the suppression of disease (Emery, 1980; Kloepper et al., 1980a). There seems to be no clear information on how important siderophores are in the general soil population or amongst potential biocontrol agents. There may be a danger of a circular argument: for example, pseudomonads are good

Table 1. Range and means of values for soil characteristics from the sampling sites.

		pH	PO_4 ppm	K ppm	Mg ppm	Ca ppm	%N	% Organic matter	% sand	% clay	% silt
General sites:	range	6.2-8.4	12-98	66-750	33- 440	650-7875	0.21-0.62	5.3-17.0	21-78	13-39	9-56
	mean	7.8	42	273	94	3973	0.32	9.0	39	25	35
Site yielding greatest number of antagonists		8.3	69	144	65	7875	0.62	17.0	46	22	32
Sand used in disease screen		7.9	4	48	48	1749	0.02	0.0	100	0	0
High iron soils:	range	5.3-6.9	18-32	132-304	48-108	2259-3630	0.46-0.60	13-17	50-53	11-15	30-36
	mean	6.1	27	230	77	2920	0.52	15.0	52	13	35

	total iron	available iron (as Fe^{+++})
range	2.2-3.7%	68-70 ppm
mean	3.1%	69 ppm

biocontrol agents, pseudomonads often produce siderophores, therefore siderophores are important in disease control.

The aim of this study was to try to separate some of these overlapping factors. Microorganisms were selected for their ability to control the take-all disease on wheat plants. The proportion of these selected organisms which were able to produce siderophores and those achieving disease control by other means, such as antibiotic production, was estimated. The frequency of siderophore-producers in the general rhizosphere population was also measured for comparison. Is siderophore production more common amongst selected biocontrol agents or PGPR's? How does the *in vitro* production of siderophores and antibiotics reflect the ability to control plant disease or promote plant growth?

MATERIALS AND METHODS

Sites for sampling were selected on a number of criteria including the geographical spread within the United Kingdom and the variation in the soil type, but many fields with long-term cereals were used (up to 35 years continuous cereals). The yield of grain from the selected sites was high and reported levels of root disease, especially take-all, were low. These were sites where the disease might be expected to be present but where it did not occur because of take-all decline (Hornby, 1979) or because of their other microbiological characteristics, though other soil factors could also contribute to lowered disease levels. A small proportion of the sites were grassland with clover, which may also reduce take-all in subsequent cereal crops (Grosman, 1967; Wong, 1981). One site was specially selected for the high iron levels in the soil (Parc 612a soil association as defined by Rudeforth et al., 1984 and personal communication Rudeforth) and a lower pH, since siderophores should not be so effective in the presence of high levels of available iron. The characteristics of the sites are summarized in Table 1.

Plants were removed at random from the fields and transported to the laboratory where the root surface and the rhizosphere microflora were isolated as soon as possible. The roots were shaken free of excess soil and placed in 20 ml sterile water with glass beads. After the roots had shaken for 10 min on a rotary shaker a dilution series was prepared in water. Aliquots (200 μl) were plated onto a variety of media to isolate as many heterotrophs as possible; the media included malt agar (Oxoid CM59); 1/10 strength tryptic soy agar (3 g/l tryptic soy broth, Difco 0370-17, plus 20 g/l agar); diagnostic sensitivity test agar (DST, Oxoid CM126); King's medium (King et al., 1954); rose bengal agar (Martin, 1950): potato dextrose agar (Oxoid CM139) plus 0.1% v/v triton N 100 (Fluka AG) and streptomycin 30 μg/ml; and starch casein agar (Kuster and Williams, 1964). The plates were incubated at 25°C for seven days then a representative sample of individual colonies was subcultured onto 1/10 tryptic soy agar and kept as stock cultures. When required for screening as potential biocontrol agents the isolates were grown as shake cultures in 1/10 tryptic soy broth at 20°C for 3 days.

Gaeumannomyces graminis var. *tritici*, for infecting the test plants was grown in a mixture of sand:wheat straw (100:3 w/w) at 20°C for 21 days. Plant pots (5 cm diam.) were filled approximately 1/2 full of coarse sand and gravel (see Table 2) and approximately 3 g of the *G. graminis* inoculum placed on top, followed by a further layer of sand. Wheat seeds (cultivar Avalon) were germinated on wet paper and, when 2 days old, sown into the pots prepared as above with two seeds per pot. Two days later 5 ml of the shake culture of the potential biocontrol agents were added to each pot. There were 3 replicate pots for each

Table 2. Results of isolation and primary screening in pots.

Primary screen of 1426 isolates from
25 different farms and
41 different sites

	Antagonistic to *G. graminis* only	PGPR only	Antagonistic to *G. Graminis* and PGPR	Total 'useful' organisms
GENERAL SITES (1319 isolates)				
Number of isolates	23	44	5	72
% isolates from these sites	1.7	3.3	0.4	5.4
Fungi only	2	5	0	7
Bacteria only	21	39	5	65
Actinomycetes only	0	0	0	0
HIGH IRON SOILS (107 isolates)				
Number of isolates	4	4	0	8
% isolates from these sites	3.7	3.7	0	7.5

isolate plus control infected and healthy plants. The microorganisms were also tested for possible plant growth promotion by inoculating into healthy plants, not infected with *G. graminis*

The seedlings were grown in a glasshouse for 4 weeks, then the roots were washed free of sand and scored on arbitrary scales for disease and growth enhancement.

Microorganisms used in the above screening were also examined for *in vitro* antagonism against *G. graminis* on DST agar plates. *In vitro* production of siderophores was assessed by adding 100 μM iron (as ferric chloride) to DST agar and to tris-glucate medium supplemented with 1 μM magnesium sulfate (Leong and Neilands, 1982), and examining the plates for loss of inhibitory activity. *G. Graminis* was inoculated centrally as a 6 mm agar disc, allowed to grow for 2 days and then 4 stabs of the microorganism under test were placed around the periphery of the petri plate. There were 3 replicate plates for each isolate. The plates were incubated at 20°C for 7-10 days then colony radii were measured for the *G. graminis* and for the test isolate. The width of any inhibition zones was also measured and recorded in 5 groups; no gap, >0.1-2.0 mm, >2.0-5.0 mm, >5.0-8.0 mm, and >8.0 mm. This was done for plates with and without added iron.

A randomly selected number of isolates from the screened populations, and all those found to be potential biocontrol agents were inoculated into glucose-tris broth (supplemented with 1 μM magnesium sulfate). The cultures were shaken at 20°C for 5-7 days and the culture supernatants then tested for catechols and hydroxamates using the tests of Arnow (1937) and Leong and Neilands (1982) respectively.

RESULTS AND DISCUSSION

The general sites selected cover a range of soil types (Table 1), but being mainly arable soils used for cereals, the nutrient status and the pH tend to be high following the application of fertilizers and sometimes lime. Organic matter levels are generally low. The high pH and the calcium levels would reduce the availability of iron, so competition for available iron may be important and organisms producing siderophores could be at an advantage.

The site yielding the greatest number of antagonists to take-all had high levels of nitrogen and calcium, a high pH and a high organic matter content. This site was different to all others, being a permanent pasture with clover. The possible effects of grasses and legumes on the suppression of take-all in subsequent cereal crops has already been noted (Grossman, 1967; Wong, 1981).

The sand used in the disease assessment screen had a low nitrogen and phosphorous content (Table 1), which would encourage take-all and make the test for biocontrol agents more severe. The iron concentration was also very low (1.4% total; 7 ppm available ferric iron) which would not discourage the selection of siderophore-producing microorganisms.

The high iron soils had a lower pH and lower calcium than the general sites, with normal or slightly elevated levels of total iron (Table 1). Levels of available iron as extracted by the method of Lindsay and Norvell (1978) were high. In the soils tested by Lindsay and Norvell the range of values for extractable iron was 1.2 to 20.2 ppm with a mean value of 6.4 ppm. The mean available iron in our soils was 69 ppm (Table 1): it would seem unlikely that iron was limiting, so selection pressure for the production of siderophores would be slight.

The screening method selected a surprisingly large number of isolates (5.4% of all isolates; Table 2). Most of the selected organisms were bacteria though this may reflect some aspects of the screening procedure rather than the relative importance of the different groups of organisms in the soil. Fungi have been reported as antagonists to *G. graminis* (Wong, 1981) but most of the reported work has been on bacteria (Cook and Baker, 1983: Cook and Rovira, 1976; Weller and Cook, 1983). We did not select any actinomycetes as biocontrol agents, although some were isolated. Actinomycetes have been reported as antagonistic to *G. graminis* (Sivasithamparam and Parker, 1978). The high iron soils (Table 2) produced a similar or slightly higher proportion of antagonists although the sample size was small: all eight isolates selected from the iron soils were bacteria.

Pseudomonas fluorescens has been much used as a biocontrol agent and siderophore-producer (Geels and Schippers, 1983a; Cook and Rovira, 1976; Howell and Stipanovic, 1979; Weller and Cook, 1983). Only 19% of the potential biocontrol agents selected by this primary screening procedure from the general soils (Table 2) were fluorescent pseudomonads. However, the remaining isolates which did not reduce disease in the screening test (the residual population of Tables 3 and 4), did not include any fluorescent pseudomonads. The *in vivo* screen against take-all infected plants strongly selects for fluorescent pseudomonads amongst the biocontrol agents (note however that 81% of the isolates were not fluorescent pseudomonads).

The total isolates from the iron soils contained only 2.8% fluorescent pseudomonads but none of these were selected as potential biocontrol agents (shown in Table 2) by the screening. This suggests that fluorescent pseudomonads, and their siderophores presumably, are not important biological control agents in soils without iron limitation.

The *in vivo* screen seems to select for microorganisms which also show considerable inhibition zones in the traditional agar plate test for the production of antibiotics (Table 3). For example 38.1% of the selected biocontrol agents produced inhibition zones of more than 5 mm but only 9.4% of the residual population produced such wide zones (Table 3). The corollary of this is that 40.6% of the residual population produced no inhibition of *G. graminis* but only 4.8% of the selected organisms failed

Table 3. Plate tests for antagonism to *G. graminis* on DST agar, % of isolates in each inhibition group

Inhibition group, zone width mm.	Biocontrol agents	Residual population	High Iron soil isolates
0	4.8	40.6	65.4
0 to 2	33.3	28.1	19.3
2 to 5	23.8	21.9	12.8
5 to 8	33.3	4.7	2.6
>8	4.8	4.7	0
Numbers in test	21	64	78

Table 4. Influence of additional iron (100 μM) in agar on the degree of antagonism

	DST agar			Tris-glucose agar
	Biocontrol agents	Residual population	Iron soil isolates	Biocontrol agents
% isolates not affected by added iron	90.5	95.3	92.3	81.3
% isolates with antagonism significantly reduced by Fe	4.8	1.6	2.6	6.3
% isolates with antagonism significantly increased by Fe	4.8	3.1	5.1	12.5
Numbers in test	21	64	78	16

to show inhibition under the conditions of this test. Even though agents selected *in vivo* are often inhibitory *in vitro* there is no reason why organisms selected on the basis of *in vitro* antagonism should colonize roots and be effective under soil conditions. We would suggest that it is better to select organisms using whole plants growing under at least semi-natural conditions.

A random sample of the isolates from the iron soils showed few or very small inhibition zones (Table 3). There were 65.4% of the isolates non-inhibitory and only 2.6% producing inhibition zones of more than 5 mm. It is not clear why antagonists producing antibiotics, which this test measures, should be relatively uncommon in high iron soils.

Additional iron in the DST agar (Table 4) did not alter the antagonism of most isolates to any great degree, though a slightly higher percentage of the biocontrol agents had their antagonism significantly reduced, presumably because the siderophores they produce were no longer effective in the presence of excess iron. Increases in antagonism also occurred in the presence of additional iron but as yet we have no explanation of this.

The use of tris-glucose agar (very low iron, 16 μM) for testing the potential biocontrol agents resulted in no substantial changes in the amount of inhibition compared with that in the presence of added iron (Table 4). There were slightly higher percentages of isolates whose inhibition was significantly reduced or increased. The actual isolates showing reduced antagonism on DST with added iron were not the same as those showing this phenomenon in tris-glucose medium. This implies that the choice of nutrient base could influence isolate activity, as well as the level of available iron. Two out of the 21 strains grew more abundantly in the presence of excess iron on tris-glucose and these were not included in the analysis.

It appears that additional iron has little effect upon the antagonism exerted by the isolates in agar culture. Those isolates whose inhibition was reduced under high iron levels produced neither hydroxamates nor catechols in iron limited shake culture.

Some strains from the residual population and from the iron soils, and all the selected biocontrol strains were grown in shake culture to test for hydroxamate and catechol production. Almost 40% of the biocontrol organisms produced either catechol or hydroxamates, and this included almost all of the fluorescent pseudomonads. Only 13% of the residual population and 15% of the iron soil isolates produced detectable amounts of these metabolites. The isolates which did produce siderophore-like compounds were not detected in the agar tests. The higher percentage of biocontrol microbes producing hydroxamates or catechols were selected by the *in vivo* screen under the limiting iron conditions in the sand. The absence of any positive or reliable test using the agar plate method may be due to the *G. graminis* test organism used. It may be that *G. graminis* competes strongly for iron in agar culture, by producing its own siderophore with a high stability constant. This possibility is at present being assessed.

CONCLUSIONS

The *in vivo* screening procedure with growing plants selected quite a high proportion of potential biocontrol agents and PGPR's. Many of these isolates showed strong antagonism to *G. graminis* in subsequent agar plate tests and they also included many fluorescent pseudomonads and organisms producing siderophores. The plate tests for siderophores did not detect siderophore activity, even though some isolates were forming hydroxamates and catechols in iron limited shake culture.

In agar plates the main inhibition of *G. graminis* is by the production of antibiotics and not by siderophores, but *in vivo* both mechanisms may operate as well as others such as lysis. In selecting potential biocontrol agents it is important to have a screening system which uses the overall beneficial effect on the plant in the presence of the pathogen, regardless of the method(s) of action of the microorganisms. This will simultaneously test for antagonism of any sort, colonization of the roots and the ability to survive and grow in the soil environment.

We have no direct evidence as to how important siderophores are in the control of plant diseases and the promotion of growth. However, many of our isolates, which were selected only for biocontrol and as PGPR's, did have siderophores when subsequently tested. The residual population, which did not control disease or promote growth, had fewer siderophore-producers.

Fluorescent pseudomonads were approximately 20% of the selected biocontrol organisms and were the most antagonistic in agar plates. There were however four times as many selected organisms which were not fluorescent pseudomonads. These latter biocontrol agents often did not show large inhibition zones in agar cultures and may not have inhibition zones in agar cultures and may not have been selected in a screening system mainly based on plate tests.

Pseudomonads and siderophores do not appear to be important in soils high in iron.

ACKNOWLEDGEMENTS

We are most grateful to ICI plc., Agricultural Division, Jealott's Hill, Bracknell, Berkshire, England, for funding this work, for locating some of the sampling sites and for soil analysis facilities. In particular we thank Keith Powell and Maryling McInnes for much practical help, discussion and encouragement. We also thank the Ministry of Agriculture, Fisheries and Food, Agricultural Development and Advisory Service for help in locating sample sites.

REFERENCES

Arnow, L.E., 1937. Colorimetric determination of the components of 3,4-dihydroxyphenylalanine tyrosine mixtures. J. Biol. Chem. 118:531-537.

Burr, T.J., and Caesar, A., 1984. Beneficial plant bacteria. CRC Crit. Rev. Plant Sci., 2:1-20.

Campbell, R., and Faull, J.L., 1979. Biological control of *G. graminis:* field trials and the ultrastructure of the interaction between the fungus and a successful antagonistic bacterium. In: "Soil-borne Plant Pathogens," 603-609. (B. Schippers and W. Gams, eds.) Academic Press, London.

Cook, R.J., and Baker, K.F., 1983. "The Nature and Practice of Biological Control of Plant Pathogens," pp. 539. American Phytopathological Society, St. Paul, Minnisota.

Cook, R.J., and Rovira, A.D., 1976. The role of bacteria in the biological control of *Gaeumannomyces graminis* by suppressive soils. Soil Biol., Biochem., 8:269-273.

Emery, T., 1980. Iron deprivation as a biological defence mechanism. Nature, 287:776-777.

Geels, F.P., and Schippers, B., 1983a. Selection of antagonistic fluorescent *Pseudomonas* spp. and their root colonisation and persistence following treatment of seed potatoes. Phytopathol. Z., 108:193-206.

Geels, F.P., and Schippers, B., 1983b. Reduction of yield depressions in high frequency potato cropping soil after seed tuber treatments with antagonistic fluorescent *Pseudomonas* spp. Phytopathol. Z. 108:207-214.

Grossman, V.F., 1967. Grundungung als Pflanzen-schutzmassnahme. Z. Pflanzen-kr. Pflanzenschutz, 74:143-149.

Hornby, D., 1979. Take-all decline: a theorist's paradise. In: "Soil-borne Plant Pathogens," 133-156. Academic Press, London.

Howell, C.R., and Stipanovic, R.D., 1979. Control of *Rhizoctonia solani* on cotton seedlings with *Pseudomonas fluorescens* and with an antibiotic produced by the bacterium. Phytopathology, 69:480-482.

King, E.O., Ward, M.K., and Raney, D.E., 1954. Two simple media for the demonstration of pyocyanin and fluorescin. J. Lab. Clin. Med., 44:301-307.

Kloepper, J.W., Leong, J., Teintz, M., and Schroth, M.N., 1980a. Enhanced plant growth by siderophores produced by plant growth promoting rhizobacteria. Nature, 286:885-886.

Kloepper, J.W., Leong, J., Teintz, M., and Schroth, M.N., 1980b. *Pseudomonas* siderophores: a mechanism explaining disease-suppressive soils. Curr. Microbiol., 4:327-320.

Kuster, E., and Williams, S.T., 1964. Selection of media for isolation of Streptomycetes. Nature, 202:928-929.

Leong, S.A., and Neilands, J.B., 1982. Siderophore production by Phytopathogenic microbial species. Arch. Biochem. Biophys., 218:351-359.

Lindsay, W.L., and Norvell, W.A., 1978. Development of DIPA soil test for zinc, iron, manganese and copper. Soil Sci. Soc. Am. J., 42:421-428

Martin, J.P., 1950. The use of acid rose bengal and streptomycin in the plate method for estimating soil fungi. Soil Sci., 69:215-232.

Neilands, J.B., 1981. Microbial iron compounds. Annu. Rev. Biochem., 50:715-731.

Rudeforth, C.C., Hartnup, R.; Lea, J.W., Thompson, T.R.E. and Wright, P.S., 1984. "Soils and their use in Wales,* Soil Survey of England and Wales, Bulletin, 11, pp. 336. Lawes Agricultural Trust, Harpenden, England.

Scher, F.M., and Baker, R., 1982. Effect of *Pseudomonas putida* and a synthetic iron chelator on induction of soil suppressiveness to *Fusarium* wilt pathogens. Phytopathology, 72: 1567-1573.

Schneider, R.W., 1982. "Suppressive soils and Plant Disease," pp. 88. (R.W. Schneider, Ed.), American Phytopathological Society, St. Paul, Minnisota.

Schroth, M.N. and Hancock, J., 1981. Selected topics in biological control. Annu. Rev. Microbiol., 35:453-476.

Sivastahamparam, K., and Parker, C.A., 1978. Effects of certain isolates of bacteria and actinomycetes on *Gaeumannomyces graminis* var. *tritici* and take-all of wheat. Aust. J. Bot., 26:773-782.

Weller, D.M. and Cook, R.J., 1983. Suppression of take-all of wheat by seed treatments with fluorescent pseudomonads. Phytopathology, 73: 463-469.

Wong, P.T.W., 1981. Biological control by cross-protection. In: "Biology and Control of Take-a;;." 417-431. (M.J.C. Asher and P.J. Shipton, Eds.), Academic Press, London.

HERBICIDE-INDUCED INTERACTIONS BETWEEN CEREAL ROOTS AND FLUORESCENT *PSEUDOMONAS* SPP

M.P. Greaves[1], J.A. Sargent[2] and J.M. Whipps[3]

Long Ashton Research Station[1]
Weed Research Division, Begbroke Hill
Yarnton, Oxford, UK.

Hexlands Ltd.[2]
East Challow, Wantage, Oxon, UK.

Glasshouse Crops Research Institute[3]
Littlehampton, Sussex, UK.

INTRODUCTION

In order for any microbial inoculant, developed to promote plant growth or combat soil-borne diseases, to be of practical value it must be integrated into existing weed and pest management systems. It must, therefore, not interact adversely with the sequence of insecticides, fungicides, herbicides or nematicides that may be used routinely in the crop. On the other hand, there may be beneficial synergistic interactions between the control agent and chemicals which can be exploited (Scheepens, 1979).

Interactions between herbicides and micro-organisms are particularly well documented. This is, in no small part, a result of the frequent observations of increased incidence and virulence of plant diseases following herbicide use (Altman and Campbell, 1979; Katan and Eshel, 1973; Papavizas and Lewis, 1979). Among the diseases involved are the foot- and root- rots caused by soil-borne micro-organisms. The reasons cited for these effects on plant-disease include herbicide-mediated changes in plant biochemistry, physiology and morphology. Much of the work described in this paper is a synthesis of an extensive research programme done at the Weed Research Organization, Oxford (now the Long Ashton Research Station, Weed Research Division) in recent years. It serves to illustrate the complex changes in plant physiology, root morphology and root-microflora interactions which can occur following herbicide use.

Although herbicides are designed to be tolerated by crops, damage can and does occur, particularly as a result of overdosing or inappropriate timing of application. Damage can occur to both leaves and roots.

It has been known for many years that certain herbicides can interfere with meristematic activity in roots and inhibit root growth (Audus, 1948). Subsequently many research reports have confirmed the early findings. These are reviewed by Currey and Teem, (1976).

Table 1. Effect of mecoprop at 10 kg a.i. ha^{-1} on relative growth rates (RGR) of spring wheat Cv Maris Dove. Sprayed 4 weeks after planting. (Whipps and Greaves, 1985)

Growth interval after treatment	2-9d		9-16d	
Treatment	Control	Mecoprop	Control	Mecoprop
Shoot	0.051 ± 0.012	0.036 ± 0.001*	0.081 ± 0.010	0.059 ± 0.021
Root	0.076 ± 0.009	0.088 ± 0.008	0.096 ± 0.009	0.040 ± 0.009*
Total Plant	0.065 ± 0.009	0.066 ± 0.004	0.089 ± 0.009	0.039 ± 0.009*

Values are means of 3 reps ± SEM. Differences tested by Student's 't' test
* = significantly different from control, P = <0.05

$$RGR = \frac{\ln F - \ln I}{t}$$ where I = Initial dry wt. (mg), F = Final dry wt (mg) and t = time in days

The effects of the phenoxy-group of herbicides have been studied more widely than any other group (Friesen et al., 1964; Maas, 1968; Nilsson, 1973a; Nilsson, 1973b; Skuterud, 1975, Tottman and Ellis, 1978). One of them, mecoprop-(2-methyl-4-chloro) phenoxy propionic acid, is the subject of the research reported here. It must be stressed at the outset that the findings of this research do not necessarily single mecoprop out as a bad herbicide. Indeed, it is of proven value and has been used successfully for many years. In the context of this paper it is used solely as an experimental tool to help elucidate some root-microorganism interactions.

EFFECTS ON PLANT GROWTH AND ROOT MICROBIOLOGY

Whipps and Greaves (1985) have described the changes in shoots and roots of wheat following treatment with mecoprop. Their results were for a herbicide application rate of 10 kg a.i. ha^{-1} (4 x field rate) but similar, though less pronounced effects have been measured after treatment at normal field rate (Greaves, unpublished data), especially with wheat grown in sand. In summary, they showed that mecoprop treatment resulted in slight scorching of some leaves followed by leaf tip chlorosis. These visible symptoms were accompanied by a decrease in leaf area and an increase in dry weight cm^{-2} of the leaves. The relative growth rates of the shoots were significantly decreased during the first 9 days after treatment (Table 1). The effects on leaf growth rate were not confined to those leaves that came into direct contact with the herbicide but occurred also in leaves that were still sheathed or were present as primordia at the time of spraying. (Greaves, unpublished data).

The major changes in growth and morphology occurred in the roots. Tottman and Ellis, (1978) have described these effects and our observations confirm them exactly (Greaves and Sargent, 1985; Whipps and Greaves, 1985). Roots of treated plants quickly (by 7 days after treatment) produce characteristic abnormal lateral roots, often in large numbers and arising close to the meristem of main axes. (Fig.1). Root hairs often develop close to the meristem (Fig 1). This symptom was marked by day 15 and the affected roots became thicker in diameter. Some recovery usually occurred after this time and the roots resumed normal growth and branching. Even though these morphological changes were pronounced there was no significant effect of mecoprop on root dry weight. However, the relative growth rates of the roots were decreased between 9 and 16 days after treatment, that is later than the effect on shoot RGR (Table 1).

Although these changes in plant morphology are significant, even greater significance may be attached to changes at the cellular level as seen in the electron microscope. These changes have been described in detail by Greaves and Sargent (1985) and subsequent observations in our laboratory have confirmed and extended the results. In controls the ribosomes are not particularly aggregated into polysomes and the endomembrane system is relatively quiescent, indicating low secretory activity (Fig. 2). In contrast, within 4 days of mecoprop application to the foliage, root cortex cells show pronounced changes. In particular, starch grains are evident and the endoplasmic reticulum is dilated. Dictyosomes are producing secretory vesicles and ribosomes are aggregated into polysomes (Fig. 2b,c).

Root cells of untreated plants showed no change during the remaining period of examination (up to 25 days after spray date). It is worth recording that sections of control roots showed only a few signs of microbial colonization. When this was observed it was confined to the epidermis or to the interior of dead surface cells and was sparse (Fig 2.d). Occasionally, bacteria were seen to be digesting the cell wall.

By 11 days after treatment with mecoprop bacterial colonization of

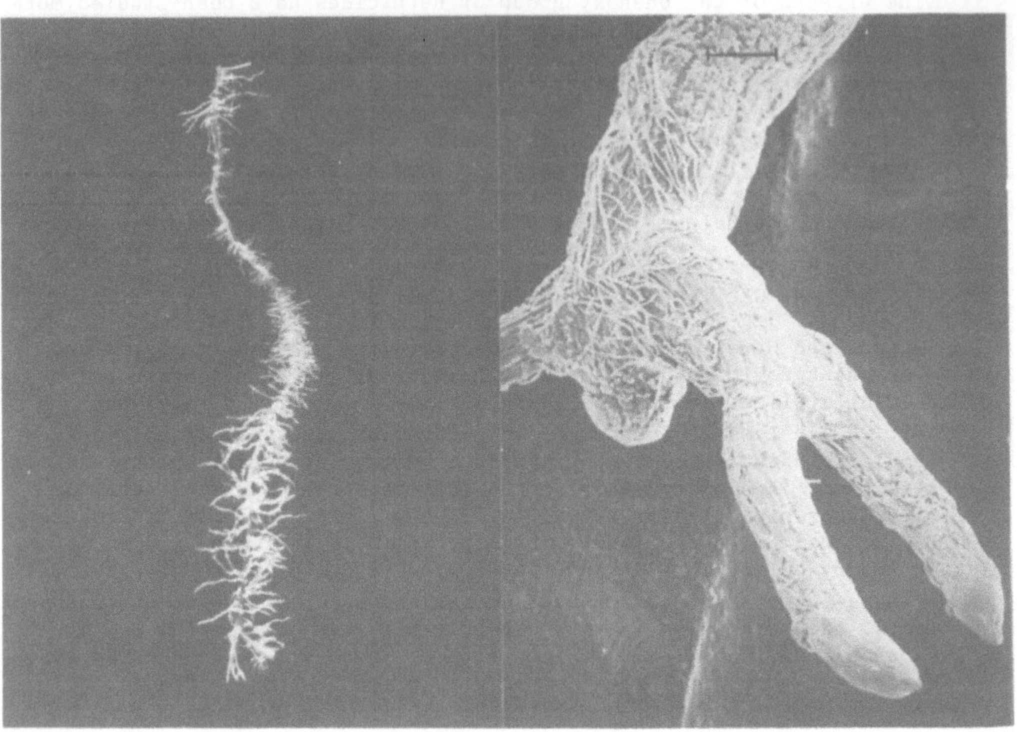

Figs. 1a,b. Main root axes showing abnormal production of lateral roots associated with mecoprop treatment.
Bar represents 0.2 mm.

the epidermis was extensive (Fig. 2e). The bacteria were almost exclusively embedded in a pronounced layer of material, the mucigel, and were generally surrounded by clear zones. These zones might indicate dissolution of the surrounding material, or more likely, production of electron-transparent material by the bacteria. More significantly, at this time, very large numbers of bacteria were present within the cortex (Fig. 2f). Generally, they were confined to intercellular regions, where their large numbers caused dilation and distortion of the spaces accompanied by changes in the structure of intercellular material. The normally uniform, electron dense (polypectate) material (cf Fig. 2g) became more granular and showed considerable areas of electron transparency. As reported by Greaves and Sargent (1985) many cells were also ruptured and colonized by bacteria, the cell contents apparently being digested. The rupture of the cells was frequently a result of bacterial attack on the cell walls (cf. Fig. 2d). In those regions of the cortex which were free of bacteria, there was evidence that cells were filling with secretory products and that there was much accumulation of secreted material in intercellular regions (Fig. 2g). The fibrous appearance of the cytoplasm (inset in Fig. 2g) suggests an accumulation of material similar to that in the intercellular spaces. The cells in these roots appear not to contain starch granules suggesting secretions may be at the expense of starch.

At 18 days after treatment, starch was once again visible in the cortical cells (Fig. 2h) and the cytoplasm was still fibrous in appearance. In the deeper cells of the cortex, the cytoplasm was less dense in appearance and ribosomes were present as polysomes (Fig. 2h) inset). In many of the cortical cells, at all depths, there were many large vesicles

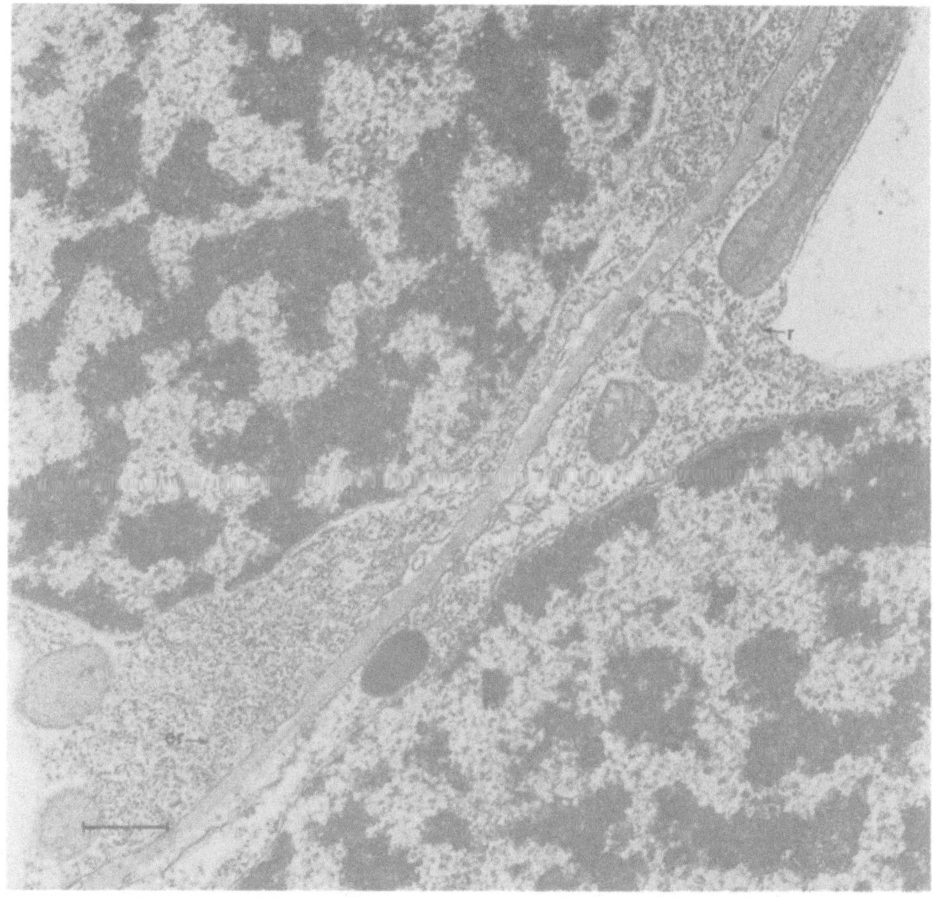

Fig. 2. Electron micrographs of sections through wheat roots at times indicated after spraying.
b, bacterium; cyt, cytoplasm; d, dictyosome; ep, epidermis; er, endoplasmic reticulum; is, intercellular space; m, mucigel: mb, mucigel boundary; ps, polysome; r, ribosome; s, secondary wall; st, starch grain; v, vesicle; w, cell wall: bars represent 0.5 μm.

Fig. 2a. Control plant, 4 days, lateral root cortex

(Fig. 2h). As noted by Greaves and Sargent (1985) many of the outer cortical cells had by now begun to show secondary thickening of the walls. This thickening, which has a laminate appearance (Fig. 2i), is most commonly seen in cells adjacent to bacteria and may, thus, represent some activation of cell defences against invasion by the bacteria. Another unusual feature associated with the presence of bacteria, was the development of defined outer boundary to the mucigel overlaying the epidermis cells (Fig. 2j). A similar structure has been reported by Greaves and Darbyshire (1972).

Little change occurred by day 25 except that, although the cortex was still colonized by bacteria, the degree of colonization was less dense than at earlier times. Cortical cells of the main axis were

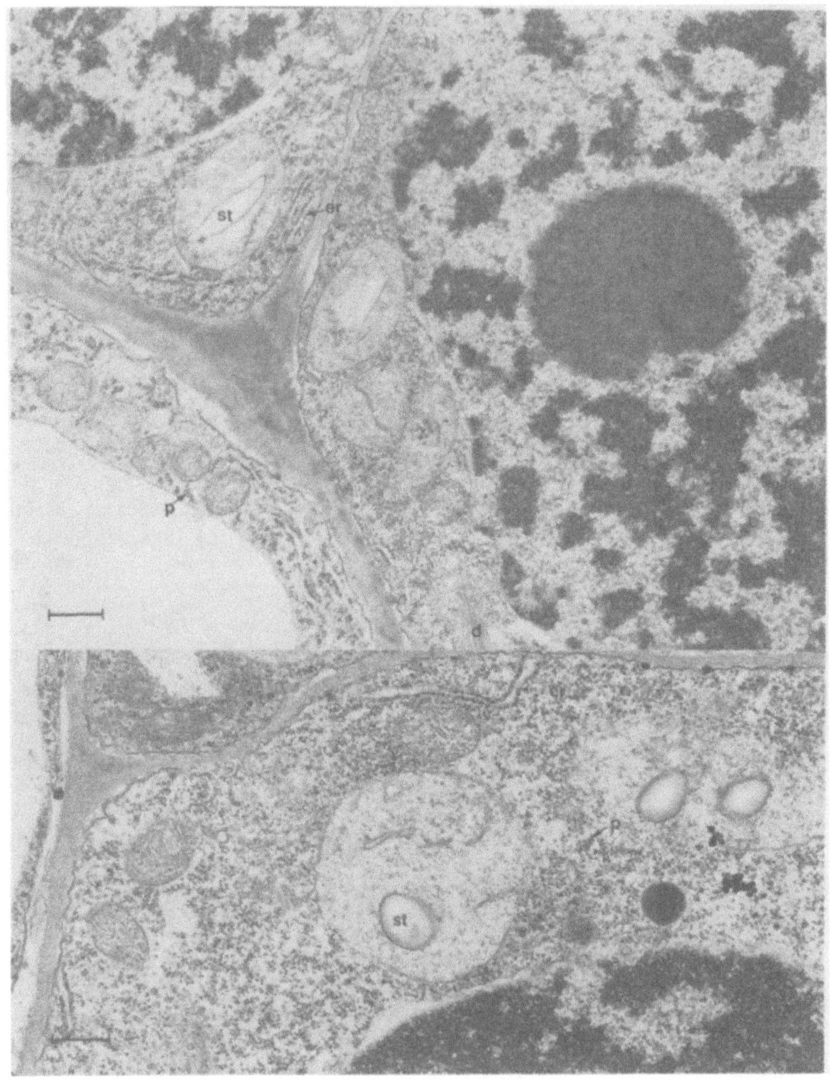

Figs. 2b, c. Mecoprop treated, 4 days, lateral root cortex

frequently empty of contents to considerable depth into the tissue and over extensive areas. Lateral root cortex cells, however, appeared to be similar to earlier samples with frequent starch grains and a fibrous appearance to the cytoplasm.

It was impossible to formally identify the mocro-organisms seen on and in root tissues in the electron microscope. There was, however, circumstantial evidence to suggest that at least a significant proportion of those observed in the work described here were fluorescent *Pseudomonas* spp. Counts of viable bacteria using the dilution-plate method with selective media have shown a massive increase in fluorescent *Pseudomonas* spp. within the cortex, reaching a maximum some 14 to 21 days after spraying with mecoprop. (Fig. 3). Other bacteria were also present in the cortex, but the increase in *Pseudomonas* spp. was by far the most pronounced. Detailed studies of these cortex populations in 1 mm root segments (Greaves, unpublished data)

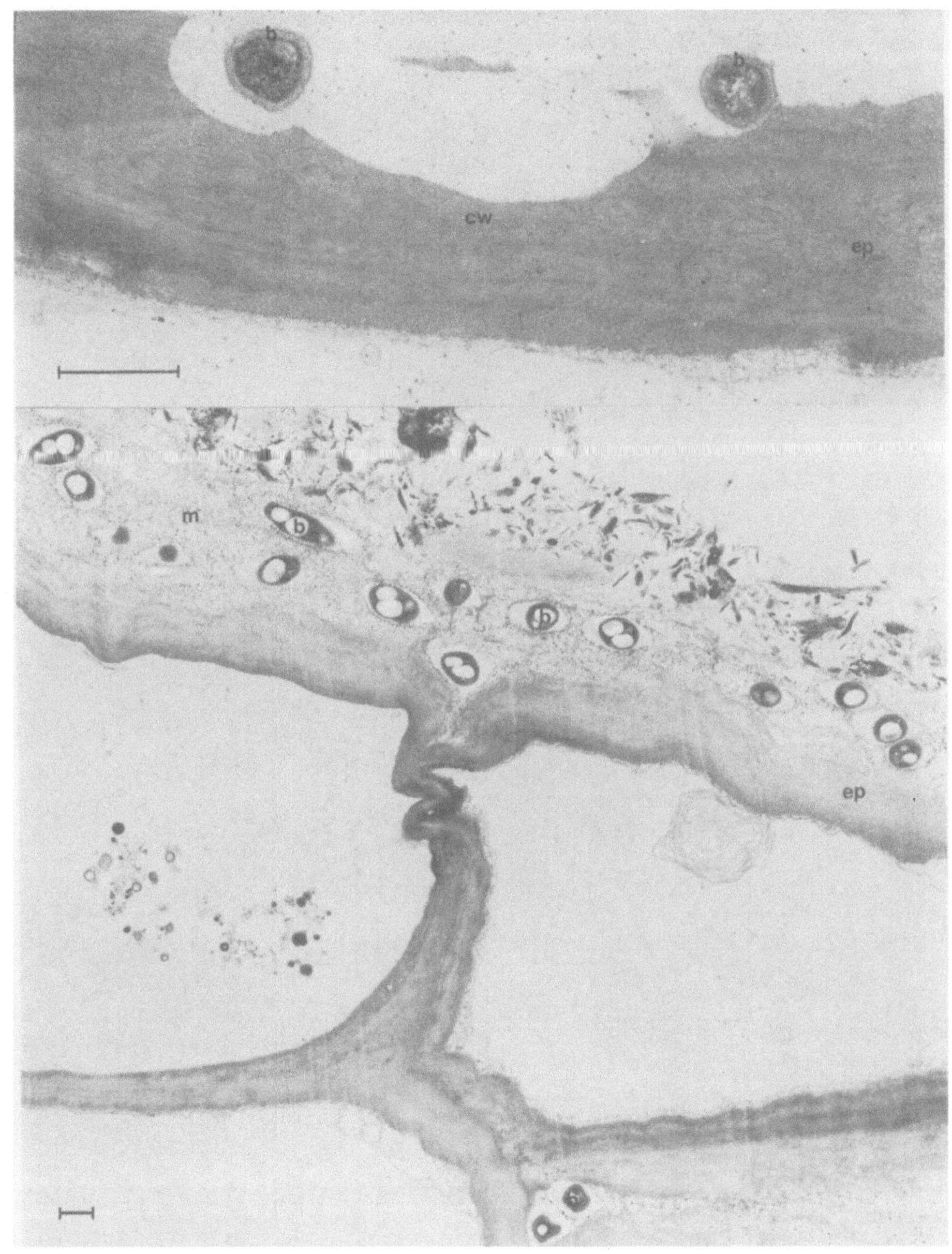

Fig. 2d. Control plant, 11 days, later root cortex.

Fig. 2e. Mecoprop treated, 11 days, lateral root epidermis.

showed that the greatest invasion of the cortex occurred adjacent to emerging laterals, where cells had been physically damaged. Nonetheless, there was also considerable invasion in regions of the root not damaged in this way. The only root region consistently found to be free of invasion was the zone of cell elongation.

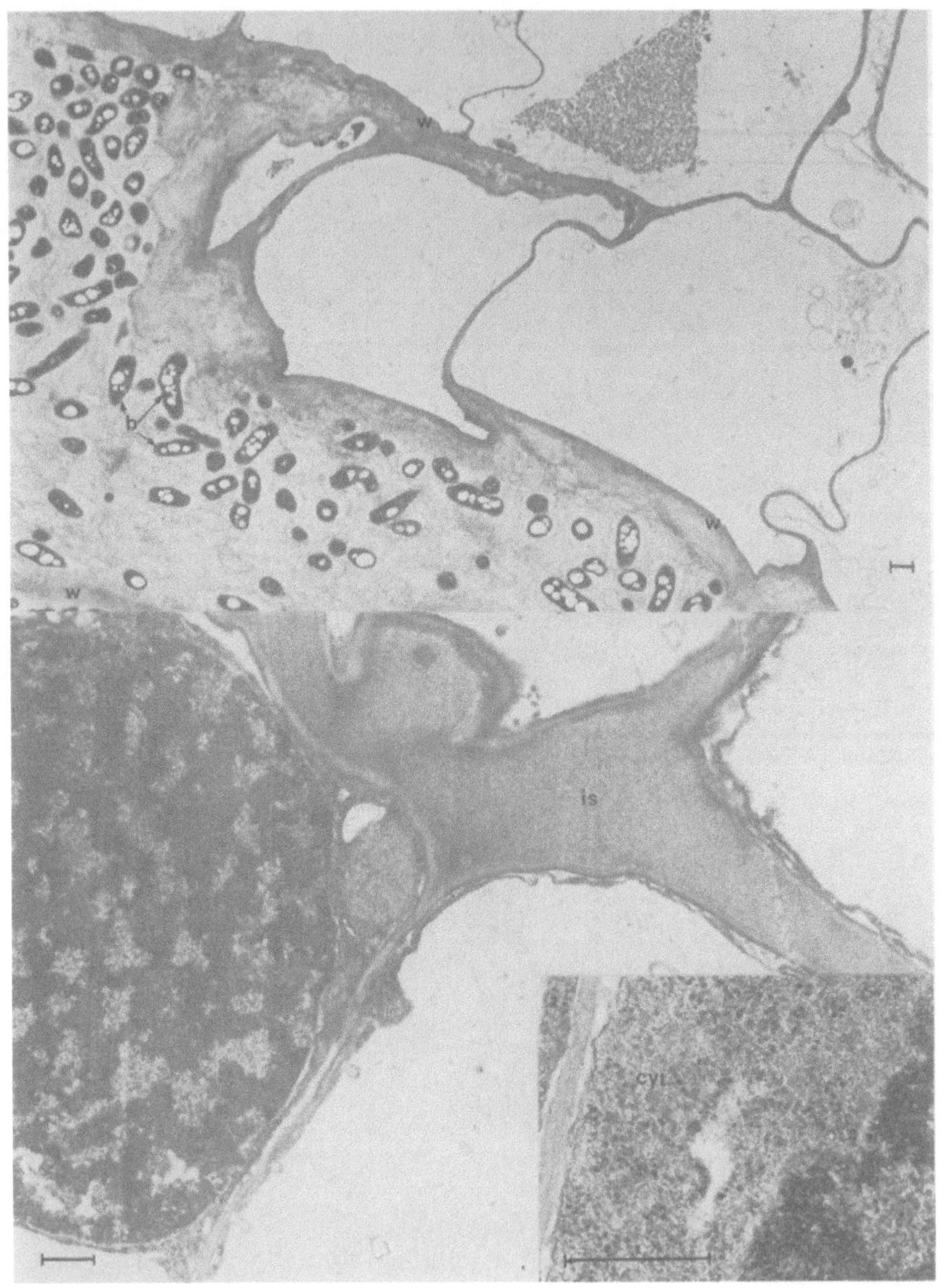

Fig. 2f. Mecoprop treated, 11 days, main root cortex.

Fig. 2g. Mecoprop treated, 11 days, lateral root cortex.

EFFECTS ON PLANT PHYSIOLOGY

Apart from the physiological changes, associated with mecoprop treatment, which are implied by the cellular changes seen in the electron

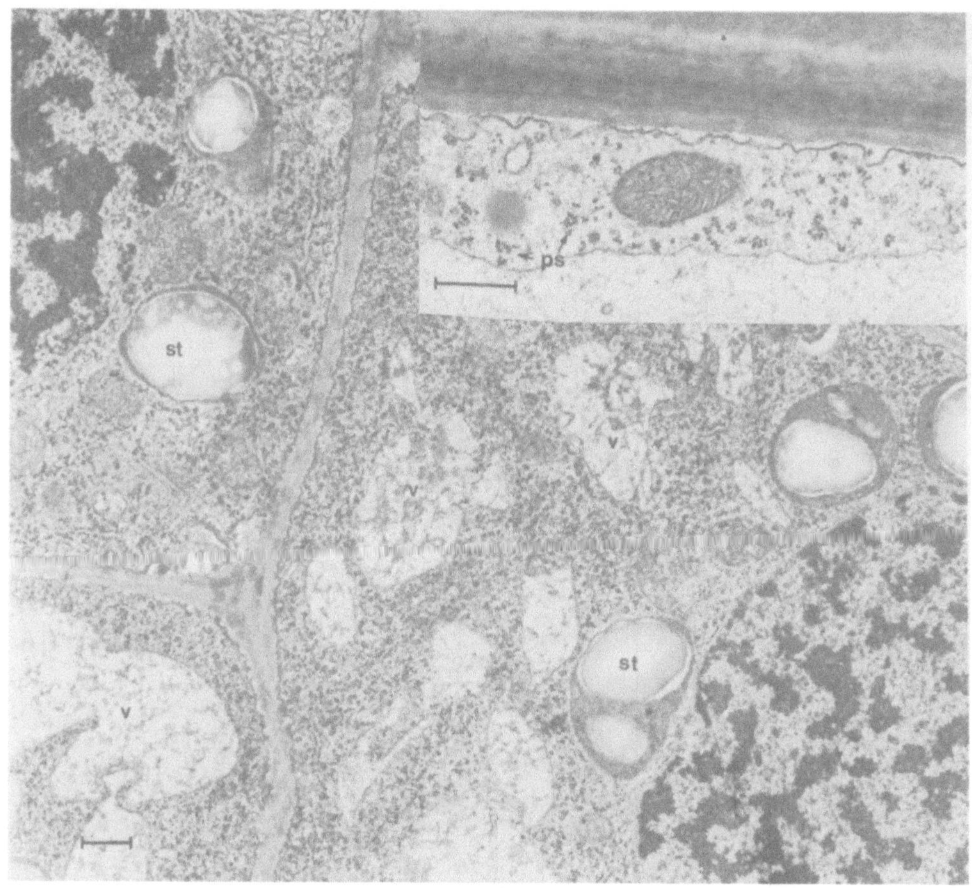

Fig. h. Mecoprop treated, 18 days, lateral root cortex

microscope, there are clear effects on photosynthesis and translocation of fixed carbon. These changes, described in detail by Whipps and Greaves (1985) and summarized in Table 2, have particular significance with respect to microbial colonization of, and function in, roots. Using the pulse-chase technique with $^{14}CO_2$, these authors showed that mecoprop treatment caused an immediate and prolonged decrease of ^{14}C fixation. Their data indicated that photosynthesis itself was affected and that the decrease in fixation was not due just to reduced plant size. There was also a decrease in the proportion of C leaving the leaves after the pulse. Mecoprop may, therefore, affect the leaf tissue such that translocation of recently fixed ^{14}C is decreased in the light. In contrast, measurement of the proportion of C translocated to the root in the chase period 9 days after herbicide treatment showed a significant increase. Equally a reduced proportion was found to have been exuded into the rhizosphere. These effects had disappeared after 16 days.

DISCUSSION

Consideration of the timing of the various events described above permits the construction of a hypothesis about the sequence and course of events occurring in the plant-microbe interactions which follow mecoprop application.

Table 2. Effects of mecoprop at 10 kg a.i. ha^{-1} on fixation and translocation of ^{14}C in spring wheat cv Maris Dove (Whipps and Greaves, 1985)

Days after treatment	1		8		15	
Treatment	Control	Mecoprop	Control	Mecoprop	Control	Mecoprop
^{14}C recovered mg^{-1} shoot (DPM x 10^{-4})	6.6 ± 1.0	5.9 ± 0.3	3.4 = 0.4	2.3 ± 0.5*	3.7 ± 0.9	2.9 ± 0.4
% ^{14}C in roots after 1h pulse	15.1 ± 1.8	4.0 ± 1.9**	19.0 ± 3.9	15.6 ± 2.0	20.4 ± 2.4	13.1 ± 3.2*
% ^{14}C in roots after 24 hour chase	34.7 ± 3.8	22.5 ± 2.9*	43.9 ± 8.3	59.9 ± 2.5*	41.9 ± 6.3	42.5 ± 5.4
Loss of ^{14}C into rhizosphere (DPM mg^{-1} d wt root)	677 ± 92	555 ± 53	554 ± 67	302 ± 29**	870 ± 133	369 ± 101**

Values are means of 3 reps ± SEM. Differences tested by Student's 't' test

*, P<0.05; **, P<0.01

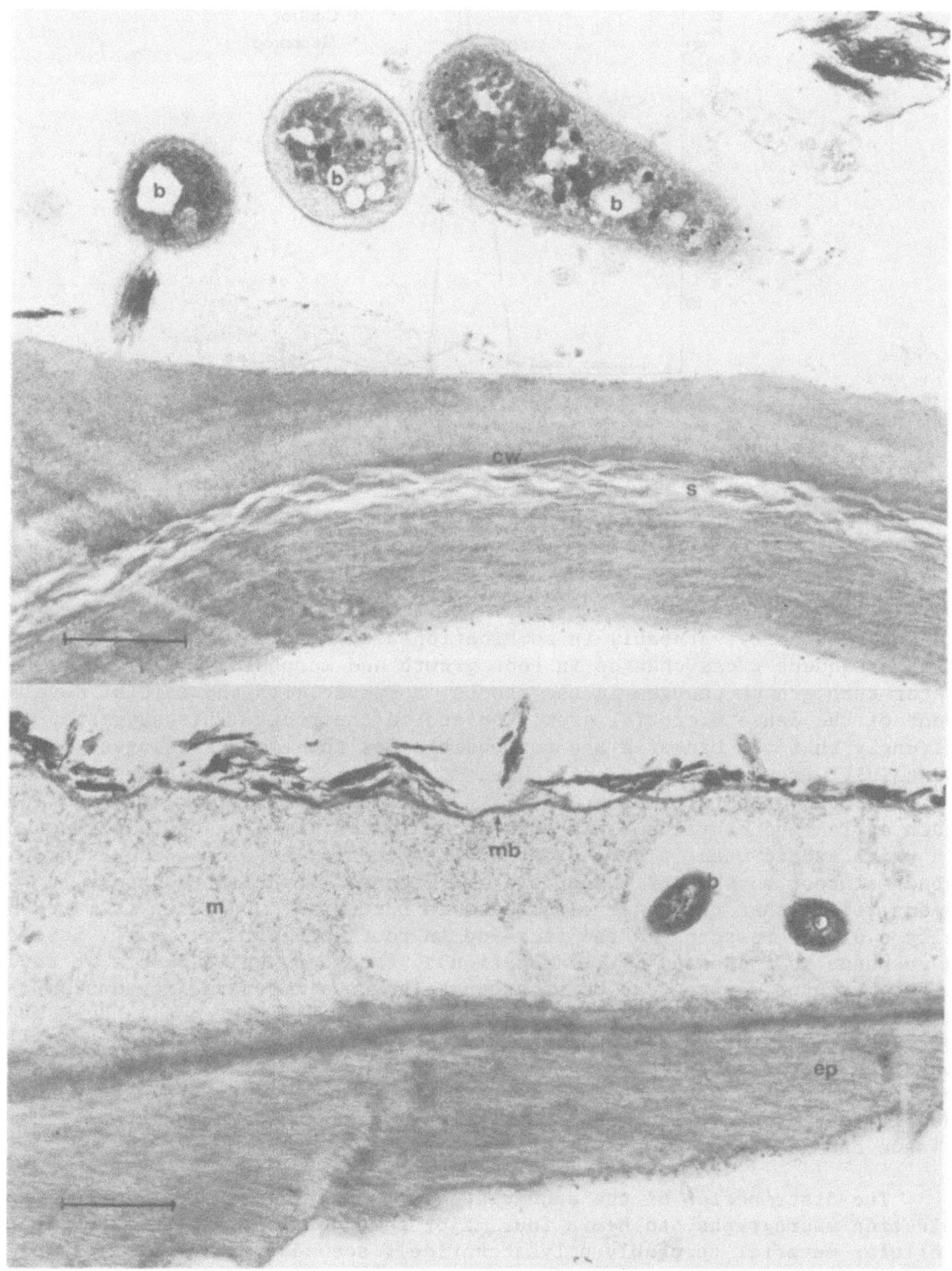

Fig. 2i. Mecoprop treated, 18 days, lateral root cortex.

Fig. 2j. Mecoprop treated, 18 days, main root epidermis.

Undoubtedly, the immediate and lasting effects on leaf growth, photosynthesis and translocation of carbon are direct effects of the herbicide. Equally, the detection of micromorphological change, and implied physiological change, in root cortex cells occurs long before any detectable change in microbial development on the root. It seems reasonable to assume that these changes are direct results of mecoprop translocated

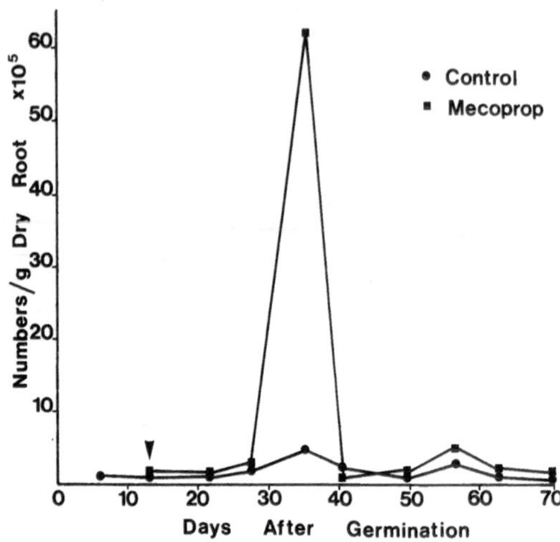

Fig. 3. Populations of fluorescent *Pseudomonas* spp. in the cortex of wheat (var. Maris Dove) treated with mecoprop (10 kg ha^{-1}). Plants were grown in sand and populations counted by plate-dilution technique after maceration of washed roots. The arrow indicates time of spraying.

to the root cells, probably in combination with changes in photosynthesis. The subsequent gross changes in root growth and morphology, which occur later than growth changes in the shoots, coincide with the initial development of the dense microbial growth on and in the roots. This suggests strongly that the bacteria are responsible for these gross changes. Certainly, *Pseudomonas* spp. which are dominant in the populations on and in the roots at this time, are well known as producers of growth regulators such as IAA and ethylene. This suggestion is confirmed by one experiment in which axenic wheat plants treated with mecoprop developed no visible abnormal root morphology (Wingfield and Greaves, unpublished data). It seems likely that the massive increase in bacterial population is initially a direct response to the increase in root secretory activity, itself a response to increased C-translocation to the root, which occurs at this time. Mecoprop may prevent this carbon being used for normal extension growth. The accumulation of secreted material within the root, and lack of exudation into the surrounding rhizosphere soil, would explain the invasion of the root tissue by the bacteria. Certainly, no increase in bacteria has been detected in the rhizosphere of these plants. The peak development of the pseudomonad population appears to be associated with tissue damage caused as the laterals emerge.

The distribution of the accumulated C in the roots appears, from the electron micrographs, to be in four major components. These are the intercellular material (probably polysaccharide), secondary thickening of cell walls, starch and microbial cells. In addition, the changes seen in the endomembrane system and formation of polyribosomes suggests increased protein synthesis which would also result in C accumulation. Possibly the early effect of mecoprop, of increasing movement of C to the roots, results in C storage as starch. Later, as the effects develop further, starch is mobilized to other sinks such as secretion polysaccharides. Alternatively, storage of C as starch, and secretion out of the cell of polysaccharides could occur simultaneously. Whether or not the presence of bacteria affects these processes is yet to be ascertained.

As well as serving to highlight the extreme complexity of herbicide-plant-microorganism interactions, these findings emphasise the need to consider carefully how to integrate microbial inoculants into the management system being applied to a crop. The data presented show two particular aspects to advantage. On the one hand, it might be possible to exploit the physiological changes induced in the plant to increase and prolong the development of an inoculated desirable *Pseudomonas* spp. on the root. Conversely, by stimulating other *Pseudomonas* spp., which obviously include deleterious strains, the inoculum of the desirable species might be overwhelmed by aggressive competition. Clearly, the usual pot experiments to determine establishment of an inoculant in the rhizosphere, will not reveal either of these possibilities.

As stated earlier, most groups of herbicides include compounds which affect root form and function. So far, only mecoprop has been investigated in depth with regard to its effects on plant-microorganism interaction. It is essential that extensive research is done in this area. Only then will it be possible to realise fully the exciting potential for manipulating plant growth and suppressing soil-borne root pathogens using microbial inoculants.

REFERENCES

Altman, J., and Campbell, C.L., 1979. Herbicides and environment: a review on stimulating and inhibiting interactions with plant diseases. Z. Pflanzenkr. Planzenschutz. 86:290-302.

Audus, L.J., 1948. Studies on the phytostatic action of 2,4-Dichlorophenoxyacetic acid and coumarin. The reversibility of root-growth inhibitions. New Phytol., 47:196-219.

Currey, W.L., and Teem, D.H., 1976. Herbicides and root development. Soil and Crop Science Society of Florida Proceedings, 36:23-28.

Friesen, H.A., Baenziger, H., and Keys, C.H., 1964. Morphological and cytological effects of dicamba on wheat and barley. Can. J. Plant Sci., 44:288-294.

Greaves M.P., and Darbyshire, J.F., 1972. The ultrastructure of the mucilaginous layer on plant roots. Soil Biol. Biochem., 4:443-449.

Greaves, M.P. and Sargent, J.A., 1985. Herbicide-induced microbial invasion of plant roots. Weed Sci. (in press).

Katan, J., and Eshel, Y., 1973. Interactions between herbicides and plant pathogens. Res. Revs., 45:145-177.

Maas, G., 1968. Damage to the anchorage roots (crown roots) of cereals after herbicide treatment. Z. Pflanzenkr. Pflanzenpathol. Pflanzenschutz, 75:139-144.

Nilsson, H.E., 1973a. Influence of the herbicide mecoprop on *Gaeumannomyces graminis* and take-all disease of spring wheat. Swedish J. Agr. Res. 3:105-113.

Nilsson, H.E., 1973b. Influence of herbicides on take-all and eye-spot disease of winter wheat in a field trial. Swedish J. Agr. Res. 3:115-118.

Papavizas, G.C., and Lewis, J.A., 1979. Side-effects of pesticides on soil-borne plant pathogens. In: "Soil-Borne Plant Pathology" 483-505. B. Schippers and W. Gams, Eds.). Academic Press, London.

Scheepens, P.C., 1979. The use of pathogenic mildews to control problem weeds. Gewasbescherming, 10:3 .

Skuterud, R., 1975. Tolerance of some varieties of cereals to phenoxy acids. Norwegian Plant Protection Institute, Department of Herbology, 132:17.

Tottman, D.R., and Ellis, E., Ll.P., 1978. The effect of herbicides on the root system of wheat plants. Ann. Appl. Biol., 90:93-99.

Whipps, J.M., and Greaves, M.P., 1985. The effect of the herbicide mecoprop on plant growth and distribution of photosynthate in wheat seedlings. Weed Res. (in press).

IRON AS A FACTOR IN DISEASE DEVELOPMENT IN ANIMALS

E.D. Weinberg

Department of Biology and Program in Medical Sciences
Indiana University
Bloomington, Indiana 47405, U.S.A.

INTRODUCTION

In nearly all forms of life, the number and diversity of enzymes that contain iron or that depend on the presence of this metal for activity is impressive. Well known groups of enzymes include electron transport proteins, iron flavoproteins, hydroperoxidases, and oxygenases. The metal plays a crucial role in DNA synthesis as a cofactor of ribonucleotide reductase. Iron is involved in notrogen and hydrogen fixation and, in animals, is essential for erythropoiesis, granulopoiesis, and synthesis of collagen, corticosteroids, and prostaglandins.

The quantitative need for the metal by cultures of diverse cells for growth and survival is remarkably uniform. For efficient growth, cells of plant, microbial, or animal species each require approximately 0.4 to 4.0 μM iron in the culture medium. This amount often may be available as a contaminant of other culture ingredients. Of course, if iron-binding compounds are present that interfere with the normal mechanism of acquisition, the quantity of iron supplied must be increased. Likewise, if the normal mechanism is damaged or absent, either an elevated amount of the metal and/or organic molecules to replace those missing must be included in the culture medium.

In most species of bacteria, iron not only controls the extent of growth but also the ability to shift into secondary metabolism of cells that have recently stopped growing (Weinberg, 1977). The range of iron concentrations that permits derepression of the genes of secondary metabolism is far narrower than the range that allows growth. Successful completion of the process of secondary metabolism is apparently essential for microbial cell survival. Moreover, such low molecular weight secondary substances as antibiotics and such protein secondary substances as exotoxins may have ecological utility for their bacterial producers.

In view of the manifold roles of iron in life processes and of the ingenious mechanisms that have by necessity been evolved by cells to acquire this unique metal, it is not surprising that numerous examples of ecological warfare exists whereby various species withhold iron from competitors or from potential parasites. During the past half century, for example, a profusion of clinical and laboratory observations have revealed that vertebrate animal hosts possess an array of mechanisms to deprive bacterial, fungal, protozoan, and neoplastic invaders of access to host iron (Weinberg, 1984a).

VERTEBRATE HOST MECHANISMS WHEREBY IRON IS WITHHELD FROM INVADERS

In advance of invasion, vertebrate hosts have stationed powerful iron-binding glycoproteins of the transferrin class at potential sites of microbial entry. These 77K molecules are present in plasma; in secretions such as milk, nasal and bronchial mucus, tears, saliva, gastrointestinal fluid, hepatic bile, synovial fluid, cervical mucus and seminal fluid; and in egg white. Two metal-binding sites are present in the polypeptide chain; three tyrosyl (or tryptophanyl) and two histidinal residues plus one carbonate ion serve as ligands in each of the two sites. Although all transition series metals that have been tested can be bound by the protein, the avidity for iron is greatest.

Consider the awareness of iron-withholding defense displayed by the egg-laying hen. She provides the potential developing embryo with a generous quantity of iron (about one milligram) in the yolk but places no iron in the white. Instead, she includes a member of the transferrin class termed conalbumin or ovotransferrin; it comprises about 12% of the egg white solids. Iron-binding activity of this protein requires an alkaline pH; accordingly, the hen has adjusted the pH of egg white to 9.6. The germ-free integrity of the hen's egg can be compromised by addling which permits the yolk iron to enter the white, or by rinsing intact eggs with dilute solutions of iron salts. In the latter case, the metal ions penetrate the shell pores and saturate the ovotransferrin. With either method of iron infiltration, such bacterial contaminants of the pores as *Pseudomonas* and *Salmonella* soon proliferate in the white.

Human plasma transferrin binds iron well at pH 7.3 but persons whose blood pH is lowered, as in an episode of diabetic ketoacidosis, may develop systemic fungal infections because of the loss of iron-withholding defense capability (Artis et al., 1982). However, one member of the transferrin class, termed lactoferrin, does chelate iron strongly at pH reactions as low as 4.0. The pH of invaded tissue generally becomes acidic due to the lactic acid released from microbial cells and/or from metabolically stimulated leukocytes. In addition to its occurence in epithelial secretions, lactoferrin is a major component of the specific granules of circulating polymorphonuclear neutrophilic leukocytes. The apoprotein is released on degranulation of the cells in a septic area; after combining with iron in the infected region, the metal-saturated protein is ingested by macrophages. The ingested iron-containing lactoferrin can then augment the release of oxygen intermediates, which increases the microbicidal action of the phagocytic cells.

At the time of invasion by microbial pathogens or neoplastic cells, vertebrate hosts quickly shift aspects of their iron metabolism. They specifically shut down intestinal assimilation of the metal as well as the recycling to transferrin of hemoglobin iron from macrophages that have ingested decayed erythrocytes. Ferritin synthesis is increased to provide intra-macrophage storage capacity for the iron that is being withheld from transferrin. In this manner, in humans, the normal serum value for iron of 12-27 μM may be reduced to a range of 2-12 μM and the normal saturation value of the metal in plasma transferrin may be lowered from a range of 25-50% to one of 10-25% (Tables 1, 2). The metabolic shift is enhanced as the clinical condition worsens and it returns to normal as the condition improves.

Table 1. Hypoferremic response to infection

	Day postexposure[1]										
	0	1	2	3	4	5	6	7	8	9	10
Mean serum iron, μM	22.3	18.0	15.5	11.2	10.0	6.4	7.3	12.3	13.0	13.2	15.3
Mean fever index, h x °F	0	1	0	17	30	12	5	2	4	0	1

[1]Four human volunteers were exposed to *Francisella tularensis* on day 0. They developed typical clinical illness as indicated by fever on days 3-8. (Data from Pekarek et al., 1969).

Table 2. Hypoferremic response to Hodgkin's disease and to non-Hodgkin's lymphoma[1]

		Hodgkin's disease				Non-Hodgkin's lymphoma
	Normal	Stage 1	Stage 2	Stage 3	Stage 4	
No. of persons	12	6	7	7	3	6
serum iron, μM	22.7 ± 1.8	16.4 ± 1.6	11.4 ± 1.5	8.6 ± 1.5	6.97 ± 0.61	10.7 ± 2.0
Transferrin saturation, %	35 ± 3.7	27 ± 3.5	23 ± 2.6	18 ± 3.5	16 ± 1.2	17 ± 3.3

[1]Values are means ± SE. Data are from Beamish et al., 1972.

The shift is triggered by a hormone that in past years has been termed leukocytic endogenous mediator but is presently named interleukin-1 (Dinarello, 1984). This 15 K polypeptide is produced and promptly released from monocytes that have been activated by invading cells or their products, by antigen-antibody complexes, or by lymphokines. In addition to its hypoferremic action, the hormone causes a shift of plasma zinc to hepatocytes which possibly require the metal in order to increase their synthesis of new RNA and protein. During invasion the rate of synthesis of at least ten "acute phase" hepatic proteins is accelerated. Moreover, the hormone stimulates the production and release of neutrophils by bone marrow, as well as their subsequent degranulation at the site of invasion, and can also cause elevation of body temperature. The ensuing fever can suppress siderophore synthesis by some gram negative bacteria and may also enhance activity of phagocytic defense cells.

Interleukin-1 also stimulates release of a T-cell lymphokine termed interleukin-2. The latter then activates T lymphocyte proliferation, at least in part, by induction of synthesis of transferrin receptor proteins by these defense cells (Neckers and Cossman, 1983). The receptor proteins enable the lymphocytes to acquire transferrin iron that will be used to catalyze ribonucleotide reductase to yield DNA. Without such iron, the cells remain quiescent and are unable to assist the host in repelling the invaders.

INVADER MECHANISMS WHEREBY IRON IS EXTRACTED FROM VERTEBRATE HOSTS

Three quarters of a century ago, at the Brown Institution of the University of London, Twort and Ingram (1912) discovered that *Mycobacterium paratuberculosis* requires a growth factor that is synthesized by other species of mycobacteria. This was the first demonstration of a bacterial growth factor; the active molecule subsequently was shown to be a siderophore, mycobactin (Snow, 1970). During the past third of a century, siderophore research has flourished; many of the key discoveries have been and are being made by participants at this meeting.

To survive and grow in the tissues of the vertebrate host, invaders must have some mechanism for acquiring iron. Often, bacterial pathogens are required not only to produce siderophores but also the proteins needed to bind and to assimilate the ferrated molecules. For example, mucoid strains of *Pseudomonas aeruginosa*, when growing in the lungs of cystic fibrosis patients, synthesize the same three outer membrane proteins (OMP) as when grown under iron-restricted conditions in laboratory medium (Brown et al., 1984). Similarly cells of *Vibrio cholerae*, grown in the gut fluids of infant rabbits (Sciortino and Finkelstein, 1983) or infant mice (Sigel et al., 1985), express novel OMPs that are similar to those observed in cells grown under conditions of iron depletion. On the other hand, cells of *V. cholerae* associated with the mucosal surface of the gut apparently obtain sufficient iron from the mucin to repress synthesis of these OMPs (Sciortino and Finkelstein, 1983; Sigel et al., 1985). Cells of some but not all strains of *Escherichia coli, Klebsiella pneumoniae,* and *Proteus mirabilis,* when growing in human urinary tracts, express at least two OMPs identical to those formed by the same organisms when grown in iron-restricted but not in iron-sufficient media (Lam et al., 1984; Shand et al., 1985).

By no means do all microbial pathogens make use of hydroxamate or catechol siderophores in acquisition of host iron. *Proteus mirabilis,* for example, has been reported to produce and utilize α-hydroxyisolvaleric acid (Evanylo et al., 1984) and *Corynebacterium diphtheriae* employs corynebactin (Russell et al., 1984). Although such anerobic species as *Bacterioides* and *Clostridium* generally are assumed to acquire ferrous rather than ferric iron, cells of *E. coli* grown anaerobically have been observed to employ the same siderophores as used under aerobic conditions (Lodge and Emery, 1984). However, the microaerophilic anerobe, *Bifidobacterium bifidum,* has been shown to acquire ferrous rather than ferric iron; a substance released into the medium by the cells markedly stimulated ferrous iron acquisition (Bezkorovainy, 1984). *Neisseria gonorrhoeae* can produce iron-repressible OMPs and acquire iron from transferrin, lactoferrin, hemin, or hemoglobin without forming detectable siderophores (Mietzner et al., 1984; West and Sparling, 1985). Likewise, *N. meningitidis* (DeVoe, 1982) and *Hemophilus influenzae* (Herrington and Sparling, 1985) can obtain iron directly from transferrin, and *Yersinia* (Perry and Brubaker, 1979) can do so from hemin.

Although *in vivo* growth of such pathogenic protozoa as *Entamoeba, Naegleria, Plasmodium,* and *Trypanasoma* is dependant on a host source of iron (Weinberg, 1984a), the existence of protozoan siderophores has not been established. *Naegleria,* for example, cannot use (and is inhibited by) bacterial and fungal siderophores, nor is it known to produce its own iron chelators (Newsome and Wilhelm, 1983). The mechanism whereby it secures the metal from its human host has not been investigated. In its free-living stage the pathogen flourishes in freshwater bodies of organically polluted water that contain high levels of iron, carbon dioxide, and acidity, with little or no dissolved oxygen. Such conditions suppress siderophore synthesis by competitors and permit an adequate supply of ferrous iron for growth of the protozoan.

Plasmodium falciparum apparently utilizes none of the 20,000 μM intra-erythrocytic heme for its iron supply, but rather obtains the metal from extra-erythrocytic transferrin (Pollack and Fleming, 1984). Though mature red blood cells have no transferrin receptors (Seligman, 1983), the parasite can insert its own proteins into the erythrocyte membrane; a transferrin receptor could be among these. Some of the numerous human transferrin subtypes may be less accessible sources of iron than other subtypes for *P. falciparum* and thus may provide a survival advantage to hosts possessing these types in endemic areas (Pollack and Fleming, 1984). Strains of *P. vivax* on the other hand, must parasitize reticulocytes rather than mature erythrocytes perhaps because of their possible inability to synthesize transferrin receptors. The remarkable amount of plasmodial growth in placental tissue has been attributed to a decrease in cell mediated immunity during pregnancy (Weinberg, 1984b) but may also result from the large quantity of transferrin iron delivered across the placenta during the latter stages of gestation (Bothwell et al., 1958).

Formation of at least two kinds of molecules may assist neoplastic cells in obtaining growth-essential iron from host iron-restricted tissue: (i) low molecular weight peptides that have specific siderophore function and (ii) elevated amounts of the 90 K cell surface glycoprotein that binds iron-transferrin. Cells that form siderophores do not need exogenous transferrin; such independence may constitute an important characteristic of the neoplastic state. Indeed, this type of self-stimulation and regulation of iron transport may also be important in the earliest stages of embryogenesis in which the embryo has no circulatory or glandular endocrine system to sustain it (Fernandez-Pol, 1983).

Many research groups have observed that, in various lines of transformed or neoplastic cells as well as in multiplying normal cells, the number of cell surface iron-transferrin receptor binding sites varies with the rate of multiplication (cf. Weinberg, 1984a). For example, transformed lymphoid cells had as many as 1,000 times the number of receptors found on normal resting lymphocytes (Larrick and Cresswell, 1979). The appearance of the receptor anticipates nuclear changes and is one of the important programmed events necessary for the cells to progress to the S phase of the cell cycle in which DNA is synthesized (Sutherland et al., 1981). A loss of receptor sites occurs as the cells enter the stationary phase and begin to differentiate (Yeh et al., 1982).

METHODS OF ENHANCEMENT OF IRON WITHHOLDING

During the past two decades, several dozen reports have been published concerning enhancement of iron withholding to reduce the incidence and intensity of infection and neoplasia. Methods of enhancement (Table 3) include (i) diets low in iron, (ii) diets low in copper, (iii) elicitation of hypoferremia by inducers of interleukin-1 or, directly, by interleukin-1 itself, (iv) elevation of temperature of invaded tissue, and (v) administration of an iron chelator or iron-binding protein that cannot be used for iron acquisition by the invading microorganism or cancer cell.

Among the recent publications on this topic, several have dealt with the administration to infected animals of low molecular weight iron chelators. For example, hepatic iron stores in mice were lowered by feeding a diet low in the metal plus injecting intraperitoneally five mg desferrioxamine (DF) on the fifth and again on the sixth day following infection with *Trypanasoma cruzi* (Lalonde and Holbein, 1984). Of the animals whose iron burden was lowered, none died, whereas the mortality rate of controls was 23% and that of mice stressed with additional iron was 50% (Table 4).

Table 3. Methods for enhancement of iron withholding defense[1]

Method	Decreased morbidity of Infection	Decreased morbidity of Neoplasia
1) Moderate restriction of iron in diet	Yes[2]	No data
2) Dietary restriction of copper (to lower ceruplasmin level, thus preventing recycling of ferric iron to transferrin)	Yes	No data
3) Hypoferremia induced by injection of materials that stimulate synthesis of IL-1	Yes[2]	Yes[3]
4) Hypoferremia induced by injection of IL-1	Yes	No data
5) Elevation of temperature of invaded tissue	Yes[2]	Yes[3]
6) Administration of an iron chelator that cannot be used by invader	Yes	Yes
7) Administration of host iron binding protein	Yes[2]	No data

[1]References are contained in Weinberg (1984a).

[2]Beneficial anti-infective effect reversed by excess iron.

[3]Possible involvement of iron nonavailability in mechanism whereby procedure functions have not yet been investigated.

Table 4. Cumulative mortality of mice infected with 10^4 epimastigotes of *Trypanasoma cruzi*[1]

Days after infection	Mortality (%) Desferrioxamine[2] (n = 28)	Control (n = 57)	Iron dextran[3] (n = 40)
40	0	9	15
50	0	18	28
60	0	21	38
70	0	23	45
80	0	23	50

[1]Data derived from Fig. 2B, Lalonde and Holbein, 1984.

[2]Injected intraperitoneally (5 mg on fifth and sixth day after infection).

[3]Injected intraperitoneally (5 mg iron on fifth day after infection).

Of a series of 45 iron chelators that were trypanocidal in *in vitro* tests, thirteen cured murine infections of *T. brucei* (Shapiro et al., 1982). Most highly active (100% cured following a single dose of 100 mg/kg) were 3,4-dihydroxycinnamic acid (caffeic acid): 2,9-dimethyl-1,10 phenanthroline (neocuproine); and 2-pyridine-carboxylaldehyde-2-pyridyl hydrazone. The therapeutic action of these chelators was reversed by excess iron.

Unfortunately, specific iron chelators sometimes can promote rather than suppress iron acquisition by some types of invaders. For example, the median lethal dose (LD_{50}) of *Yersinia enterocolitica* was reduced ten-fold in mice stressed with iron dextran as compared with controls. However, in mice injected with desferrioxamine, the dose was lowered by >100,000 fold (Robins-Browne and Prpic, 1985). Accordingly, it is essential that cells of the invader be isolated from the patient, in advance of chelator use, so that *in vitro* tests can be employed to determine which chelators might assist rather than hamper host defense.

Iron chelators increasingly are being examined as possible anti-cancer agents. For example, in *in vitro* tests, 100 μM desferrioxamine blocked cell cycle progression at early S phase of lymphoblastoid cell lines; the action was reversible by equimolar concentrations of iron (Lederman et al., 1984). Unfortunately, the drug was as effective against normal mitogen-stimulated B and T lymphocytes isolated from peripheral blood as against the neoplastic cell lines. A series of spermidine catecholamide chelators inhibited growth of cultured L1210 leukemia cells at 2-14 μM (Bergeron et al., 1985). Their activity correlated with the iron formation constants (10^{36} to 10^{48} moles/l) and with their lipophilicity.

To assist the host in withholding iron from neoplastic cells, an alternative method to using chelators could be that of replacing the metal with one that is more toxic, such as gallium (Rasey et al., 1982). The cytotoxicity of gallium for mouse sarcoma cells was enhanced by transferrin, which facilitated uptake of the metal. However, excess iron was preferentially transported by the protein and thus could protect the tumor cells from inhibition by gallium.

Still another potential antineoplastic strategy would be the use of monoclonal antibodies to transferrin receptor proteins (Taetle et. al., 1983). In *in vitro* growth tests, ID_{50} values of <5 μg/ml were obtained against both non-malignant and some strains of malignant cells. No overall selectivity against malignant strains was observed, although tumor cells from one patient were more susceptible than that person's normal cells. In such patients, it might become possible to enhance selectivity by delivering the antibodies directly to the tumor site. Because of the presence of the tumor cells in such sites, transferrin binding activity is much higher than in normal tissue (DeSousa and Potaznik, 1984).

In development and application of these various strategies for assisting the host to withhold iron from invaders, caution is required lest a condition of true iron-deficiency be inadvertently created. If hosts were to become markedly iron-deficient, susceptibility to infection (Weinberg, 1978; Gross and Newberne, 1980) and to neoplasia (Vitale et al., 1978) could be intensified presumably because the metal is needed to catalyze several aspects of the humoral and cell-mediated immune systems.

CONSEQUENCES OF IRON OVERLOAD

During the past three decades, scores of reports have shown that added iron enhances the ability of designated microbial strains to grow in host fluids, cells, or tissues. The genera thus far examined are contained in Table 5. Not all strains of a genus respond equally well to added iron because of the considerable variation in their ability to produce and/or utilize siderophores and iron uptake proteins. Generally, an amount of iron sufficient to double the saturation value of the iron-binding host-defense protein in the system has a marked effect on the growth of cells of responsive strains. The minimal size of the inoculum required for microbial growth and disease production in the controls is commonly lowered by several log units in the presence of excess iron. For example, the LD_{50} of *Pasteurella multocida* in mice was lowered 320 fold by intraperitoneal (i.p.) injection of 100 g iron (as sulfate) per test animal (Flossman and Müller, 1980). Even more dramatically, the LD_{50} of *Vibrio vulnificus* was lowered by i.p. injection of 80 g per mouse from 6 x 10^6 to 1 x 10^0 organisms (Wright et al., 1981). Moreover, with an i.p. injection as small as 5 μg iron per mouse (either as ferric ammonium citrate, hemoglobin, or heme), the LD_{50} of *V. vulnificus* was lowered 1000 fold (Helms et al., 1984). In healthy human neonates, a seven-fold increase in septicemias and meningitis caused by *Escherichia coli* and other environmental gram negative bacteria occurred within one week following intramuscular injection of approximately 30 mg iron (as dextran) per child (Barry and Reeve, 1977; Becroft et al., 1977).

A variety of sources of iron are effective in enhancing infections; ferric and ferrous inorganic salts, ferric ammonium citrate, iron sorbitol, iron dextran, heme, hematin, hemin, hemoglobin, and ferritin as well as highly iron-saturated transferrin. On the other hand, organic compounds that lack iron (such as ammonium citrate, dextran, apotransferrin, and hematoporphyrin) are devoid of activity. Blockage of the reticuloendothelial system (RES) is unnecessary for enhancement of infection; low-molecular weight iron salts are active but do not block the RES, whereas carbon particles and dextran (each of which is a blocker) have no activity in the absence of iron. Moreover, iron enhancement of pathogen growth is readily demonstrated in such body fluids as serum that lack elements of the RES. Nor does iron function by neutralizing complement, inasmuch as enhancement of microbial growth by the metal occurs equally well in control sera and in sera whose complement has been inactivated at 56°C (Weinberg , 1984a).

Not surprisingly, the clinical observations in humans burdened with excess iron confirm the results obtained with experimental laboratory models. Iron overload of human fluids, cells, or tissues can occur from (1) exogenous sources via ingestion, injection, or inhalation; (ii) destruction of

Table 5. Microbial genera with strains whose growth in body fluids, cells, tissues, and/or intact vertebrate hosts is stimulated by excess iron.[1]

Gram-negative bacteria		Gram-positive and acid-fast bacteria	Fungi and protozoa
Acinetobacter	Neisseria	Bacillus	Candida
Aeromonas	Pasteurella	Clostridium	Cryptococcus
Alcaligenes	Proteus	Corynebacterium	Entamoeba
Enterobacter	Pseudomonas	Erysipelothrix	Histoplasma
Escherichia	Salmonella	Listeria	Naegleria
Klebsiella	Shigella	Mycobacterium	Plasmodium
Legionella	Vibrio	Staphyloccus	Torulopsis
Moraxella	Yersinia	Streptococcus	Trichophyton
			Trichosporun
			Trypanasoma

[1]References are contained in Weinberg, 1984.

iron-storage cells (as in hepatitis): (iii) excessive destruction of erythrocytes (as in clinical episodes of various hemoglobinopathies, malaria, bartonellosis, leukemias, and lymphomas): and (iv) decreased synthesis of transferrin (as in Kwashiorkor and jejunoileal bypass). Persons with these underlying conditions have often been reported to be at increased risk of developing infection (references cited in Weinberg, 1984a).

In some reports, the hyperferremic plasma obtained from patients in the first three categories (and the plasma with little iron-binding capacity from patients in the fourth category) were shown to support excessive microbial growth that could be prevented by addition of apotransferrin. Of course, some individuals are able to evade infection: patients with these underlying conditions who can minimize their exposure to potentially virulent microbial strains and/or those who are in remission and thus have iron levels close to normal. Conversely, the patients undergoing clinical episodes of these diseases may simultaneously be compromised in some aspect(s) of humoral or cell-mediated immunity; they would therefore have even less resistance to environmental pathogens than those deficient only in the ability to withhold iron.

Mice injected with L1210 leukemia cells and stressed with iron dextran developed a 4.6-fold greater tumor load and survived 25% fewer days than did controls who received tumor cells but no iron (Bergeron et al., 1985). The development of sarcomas at the site of injections of iron dextran in rodents, rabbits, and some humans has been reported frequently (summarized in Weinberg, 1981). Numerous cases of primary hepatocellular carcinoma (PHC) as well as other neoplasms occur in persons who have developed siderosis because of excessive ingested iron and/or inordinate intestinal absorption of the metal. PHC is also more likely to develop in hepatitis B carriers who have elevated plasma iron (Blumberg et al., 1981) or elevated iron stores (Stevens et al., 1983).

Although the most obvious and most frequently demonstrated mechanism whereby excess iron enhances infection and neoplasia is that of serving as a nutrient for the invading microbial or neoplastic cells, additional modes of action may also contribute to a weakening of host defense. Excess iron can suppress such key functions of host defense as T cell mitogen responses, natural killer cell activity, and phagocytic function of monocytes and granulocytes. For example, a marked decrease in phagocytic capacity of monocytes ($p<0.015$) and polymorphonuclear leukocytes ($p<0.03$), as well as bactericidal activity of the former ($p<0.05$) and chemotactic response of the latter ($p<0.025$), were observed in 10 of 16 patients who had various iron overload diseases (van Asbeck et al., 1984). Reversal of overload by phlebotomy was associated with a return to normal phagocytic function.

Iron overload increases risk not only to infection and neoplasia but also to cardiac dysfunction. Indeed, a common cause of death in persons stressed with excess iron is heart failure; autopsy reveals prominent myocardial deposits of the metal (Bothwell and Charlton, 1979; McLaren et al., 1984). Cardiac muscle has a greater affinity for the metal than does skeletal or smooth muscle. Unfortunately, the "iron heart is not a strong heart but a weak one" (Biya and Roberts, 1971). Thus procedures for preventing the accumulation of excess iron in a human or animal population would be expected to lower the incidence not only of infection and neoplasia but also of chronic cardiac failure. For example, adult men and postmenopausal women whose meat-rich diets in affluent societies predispose to iron overload have a high risk of cardiomyopathy. This risk might be reduced by annual donation of three units of whole blood by each person (Sullivan, 1981).

Persons with iron overload frequently experience destruction of hepatic and pancreatic as well as cardiac tissue. Iron has long been known to catalyze lipid peroxidation with consequent membrane damage (reviewed in McLaren et al., 1984). More recently, attention has been focused on the ability of the metal to promote formation of hydroxyl radicals from superoxide anion radicals and hydrogen peroxide (Anonymous, 1985). Additionally, accumulation of excess iron within liposomes might lead to disruption of these organelles, thus hastening cell destruction (McLaren et al., 1984).

The ability of asbestos to catalyze lipid peroxidation and hydroxyl radical formation has been attributed to the iron component of its various fibrous silicates; for example, amosite contains 28% iron; crocidolite, 27%; and chrysotile, 2.6% (Weitzman and Graceffa, 1984).

Inflammation induced in animals by a variety of agents likewise can be suppressed by desferrioxamine (Blake et al., 1983). In humans with rheumatoid disease, excess iron is deposited in synovial membranes. Unfortunately, attempts to remove the metal by desferrioxamine had to be halted because of serious side effects (Anonymous, 1985). Research is needed to develop novel iron chelators that would prevent the catalytic accumulation of hydroxyl radicals, could be administered orally, and which would be safer than desferrioxamine (Blade et al., 1983).

REFERENCES

Anonymous, 1985, Metal chelation therapy, oxygen radicals, and human disease, The Lancet, 1:143.

Artis, W.M., Fountain, J.A., Delcher, H.K., and Jones, H.E., 1982, A mechanism of susceptibility to mucormycosis in diabetic ketoacidosis: transferrin and iron availability, Diabetes, 31:1109.

Barry, D.M.J., and Reeve, A.W., 1977, Increased incidence of gram-negative neonatal sepsis with intramuscular iron administration, Pediatrics, 60:908.

Beamish, M.R., Jones, P.A., Trevett, D., Evans, I.H., and Jacobs, A., 1972, Iron metabolism in Hodgkin's disease, Br. J. Canc., 26:444.

Becroft, D.M.O., Dix, M.R., and Farmer, K., 1977, Intramuscular iron-dextran and susceptibility of neonates to bacterial infection, Arch. Dis. Child., 52:778.

Bergeron, R.J., Streiff, R.R., and Elliott, G.T., 1985, Influence of iron on *in vivo* proliferation and lethality of L1210 cells, J. Nutr., 115:369.

Bezkorovainy, A., 1984 Iron uptake by the microaerophilic anaerobe *Bifidobacterium bifidum* var. *pennsylvanicus,* Clin. Physiol. Biochem., 2:291.

Biya, L.M., and Roberts, W.C., 1971, Iron in the heart, Am. J. Med., 51:202.

Blake, D.R., Hall, N.D., Bacon, P.A., Dieppe, P.A., Halliwell, B., and Gutteridge, J.M.C., 1983, Effect of a specific iron chelating agent on animal models of inflammation, Annals Rheum. Dis., 42:89.

Blumberg, B.S., Lustbader, E.D., and Whitford, P.L., 1981, Changes in serum iron levels due to infection with hepatitis B virus, Proc. Natl. Acad. Sci., USA, 78:3222.

Bothwell, T.H., and Charlton, R.W., 1979, Current problems of iron overload, Recent Results Canc. Res., 69:86.

Bothwell, T.H., Pribilla, W.F., Mebust, W., and Finch, C.A., 1958, Iron metabolism in the pregnant rabbit; iron transport across the placenta, Am. J. Physiol., 193:615.

Brown, M.R.W., Anwar, H., and Lambert, P.A., 1984, Evidence that mucoid *Pseudomonas aeruginosa* in the cystic fibrosis lung grows under iron-restricted conditions, FEMS, Microbiol. Letts., 21:113.

DeSousa, M., Potaznik, D., 1984, Proteins of the metabolism of iron, cells of the immune system, and malignancy in: "Vitamins, Nutrition, and Cancer," Prasad, ed., Karger, Basel.

DeVoe, I., 1982, The meningococcus and mechanisms of pathogenicity, Microbiol. Revs., 46:162.

Dinarello, C.A., 1984, Interleukin-1, Revs. Infec. Dis., 6:51.

Evanylo, L.P., Kadis, S., Maudsley, J.R., 1984, Siderophore production by *Proteus mirabilis,* Can. J. Microbiol., 30:1046.

Fernandez-Pol, J.A., 1983, Siderophore-like growth factor synthesis by transformed cells, Microbiology-1983, P. 313.

Flossman, K.-D., and Muller, G., 1980, Einfluss von Eisen auf die Virulenz von *Pasteurella multocida,* Acta Biol. Med. Germ., 39:327.

Gross, R.L., and Newberne, P.M., 1980, Role of nutrition in immunologic function, Physiol. Revs., 60:188.

Helms, S.D., Oliver, J.D., and Travis, J.C., 1984, Role of heme compounds and haptoglobin in *Vibrio vulnificus* pathogenicity, Infec. Immun., 45:345.

Herrington, D.A., and Sparling, P.F., 1985, *Haemophilus influenzae* can use human transferrin as a sole source for required iron, Infec. Immun., 48:248.

Lalonde, R.G., and Holbein, B.E., 1984, Role of iron in *Trypanasoma cruzi* infection of mice, J. Clin. Invest., 73:470.

Lam, C., Turnowsky, F., Schwarzinger, E., and Neruda, W., 1984, Bacteria recovered without subculture from infected urines expressed iron-regulated outer membrane proteins, FEMS, Microbiol. Letts., 24:255.

Larrick, J.W., and Cresswell, P., 1979, Modulation of cell surface iron transferrin receptors by cellular density and state of activation, J. Supramol. Struct., 11:579.

Lederman, H.M., Cohen, A., Lee, J.W.W., Freedman, M.H., and Gelfand, E.W., 1984, Deferoxamine: a reversible S-phase inhibitor of human lymphocyte proliferation, Blood, 64:748.

Lodge, J.S., and Emery, T., 1984, Anaerobic iron uptake by *Escherichia coli,* J. Bacteriol., 160:801.

McLaren, G.D., Muir, W.A., and Kellermeyer, R.W., 1984, Iron overload disorders: natural history, pathogenesis, diagnosis, and therapy, CRC, Crit. Rev. Clin. Rev. Lab. Sci., 19:205.

Mietzner, T.A., Luginbuhl, G.H., Sandstrom, E., and Morse, S.A., 1984, Identification of an iron-regulated 37,000-dalton protein in the envelope of *Neisseria gonorrhoeae,* Infec. Immun., 46:410.

Neckers, L.M., and Cossman, J., 1983, Transferrin receptor induction in mitogen-stimulated lymphocytes is required for DNA synthesis and is regulated by interleukin-2, Proc. Natl. Acad. Sci., USA, 80:3494.

Newsome, A.L., and Wilhelm, W.E., 1983, Inhibition of *Naegleria fowleri* by microbial iron-chelating agents: ecological implications, Appl. Environ. Microbiol., 45:665.

Pekarek, R.S., Bostian, K.A., Bartelloni, P.J., Calia, F.M., and Beisel, W.R., 1969, The effects of *Francisella tularensis* infection on iron metabolism in man, Am. J. Med. Sci., 258:14.

Perry, R.D., and Brubaker, R.R., 1979, Accumulation of iron by *Yersinia,* J. Bacteriol., 137:1290.

Pollack, S., and Fleming, J., 1984, *Plasmodium falciparum* takes up iron from transferrin, Br. J. Haematol., 58:289.

Rasey, J.S., Nelson, N.J., and Larson, S.M., 1982, Tumor cell toxicity of stable gallium nitrate: enhancement by transferrin and protection by iron, Eur. J. Oncol., 18:661.

Robins-Browne, R.M., and Prpic, J.K., 1985, Effects of iron and deferrioxamine on infections with *Yersinia enterocolitica,* Infec. Immun., 47:774.

Russell, L.M., Cryz, S.J., Jr., and Holmes, R.K., 1984, Genetic and biochemical evidence for a siderophore-dependent iron transport system in *Corynebacterium diphtheriae*, Infec. Immun., 45:143.

Sciortino, C.V., and Finkelstein, R.A., 1983, *Vibrio cholerae* expresses iron-regulated outer membrane proteins *in vivo*, Infec. Immun., 42:990.

Seligman, P.A., 1983, Structure and function of the transferrin receptor, Prog. Hematol., 13:131.

Shand, G.H., Anwar, H., Kadurugamuwa, J., Brown, M.R.W., Silverman, S.H., and Melling, J., 1985, *In vivo* evidence that bacteria in urinary tract infection grow under iron-restricted conditions, Infec. Immun., 48:35.

Shapiro, A., Nathan, H.C., Hutner, S.H., Garofalo, J., McGlaughlin, S.D., Rescigno, D., and Bacchi, C.J., 1982, *In vivo* and *in vitro* activity by diverse chelators against *Trypanasoma brucei*, J. Protozool., 29:85.

Sigel, S.P., Stoebner, J.A., and Payne, S.M., 1985, Iron-vibriobactin transport system is not required for virulence of *Vibrio cholerae*, Infec. Immun., 47:360.

Snow, G.A., 1970, Mycobactins: iron-chelating growth factors from mycobacteria, Bacteriol. Revs., 34:99.

Stevens, R.G., Kuvibidila, S., Kapps, M., Friedlaender, J., and Blumberg, B.S., 1983, Iron-binding proteins, hepatitis B virus, and mortality in the Solomon islands, Am. J. Epidemiol., 118:550.

Sullivan, J.L., 1981, Iron and the sex difference in heart disease risk, The Lancet, 1:1293.

Sutherland, R., Delia, D., Schneider, C., Neuman, R., Kemshead, J., and Greaves, M., 1981, Ubiquitous cell-surface glycoprotein on tumor cells is proliferation-associated receptor for transferrin, Proc. Nat. Acad. Sci., USA, 78:4515.

Taetle, R., Honeysett, J.M., and Trowbridge, I., 1983, Effects of anti-transferrin receptor antibodies on growth of normal and malignant myloid cells, Internat. J. Canc., 32:343.

Twort, F.W., and Ingram, G.L.Y., 1912, A method for isolating and cultivating the Mycobacterium enteritidis chronicae pseudotuberculosis bovis, Johne, and some experiments on the preparation of a diagnostic vaccine for pseudo-tuberculous enteritis of bovines, Proc. Roy. Soc. Ser. B Biol. Sci., 84:517.

van Asbeck, B.S., Marx, J.J.M., Struyvenberg, A., and Verhoef, J., 1984, Functional defects in phagocytic cells from patients with iron overload, J. Infec., 8:232.

Vitale, J.J., Broitman, S.A., Vavrousek-Jakula, E., Reddy, P.W. and Gottlieb, L.S., 1978, The effects of iron deficiency and the quality of fat on chemically induced cancer, Adv. Exp. Med. Biol., 91:229.

Weinberg, E.D., 1977, Mineral element control of microbial secondary metabolism, in: "Microorganisms and Minerals," E.D. Weinberg, ed., M. Dekker, Inc., New York.

Weinberg, E.D., 1978, Iron and infection, Microbiol. Revs., 42:45.

Weinberg, E.D., 1981, Iron and neoplasia, Biol. Trace Elem. Res., 3:55.

Weinberg, E.D., 1984a, Iron withholding: a defense against infection and neoplasia Physiol. Revs., 64:65.

Weinberg, E.D., 1984b, Pregnancy-associated depression of cell-mediated immunity, Rev. Infec. Dis., 6:814.

Weitzman, S.A., and Graceffa, P., 1984, Asbestos catalyzes hydroxyl and superoxide radical generation from hydrogen peroxide, Arch. Biochem. Biophys., 228:373.

West, S.E.H., and Sparling, P.F., 1985, Response of *Neisseria gonorrhoeae* to iron limitations; alterations in expression of membrane proteins without apparent siderophore production, Infec. Immun., 47:388.

Wright, A.C., Simpson, L.M., and Oliver, J.D., 1981, Role of iron in the pathogenesis of *Vibrio vulnificus* infections, Infec. Immun., 34:503

Yeh, C,-J., Papamichael, G.M., and Faulk, W.P., 1982, Loss of transferrin receptors following induced differentiation of HL-60 promyelocytic leukemia cells, Exp. Cell Res., 138:429.

STIMULATION OF DISEASE DEVELOPMENT BY SIDEROPHORES AND INHIBITION BY CHELATED IRON

T.R. Swinburne

East Malling Research Station

Maidstone, Kent, U.K.

INTRODUCTION

Most of those contributions to this Symposium which deal specifically with iron as a factor in the competition between saprophytic and parasitic microorganisms, conclude that iron withholding mechanisms have adverse effects on fungal pathogens and can thereby achieve biological control of plant disease. This review will consider contrary evidence and will seek to show that iron chelating ligands, including siderophores can stimulate the development of fungal pathogens and the diseases which they incite, at least for pathogens of stem, leaf and fruit.

The chemical environment of the plant surface is derived from two main sources; substances leached from the plant tissues and metabolites produced by the epiphytic microflora (Preece and Dickinson, 1971). These, together with substances which are carried in or on fungal spores interact to influence three distinct phases of pathogen development; germination, elongation of germ tubes and the formation of infection structures such as appressoria. The latter phase must be extended to include infection *per se*. Chemicals which can control these processes have been extensively studied, particularly those with nutrient properties. The presence of relatively high concentrations of nutrients promotes germ tube and hyphal growth, usually to the detriment of appressorial development and consequently to infection. This has led to the assumption that appressoria initiation is a response to nutrient starvation (Emmet and Parbery, 1975), although it is also recognised that this initiation is more demanding in terms of chemical specificity than germination (Preece and Dickinson, 1971).

Over the last ten years or so evidence has been obtained to suggest that some of the chemicals present on plant surfaces which stimulate both germination and appressorium formation exert their effect at concentrations too low to be considered in terms of nutrition (Swinburne, 1976). Anthranilic acid found in banana fruit leachates, p-coumarylquinic, chlorogenic and caffeic acids found in apple fruit leachate stimulate both *Colletotrichum musae* and *Diaporthe perniciosa* at optimal concentrations of 10^{-3}-10^{-4} M (Brown and Swinburne, 1978). The property which these, and a wide range of synthetic compounds share is their capacity to chelate iron or to be rapidly converted to iron chelating agents (Harper et al., 1980). Similarly, siderophores produced by epiphytic bacteria also stimulate both germination and appressoria formation (McCracken and Swinburne, 1979).

Thus many of the non-specific stimulants present on plant surfaces apparently involve iron in their mode of action.

GERMINATION OF FUNGAL SPORES

Conidia of a number of fungi germinate poorly in water but do so readily in plant leachates. For *C. musae* the active compound in banana fruit leachate was shown to be anthranilic acid (Swinburne, 1976) by virtue of rapid conversion to 2,3-dihydroxybenzoic acid (DHBA) by the ungerminated spore (Harper and Swinburne, 1979). DHBA in common with other phenolic compounds including caffeic and chlorogenic acids present in apple leachates (Brown and Swinburne, 1978) can complex iron between two adjacent hydroxyl groups as described by Hider in this volume. Catecholate siderophores, such as pseudobactin and the pyoverdines (see Abdallah and Myers, this volume) are complex molecules in which catechol units are conjugated with amino acids. One such siderophore, related to pseudobactin (Teintze et al., 1981) produced by *Pseudomonas fluorescens* Biovar V (Slade, 1985) was shown to be highly stimulatory to both germination and appressorium formation in a number of fungi including *C. musae* and *Botrytis cinerea* (McCracken and Swinburne, 1975); Slade, 1985). This siderophore, referred to as SA, produced maximal response in *C. musae* at a concentration of approximately 5 x 10^{-5} M, the concentration used for most subsequent experiments.

In addition to this evidence it was also shown that conidia produced by *C. musae* in a liquid medium containing as little as 10 $\mu g\ ml^{-1}$ iron germinated freely without the addition of iron chelating ligands (Harper et al., 1980). As the iron content of the medium was increased, germination of the resultant conidia in water decreased, except where iron chelating ligands were added to the germination droplet. This suggested that at least some of the iron in conidia from iron replete media was associated with a biochemical site which inhibited germination and that this inhibition was relieved by the addition of iron chelating ligands. Iron chelating ligands complexed with iron before addition to conidia are not stimulatory and in some circumstances can be inhibitory, as will be discussed later. Thus this process of stimulating germination apparently involves the removal of iron from some site within the conidia. By contrast, iron uptake has been shown to be a prerequisite for germination of ascospores of *Neurospora crassa* (Horowitz et al., 1976). Iron replete and iron deficient conidia of *C. musae* absorbed substantial quantities of iron from their environs in the absence of added chelating ligands, whose presence did not alter uptake (Graham and Harper, 1983). Similarly, in double radio-labelled experiments (Graham, 1981) it was shown that uptake of DHBA (from anthranilic acid) was not affected by the presence of iron. These experiments also showed that the addition of iron chelating ligands to replete conidia containing radio-labelled iron did not promote release of iron into the surrounding medium.

The essential questions concerning the location and function of iron associated with germination remain largely unanswered, but two hypotheses warrant further investigation. Graham and Harper (1983) found that the organelle fraction of iron-replete conidia accumulated 15 times as much EDTA as did that from iron deficient conidia, implicating ribosomes or mitochondria. As ribosomal translation is controlled by iron which is additionally involved in mitochondrial metabolism (Downer et al., 1970); Zahringer et al., 1976) either of these organelles could be the location. RNA transcription and *de novo* synthesis of proteins are the first detectable biochemical changes during spore germination (Barash et al., 1967; Shepherd et al., 1980), which suggests that ribosomes should be considered first in further studies. Another possible location for the iron is the conidial wall, which comprises 30% of the total iron present in replete conidia (Graham and Harper, 1983). Winkelmann has suggested at this

Symposium that catecholate siderophores, such as pseudobactin and SA are not transported by fungi, which employ hydroxamate-siderophores (Wiebe and Winkelmann, 1975). If this is so the high affinity of catecholate siderophores for iron must impose an iron withdrawing effect on conidia, affecting the wall-bound iron first. It would be interesting to repeat the double labelled EDTA/Fe experiments with SA. Under these conditions iron may be lost from conidia, rather than remain complexed within the cell, as was found with EDTA.

GERM TUBE ELONGATION AND APPRESSORIUM FORMATION

Apart from their effects on germination iron chelating agents including siderophores reduce germ tube length and accelerate the formation of appressoria in *C. musae* (Swinburne, 1976) and *B. cinerea* (Slade, 1985). Germ tube growth and appressorium formation are mutually exclusive processes and it is not possible to determine if the mode of action of the iron-chelating agents is one of inhibition or stimulation. Scher and Baker (1982) showed that iron limitation reduced germ tube growth in *Fusarium* but did not link this with appressorium formation, which is difficult to observe in that genus. If iron limitation is the cause of reduction in germ tube growth in *C. musae* then it must be assumed that iron complexed by, for example, DHBA remains unavailable to the metabolism of the fungus, in spite of its' intracellular location.

A striking feature of the appressoria of *Colletotrichum* spp. is the deposition of melanin in the thickened wall (Kubo et al., 1985). Muirhead and Deverall (1981) showed that appressoria which remained hyaline were relatively ineffective in that they either germinated laterally over the surface or penetrated so quickly that innate host resistance prevented further development. DHBA increases both germination and the proportion of resultant appressoria that become melanised (Swinburne and Brown, 1983; Slade, 1985). It is not clear if this latter process involves iron metabolism directly. Oxidation of phenolic compounds, such as catechol, is involved in the synthesis of melamin. Thus either the chelating agent as substrate or the electron transport processes of oxidation in fungi (Neilands, 1974) could be involved.

Appressoria formation is also stimulated in other fungi, such as *B. cinerea* in which the deposition of melanin is not observed (Brown and Swinburne, 1982; Slade, 1985). The effect of increasing the inoculum potential by the presence of iron chelating ligands has also been observed with *D. perniciosa* (Brown and Swinburne, 1978) and it is reasonable to conclude that other pathogens will be found to conform to this phenomenon.

INFECTION AND HOST RESPONSES

The presence of DHBA in inoculum droplets of *C. musae* on banana fruits or of SA in those of *B. cinerea* on *Vicia faba* increases apparent disease severity (Brown and Swinburne, 1983; Slade, 1985). Comparable results can be obtained with iron deficient conidia and *Glomorella cingulata* on pepper fruit added to the list of examples (Adikaram et al., 1982a). To evaluate the factors involved in such observations it is necessary to consider firstly the effect of iron chelating ligands on inoculum potential and secondly their effect on the host's response.

Colletotrichum musae infection of banana fruit

Having demonstrated the effects of DHBA on the development of conidia of *C. musae in vitro* it was of interest to know how these processes

Table 1. Germination and appressorium formation, recorded 48 h after inoculation, and progressive lesion development on banana fruits 14 days after inoculation with conidia of *Colletotrichum musae*

Inoculum	Germination (%)	Appressoria* (%)	Melanised** appressoria (%)	Progressive lesions (%)
Low-iron conidia	78[a][†]	82[b]	-	71[d]
Iron-replete conidia	75[a]	71[a]	47[c]	41[c]
Iron-replete conidia + SA	84[a]	89[bc]	78[d]	73[d]
Iron-replete conidia + Fe-SA	41[b]	66[a]	31[b]	25[b]
Iron-replete conidia + DHBA	87[a]	97[c]	86[d]	84[d]
Iron-replete conidia + Fe-DHBA	52[b]	70[a]	0[a]	0[a]

SA (10^{-4} M), catecholate siderophore produced by *Pseudomonas fluorescens* (isolate UV3); Fe - SA (10^{-4} M), iron-siderophore complex.
DHBA (5×10^{-4} M), 2,3-dihydroxybenzoic acid; Fe-DHBA (5×10^{-4} M), iron complex.
* Percentage of germinated conidia with appressoria
** Percentage of appressoria which were melanised
† values followed by the same letter in the same column do not differ significantly (P = 0.05).

would influence disease progression *in vivo.* The problems associated with the design of suitable experiments stems from interference by compounds naturally present in the infection count. To some extent these were overcome by using conidia obtained from iron depleted media, which, as has already been described, do not respond to added iron chelating agents. Iron deficient conidia of *C. musae* brought about the formation of progressive lesions on banana fruit much more rapidly than iron-replete conidia (Table 1), which is an extension of the data recorded in Swinburne and Brown (1983). DHBA and SA increased the rate of lesion formation of iron replete conidia to that of iron deficient conidia. Interestingly, these results were independent of germination, which on the banana surface were unrelated to the iron content of the conidia. Low iron status or the presence of chelating agents resulted in much less necrosis and lower levels of phytoalexin (Brown and Swinburne, 1980 and 1981) in the peel of inoculated green fruit than from iron replete conidia in water. Following the observations of Muirhead and Deverall (1981) it is reasonable to infer that as the proportion of melanised to hyaline appressoria is increased by either low iron status or the presence of chelating agents inoculations will result in less infection of green fruit and consequently when infection of ripened fruit occurs from melanised appressoria the penetration tubes and subsequent hyphae encounter less phytoalexin and colonise rapidly. The mechanisms controlling the 'germination' of melanised appressoria are unknown.

The addition of iron to either SA or to DHBA in sufficient quantities to form the complex did not merely nullify the stimulation observed with ligand alone, but inhibited germination, melanised appressoria formation and lesion development. This suggested that disease control might be achieved by enhanced levels of iron at the infection count. These possiblities are explored further and by Brown in this volume.

Glomorella cingulata infection of *Capsicum* fruits

The example of anthracnose of banana, described above, suggests that the influence which iron status and siderophores etc. might exert on disease is through their interaction with preinfectional events. That this is not the whole story was first intimated by experiments with *G. cingulata* and anthracnose of peppers (Adikaram et al., 1982a). Inoculation of undamaged fruit with conidia rarely results in lesion formation and in spite of large appressoria, as with the disease in the field, inoculations were routinely made on puncture wounds. Appressoria produced by this organism were invariably melanised. It was therefore somewhat surprising to find that iron deficient conidia or iron replete conidia in the presence of EDTA produced progressive lesions even before ripening had occurred. Moreover, such inoculations of immature fruit were not followed by the accumulation of sequiterpenoid phytoalexins, as was found following inoculation with iron replete conidia (Adikaram et al., 1982b). The presence of EDTA did not inhibit phytoalexin synthesis when these were elicited abiotically (Brown, personal communication). Therefore the fine detail of these mechanisms remains to be explained.

Botrytis cinerea infection of *Vicia faba*

The enhancement of germination in the presence of DHBA and EDTA (Brown and Swinburne, 1982) and SA (Slade, 1985) must be a major factor in their ability to promote spreading lesions of *B. cinerea* on the otherwise resistant *V. faba*. This resistance has been associated with phytoalexin accumulation by the host (Deverelle and Vessey, 1969) particularly wyerone (Rossall et al., 1979). There was some indication that during the early stages of infection by conidia less wyerone was accumulated in the presence of SA than in water alone (Slade, 1985). Such experiments with live fungus are difficult to interpret because wyerone is formed in progressing lesions and it is necessary to equate this with the number of reacting cells. This can be avoided with experiments made with abiotic elicitors of phytoalexin such as heat-killed conidia or damage from burning (Slade, 1985, Table 2). From such experiments it was apparent that siderophore SA reduced wyerone accumulation through a direct effect on the host. Ferri-siderophore had no such effect. From the dependence of many enzyme systems on iron cofactors (Lehninger, 1970) one can speculate that removal of iron by SA could inhibit synthesis of phytoalexin. This remains to be fully explained.

Colletotrichum acutatum infection of strawberry

The effect of SA on the development of conidia of *C. acutatum* and subsequent disease development on strawberry (Slade et al., 1985) corresponds broadly with observations made with *C. musae*. Thus both germination and appressorium formation by *C. acutatum* are increased by SA or the presence of cells of *P. fluorescens* UV3. Likewise conidia produced in iron deficient media germinated readily without the need of added stimulants. Interestingly these experiments produced evidence for the acquisition of an iron-uptake mechanism by such pathogens, as was suggested earlier by Graham and Harper (1983). Three iron-binding compounds were detected after growth of *C. acutatum* in iron deficient media, and whilst these were not identified, their ability to stimulate germination and appressorium formation in iron-replete conidia was demonstrated. Their role in disease development warrants further work.

Table 2. Effect of SA and Fe-SA on wyerone accumulation in *Vicia faba* cotyledons elicited by burning surface cells or in response to heat-killed conidia of *Botrytis cinerea*. (Data from Slade, 1985).

Treatment	Mean wyerone concentration ($\mu g\ g\ FW^{-1}$)	
	Burnt cells	Heat-killed conidia
Unelicited control	-	26
H_2O	868	232
SA (5×10^{-5}M)	480	130
Fe-SA (5×10^{-5}M)	860	-
LSD (P = 005)	129	43

SA and Fe-SA, see footnote to Table 1.

The enhancement of apparent aggression of *C. acutatum* by SA, shown by enhanced lesion development on ripe strawberry fruits, stolons and leaves could be explained by the increase in inoculum potential which follows stimulation of germination and appressorium formation. Thus, in general, fewer conidia were needed to produce a given rate of lesion formation in the presence of SA than in water alone. However, conidia applied in water to green, unripe fruit produced only slight necrotic flecking of the underlying cells which did not expand into progressive lesions, even after fruit ripening, whereas inoculations made in SA developed into expanding lesions when fruits ripened. Little is known of the infection processes which lead to black rot of strawberry, but these may resemble those described above for *C. musae* on banana. Phytoalexins have been detected in strawberry (Mussell and Staples, 1971; Slade, 1985) but their role in quiescent infections is so far unexplained. If melanisation of appressoria is an important factor in such infections, or indeed to survival in soil, then the presence of iron-chelating ligands, especially bacterial siderophores, may be essential components of the process.

Suppression of infection by iron-complexes

Germination of iron-deficient conidia of *C. musae* is inhibited by the addition of iron to the germinating medium (Swinburne, 1981). This suggested that it might be possible to suppress disease using compounds which were not inherently biocidal, and this has been explored recently (see also Brown, this volume).

Iron-DHBA complex was slightly inhibitory to the germination of conidia of *C. musae* but dramatically reduced the maturation of appressoria (Swinburne and Brown, 1983; see Table 1). The presence of such complexes in inoculum drops effectively controlled disease development. The first assumption was that ferric-ions were involved, especially in view of the assumed specificity of siderophores for Fe^{+++}. However, polarographic data indicated that at pH 4.5-5.0, that used in the assays, the dominant ion in the DHBA complex was Fe^{++} (Brown and Sharma, 1985). Ferrous complexes of other phenolic compounds, notably gallic acid and other chelating agents such as 2,2'-dipyridyl and 1,10-phenanthroline not only suppressed maturation of appressoria in *C. musae* but also totally suppressed spore germination of several plant pathogenic fungi both *in vivo* and *in vitro* (Brown, this volume).

Fe-Sa also suppressed melanisation of appressoria of *C. acutatum*, with a consequent reduction in disease development (Slade et al., 1985). This result, coupled with the assumed specificity of siderophores for ferric ions seems to contradict the results obtained with ferric complexes of EDTA and EDDHA, which had no effect what so ever! However, as Hider shows in his review in this Symposium, at the acidity of the Fe SA complex (pH 3.5) a large proportion of the iron bound to catecholate siderophores (such as pseudobactin and therefore presumably SA) is in the ferrous state, as a consequence of internal redox reactions. The low pH of SA-Fe complexes, prepared with Fe^{+++} salts, was not in itself inhibitory to germination or appressorium formation (Slade, 1985). Thus the Fe-SA and Fe-DHBA complexes probably have a similar mode of action and indicate that the effective inhibitory ion is Fe^{++}. This leaves the valency of iron associated with the site of inhibition of germination in spores, which is so efficiently removed by SA at pH 5.0, open to speculation. Similarly it is not possible to be certain that inhibition of germination and subsequent development by ferrous ions is directly linked with the stimulation observed in the presence of iron chelating ligands.

Although SA inhibited phytoalexin accumulation in *Vicia faba* Fe SA had no effect (Table 2), which might indicate that either the host cells are unable to utilise iron supplied in this way or that the normal cell contains sufficient iron for this not to be rate-limiting. Thus it is unlikely that added iron could control disease through enhancement of resistance.

CONCLUDING DISCUSSION

Iron chelating ligands, such as phenolic compounds of host origin and siderophores of bacterial origin have been shown to stimulate the development of several fungal pathogens and of the diseases they incite by direct actions on the pathogen and on the host. Whilst the biochemical mechanisms which govern these processes remain obscure, their biological significance cannot be ignored. Thus the innumerable factors known to predispose plants to infection should be re-examined to determine their influence on the epiphytic microflora etc. Indeed the presence of bacteria on test plants should be considered in all inoculation experiments, both before and after lesion development. The results which lead to these conclusions are mostly based on one isolate of *P. fluorescens* but the ability of virtually all gram negative and gram positive isolates from banana fruits to stimulate *C. musae* (McCracken and Swinburne, 1979) suggests that the total epiphytic population, rather than just components, are actively supportive of the infection process. The ability of so-called saprophytes to participate in infection requires much closer examination, especially as their control by biological or chemical means, could be an additional (if indirect) means of controlling disease, or at least of avoiding iatrogenic consequences of some agricultural practices. The apparent contradiction between the work reviewed here for air-borne pathogens and other studies recorded in this volume for soil-borne pathogens (e.g. Baker, Scher and Alaboun etc.) cannot be explained at this time. It is indeed curious that iron deprivation should be considered to have completely opposite effects on these two groups of pathogens. So far the epiphytic flora of stems and leaves etc. has not been screened for antagonists in the same way as the soil flora. The frequency of isolates from soil which show antagonism on, for example, the iron-limited medium of King's B is relatively low, and the interaction between non-antagonists and pathogens *in vivo* not recorded. *Ps. fluorescens* UV3 shows no antagonism with *C. musae* on King's B but this test does not reveal stimulation either (Slade and Swinburne, unpublished observation). There is a major difference between the nature of the propagules of pathogens present in soil and those which are dispersed by air. The latter are usually sexual or asexual spores whereas chlamydospores and other structures derived from the vegetative thallus predominates in the former. Stimuli for germination and subsequent development may differ widely between these groups. If this could be demonstrated it would indicate an interesting evolutionary divergence. Unfortunately very little work has been done on the effect of siderophores on the development of soil borne fungi. Whilst such *in vitro* experiments are not as easy as with fungal conidia of air borne pathogens the results would be of enormous importance to the interpretation of *in vivo* experiments and to the rationalisation of screening methods.

If it is assumed that bacterial saprophytes can stimulate fungal pathogens then other interpretations of the experiments made with soil-borne organisms are possible. For example, putative biocontrol agents could exert their effect on disease through interaction with members of the saprophytic microflora rather than through direct interaction with the pathogen.

Another possibility is that the same process of stimulation of appressorium formation which increases apparent aggression in air-borne pathogens could paradoxically have the opposite effect on soil-borne pathogens. Whereas deposition processes bring air-borne pathogens into direct contact whilst the propagule is still dormant, soil-borne pathogens must usually grow to achieve contact with the root surface. If the germ-tube of soil-borne propagules was stimulated to form infection structures too early, then contact may be prevented.

In spite of the technical difficulties it is obviously necessary to make direct observations on the effect of siderophores on soil-borne pathogens *in vivo* to resolve the apparent conflict. This is especially urgent in view of the rapid progress now being made towards commercial use of fluorescent pseudomonads as biocontrol agents.

REFERENCES

Adikaram, N.K.B., Brown, A.E. and Swinburne, T.R., 1982a. Rotting of immature *Capsicum frutescens* L. *(C. annum)* fruit by iron-depleted *Glomorella cingulata* (Stonam.). Physiol. Plant Pathol., 21:171-177.

Adikaram, N.K.B., Brown, A.E. and Swinburne, T.R., 1982b. Phytoalexin involvement in the latent infection of *Capsicum annuum* L. fruit by *Glomorella cingulata* (Stonam.). Physiol. Plant Pathol., 21:161-170.

Barash, I., Conway, M.L. and Howard, D.H., 1967. Carbon catabolism and synthesis of macromolecules during spore germination of *Microsporum gypseum*. J. Bacteriol., 93:656-662.

Blakeman, J.P., 1971. The chemical environment of the leaf surface in relation to growth of pathogenic fungy. In: 'Ecology of Leaf Surface Micro-organisms,'p.p. 255-268. (T.F. Preece and C.H. Dickinson, Eds.), Academic Press, London.

Blakeman, J.P. and Brodie, I.D.S., 1976. Inhibition of pathogens by epiphytic bacteria on aerial plant surfaces. In: 'Microbiology of Aerial Plant Surfaces,' p.p. 529-558. (C.H. Dickinson and T.F. Preece, Eds.), Academic Press, London.

Brown, A.E., 1986. Ferrous complexes and chelating compounds in suppression of fungal diseases of cereals. In: 'Iron, Siderophores and Plant Disease.' NATO Advanced Study Institute Series. (T.R. Swinburne, Ed.). Plenum Press, New York and London.

Brown, A.E. and Sharma, H.S.S., 1985. A possible role for famous complexes in fungal disease suppression: glume blotch and net blotch of cereals. Phytopathol. Z., 113:178-188.

Brown, A.E. and Swinburne, T.R., 1978. Stimulants of germination and appressoria formation by *Diaporthe perniciosa* (March) in apple leachate. Trans. Br. Mycol. Soc., 71:405-411

Brown, A.E. and Swinburne, T.R., 1980. The resistance of immature banana fruits to anthracnose [*Colletotricham musae* (Berk. & Cust) Arx.]. Phytopathol. Z., 99:70-80.

Brown, A.E. and Swinburne, T.R., 1981. The influence of iron and iron chelates on the formation of progressive lesions by *Colletotrichum musae* on banana fruits. Trans. Br. Mycol. Soc., 77:119-124.

Brown, A.E. and Swinburne, T.R., 1982. Iron-chelating agents and lesion development by *Botrytis cinerea* on leaves of *Vicia faba*. Physiol. Plant Pathol., 21:13-21.

Deverell, B.J. and Vessey, J.C. 1969. Role of a phytoalexin in controlling lesion development in leaves of *Vicia faba* after infection by *Botrytis* spp. Ann. Appl. Biol., 63:449-458.

Downer, D.N., Davis, W.B., and Byerer, B.R., 1970. Repression of phenolic acid synthesizing enzymes and its relation to iron uptake in *Bacillus subtilis*. J. Bacteriol., 101:181-187.

Emmett, R.W. and Parbery, D.G., 1975. Appressoria. Ann. Rev. Phytopathol., 13:147-167.

Graham, A.H., 1981. Studies on the role of iron in germination of conidia of *Colletotricham musae*. Ph. D. Thesis, The Queen's University of Belfast.

Graham, A.H. and Harper, D.B., 1983. Distribution and transport of iron in conidia of *Colletotrichum musae* in relation to the mode of action of germination stimulants. J. Gen. Microbiol., 129:1025-1034.

Harper, D.B. and Swinburne, T.R., 1979. 2,3-Dihydroxybenzoic acid and related compounds as stimulants of germination of conidia of *Colletotrichum musae* (Bok. & Cwt.) Arx. Physiol. Plant Pathol., 14:363-370.

Harper, D.B., Swinburne, T.R., Moore, S.K., Brown, A.E. and Graham, H., 1980. A role for iron in germination of conidia of *Colletotrichum musae*. J. Gen. Microbiol., 121:169-174.

Horowitz, N.H., Charlang, G., Horne, G., and Williams, N.P., 1976. Isolation and identification of the conidial germination factor for *Neurospora crassa*. J. Bacteriol., 127:135-140.

Kubo, Y., Suzuki, K., Furasawa, I. and Yamamoto, M., 1985. Melanin biosynthesis as a prerequisite for penetration by appressoria of *Colletotrichum lagenarium:* site of inhibition by melanin-inhibiting fungicides and their action on appressoria. Pestic. Biochem. Physiol., 23:47-55.

Lehninger, A.L., 1970. Biochemistry. Worth, New York, 1104 pp.

McCracken, A.R., and Swinburne, T.R., 1979. Siderophores produced by saprophytic bacteria as stimulants of germination of conidia of *Colletotrichum musae*. Physiol. Plant Pathol., 15:331-340.

Muirhead, I.F. and Deverall. B.J., 1981. Role of appressoria in latent infection of banana fruits by *Colletotrichum musae*. Physiol. Plant Pathol., 19:77-84.

Mussell, H. and Staples, R., 1971. Phytoalexin-like compounds apparently involved in strawberry resistance to *Phytophthora fragariae*. Phytopathology, 61:515-517.

Neilands, J.B., 1974. 'Microbial Iron Metabolism,' 597 pp. (J.B. Neilands, Ed.), Academic Press, New York and London.

Preece, T.R. and Dickinson, C.H., 1971. 'Ecology of Leaf Surface Micro-organisms. pp. . (T.R. Preece and C.H. Dickinson, Eds.). Academic Press, London.

Rossall, S., Mansfield, J.W. and Hutson, R.A., 1979. Death of *Botrytis cinerea* and *B. fabae* following exposure to wyerone derivatives *in vitro* and during infection development on leaves of *Vicia faba*. Physiol. Plant Pathol., 16:135-146.

Scher, F.M. and Baker, R., 1982. Effect of *Pseudomonas putida* and a synthetic iron chelator on induction of soil suppressiveness to Fusarium wilt pathogens. Phytopathology, 72:1567-1573.

Shepherd, M.G., Yin, C.Y., Ram, S.P. and Sullivan, P.A., 1980. Germ tube induction in *Candida albicans*. Can. J. Microbiol., 26:21-26.

Slade, S.J., 1985. The effect of iron, and of a bacterial siderophore, on fungal infection of plants. Ph.D. Thesis. Imperial College of Science and Technology/East Malling Research Station.

Slade, S.J., Swinburne, T.R. and Archer, S.A., 1986. The role of a bacterial siderophore and of iron in the germination and appressorium formation by conidia of *colletotrichum acutatum*. J. Gen. Microbiol., (in press).

Swinburne, T.R., 1976. Stimulants of germination and appressoria formation by *Colletotrichum musae* (Bok. & Cwt.) Arx. in banana lechates. Phytopathol. Z., 87:74-90.

Swinburne, T.R., 1981. Iron and iron chelating agents as factors in germination, infection and aggression of fungal pathogens. In: 'Microbial Ecology of the Phylloplane,' pp. 227-243. (J.P. Blakeman, Ed.). Academic Press, New York and London.

Swinburne, T.R. and Brown, A.E., 1983. Appressoria development and quiescent infections of banana fruit by *Colletotrichum musae*. Trans. Br. Mycol. Soc., 80:176-178.

Teintze, M., Hossain, M.B., Barnes, C.L., Leong, J., and van der Helm, D., 1981. Structure of ferric pseudobactin, a siderophore from a plant growth promoting *Pseudomonas*. Biochemistry, 20:6446-6457.

Wiebe, C. and Winkelmann, G., 1975. Kinetic studies on the specificity of chelate-iron uptake in *Aspergillus*. J. Bacteriol., 123:837-842.

Zahringer, J., Baliga, B.S. and Munro, H.N., 1976. Novel mechanism for translational control in regulation of ferritin synthesis by iron. Proc. Nat. Acad. Sci., U.S.A., 73:857-861.

A NEW SIDEROPHORE IN *AEROMONAS HYDROPHILA*:

POSSIBLE RELATIONSHIP TO VIRULENCE

B.R. Byers, D. Liles, P.E. Byers and J.E.L. Arceneaux

Department of Microbiology
University of Mississippi Medical Center
Jackson, Mississippi, 39216-4505, U.S.A.

INTRODUCTION

Aeromonas hydrophila, a freshwater gram negative bacterium, causes a fatal disease in aquatic animals and is responsible for serious economic loss in commercial fish farming operations. In Mississippi, damage to the channel catfish industry is estimated at $800,000 per year (T. Wellborn, personal communication). *A. Hydrophila* also is the etiology of serious human diseases, including wound infections, acute gastroenteritis, and septicemia (see Jand   et al 1984).

Research by many investigators, summarized recently by Weinberg (1984), reveals intense competition between a vertebrate host and an invading pathogenic microorganism for the essential nutrient iron. The host has several defense mechanisms (including the iron-binding glycoproteins transferrin and lactoferrin) that withhold iron from many microorganisms. Some pathogenic forms produce siderophores that effectively mobilize iron from transferrin for microbial use. Production of a siderophore has been co-related with the high virulent phenotype in *Escherichia coli* (Williams and Carbonetti, 1984), *Vibrio anguillarum* (Crosa, 1984) and *Corynebacterium diphtheriae* (Russel and Holmes, 1985).

Recent studies (Andrus and Payne, 1983; Byers et al., 1983) showed that two strains of *A. hydrophila* produced the well-known siderophore enterobactin. In present studies, thirteen other strains were found to produce a previously unidentified phenolate siderophore that appeared to confer resistance to transferrin inhibition on these strains. Production of the new siderophore may be a virulence factor, and future studies will determine its possible role in *A. hydrophila* infections in the catfish.

MATERIALS AND METHODS

Eight of the *A. hydrophila* strains (495A2, BAP, 193, 167, 132, 179, 209, 139), isolated locally from diseased aquatic animals, were obtained from C. Lobb of this Department; strain UMC6, also isolated locally from a human infection, was obtained from B. Mitchell, University Hospital, Jackson, Mississippi. Strains 255, 44, 25 and 294 were obtained from R. Brendon, Mount Sinai Medical Center, New York. The enterobactin producing strain UT was obtained from S.M. Payne, University of Texas, Austin.

The organisms were grown in a glucose-mineral salts medium containing (per liter):glucose (5 g), $(NH_4)_2HPO_4$ (1 g), K_2HPO_4 (2.5 g), and KH_2PO_4 (2.5 g). The medium was treated with Chelex-100 (Bio-Rad Laboratories, Richmond, California) to lower its contamination with trace metals, and then filter sterilized and supplemented with filter sterilized solutions of high purity sulfate salts (Johnson-Matthey, Inc., Seabrook, New Hampshire) of magnesium (820 μM), manganese (36 μM), and iron at various concentrations (0-18 μM). If desired, the medium was solidified by addition of agar that was previously washed with EDTA. To determine effects of transferrin on growth, essentially iron free human transferrin (Sigma Chemical Co., St. Louis, Missouri) was added to the medium. To measure inhibition of growth by the chelating agent ethylenediamine-di-(o-hydroxyphenylacetic acid)(EDDA), molten Brain-Heart Infusion agar containing 1 mg EDDA per ml was seeded with 10^4 CFU per ml and the seeded agar solidified in plates. To determine reversal of EDDA inhibition, sterile paper disks containing the desired agents were placed on the surface of the agar. All incubation was aerobic at 30°C.

Thin layer chromatography of siderophores was done using two systems. System 1 was cellulose plates (Eastman Chromagram 13255) developed with benzene:acetic acid:water (125:72:3), and system 2 was polyamide plates (Schleicher and Schuell G1600) developed with methanol:tetrahydrofuran:5% ammonium acetate (1:1:1). Phenolate siderophores were visualized by ultraviolet light and with a 1% ferric chloride spray.

RESULTS

Thirteen strains of *A. Hydrophila* were found to produce a siderophore during growth in low-iron (0.18 μM iron) medium which (unlike enterobactin and most other phenolate siderophores) could not be extracted into ethyl acetate from culture supernatant. A modification of the methods of Robinson (1979) was used to purify the *A. Hydrophila* siderophore (designated AHS) from culture supernatant of strain 495A2. The culture was grown from an inoculum of 10^4 CFU per ml in 10-l amounts in a Virtis fermentor (with Teflon coated stainless steel components; The Virtis Co., Gardiner, New York) at 30°C with aeration at 10-l per minute. Growth and siderophore production (determined by ferric chloride reactivity of the supernatant) reached maximum at 15 hours. Aeration was stopped and the culture was sparged with nitrogen. Cells were removed by centrifugation and the supernatant collected in Teflon bottles to prevent formation of the ferri-AHS chelate by leaching of iron from glass vessels. The siderophore was adsorbed to polyamide (Woelm, obtained from Universal Scientific, Atlanta, Georgia) by passing the supernatant through a 3 x 20 cm column of polyamide. The column was washed with water, and the siderophore eluted with methanol:acetone (1:1). Siderophore containing fractions were concentrated by vacuum evaporation to approximately 20 ml and lyophilized. The dried powder was dissolved in 10 ml of methanol and filtered through Teflon (0.5 μm pore diameter). The AHS was precipitated by addition of ethyl acetate and was collected and dried. Usual yield was 100-120 mg of AHS from 10-l of culture supernatant.

The purified AHS gave Rf values of 0.10 and 0.70 in thin layer chromatographic (TLC) systems 1 and 2, respectively. Partially purified siderophore (the eluate from polyamide columns) produced by all of the *A. hydrophila* strains used here, except the enterobactin producing strain UT, gave similar Rf values, suggesting that these strains also produced AHS. In TLC systems 1 and 2, authentic enterobactin gave respective Rf values of 0.40 and 0.45. As expected, partially purified siderophore produced by strain UT had Rf values identical to enterobactin.

Table 1. Reversal of EDDA Inhibition by Siderophores

	AHS^+ strains		ent^+ strain
Siderophore	495A2	BAP	UT
AHS	yes	yes	no
Enterobactin	no	yes	yes

Table 2. Effect of Human Transferrin on Growth

	Maximum growth (Klett Units)	
Addition	AHS^+ strain	ent^+ strain
None	216	225
Transferrin (30 μM)	214	6

The presence of 2,3-dihydroxybenzoic acid was qualitatively determined by hydrolyzing a 1 mg sample of AHS in 6N HCl. The 2,3-dihydroxybenzoic acid in the hydrolysate was identified by two dimensional TLC (O'Brien et al., 1970) using the authentic compound as the standard. Amino acid analysis of AHS showed that the preparation contained lysine, glycine, phenylalanine and tryptophan, suggesting that AHS may be a phenolate siderophore containing aromatic amino acids.

Addition of the chelating agent EDDA to agar medium inhibited growth of the AHS producing (AHS^+) strains and the enterobactin producing (ent^+) strain UT. Inhibition of growth was reversed by placing disks containing an iron salt on the seeded agar. Growth inhibition of an AHS^+ strain was reversed by disks containing AHS: however, this strain was unable to use enterobactin (Table 1). The ent^+ strain used enterobactin to reverse EDDA inhibition, but could not use AHS. As a specific enterobactin receptor is required for enterobactin uptake, these data suggest that a specific receptor of AHS also may be obligatory and that the ent^+ strain does not produce the AHS receptor. Only one AHS^+ strain (BAP) capable of using enterobactin was identified (Table 1). This strain may produce both the enterobactin and the AHS receptors, but only AHS was found in its culture supernatant.

Analysis of the outer membrane proteins of *A. hydrophila* 495A2 (AHS^+) by SDS-polyacrylamide gel electrophoresis (Filip et al., 1973) showed that the concentrations of at least three outer membrane proteins, with molecular weights near 68,000, were increased by growth in low-iron medium. By comparison with results obtained in other siderophore producing gram negative bacteria, one or more of these proteins may represent the AHS receptor.

In low-iron medium (0.18 μM iron), an AHS^+ strain (495A2) was not inhibited by addition of the concentration (30 μM) of human transferrin usually present in blood, while the ent^+ strain was inhibited by this level of transferrin (Table 2). Growth of the AHS^+ strain was not inhibited by transferrin concentrations as high as 100 μM. Production of AHS may confer insensitivity to transferrin inhibition and Ahs may be able to extract iron from transferrin for use by the microbe.

DISCUSSION AND SUMMARY

A previously unknown phenolate siderophore (designated AHS) was purified from a strain of *A. hydrophila,* and twelve additional strains of *A. Hydrophila* also produced AHS. Although determination of structure is still underway, this new siderophore may represent the first known phenolate siderophore containing aromatic amino acids. A siderophore is also produced by a related microorganism *Aeromonas salmonicida* that is strictly pathogenic for fish (Chart and Trust, 1983) but it is not known if this microorganism might produce AHS. Production of AHS may confer high resistance to inhibition by transferrin; therefore, AHS may be a virulence factor in *A. hydrophila.* Future studies will determine if AHS has a role in infections in the channel catfish. An enterobactin producing strain of *A. hydrophila* was inhibited by transferrin. This observation may be related to the recent demonstration that enterobactin may function poorly in the environment of the host, due to binding of enterobactin by serum albumin (Konopka and Neilands, 1984). Enterobactin producing strains may be of low virulence. Utilization of AHS probably requires a unique receptor specific for AHS, suggesting that synthesis of AHS may be coded for by an operon that also contains information for the AHS receptor. One AHS producing strain that used both AHS and enterobactin was identified. This strain may have all or part of the enterobactin operon; other AHS+ strains may have an unexpressed enterobactin operon. Genetic studies to answer these and other questions are underway with *A. hydrophila.*

ACKNOWLEDGEMENTS

We are grateful to M. Olson, University of Mississippi Medical Center, Jackson, for assistance with amino acid determinations and to K.N. Raymond, University of California, Berkeley, for collaboration on determination of the AHS structure. We thank S.M. Payne, University of Texas, Austin, and R. Brendon, Mount Sinai Medical Center, New York, for cultures. The research was supported in part by grant DE04903 from the National Institute of Dental Research, National Institutes of Health.

REFERENCES

Andrus, C., and Payne, S.M., 1983. Siderophores and iron regulated proteins in *Vibrio* and *Aeromonas* species. Abstracts Annual Meeting American Society for Microbiology, page 61.

Byers, R., Byers, P., and Arcenequx, J.E.L., 1983. Production and utilization of siderophores during iron-limited growth of *Aeromonas* species. Abstracts Annual Meeting American Society for Microbiology, page 207.

Chart, H., and Trust, T.J., 1983. Acquisition of iron by *Aeromonas salmonicida.* J. Bacteriol., 156:578-764.

Crosa, J.H., 1984. The relationship of plasmid-mediated iron transport and bacterial virulence. Ann. Rev. Microbiol., 38:69-84

Filip, C., Fletcher, G., Wulf, J.L., and Earhart, C.F., 1973. Solubilization of the cytoplasmic membrane of *Escherichia coli* by the ionic detergent sodium-lauryl sarcosinate. J. Bacteriol., 115:717-722.

Janda, J.M., Reitano, M., and Bottone, E.J., 1984. Biotyping of *Aeromonas* isolates as a correlate to delineating a species-associated disease spectrum. J. Clin. Microbiol., 19:44-47.

Konopka, K., and Neilands, J.B., 1984. Effect of serum albumin on siderophore-mediated utilization of transferrin. Biochemistry, 23:2122-2130

O'Brien, I.G., Cox, G.B., Gibson, F. 1970. Biologically active compounds containing 2,3-dihydroxybenzoic acid and serine formed by *Escherichia coli*. Biochim. Biophys. Acta., 201:453-460.

Robinson, A.V., 1979. A rapid column chromatographic method for isolation of catechol-type siderophores. Anal. Biochem. 95:364-370.

Russel, L.M., and Holmes, R.K., 1985. Highly toxinogenic but avirulent Park-Williams 8 strain of *Cornybacterium diphtheriae* does not produce siderophore. Infec. Immun., 47:575-578.

Weinberg, E.D., 1984. Iron withholding: a defense against infection and neoplasia. Physiol. Revs., 64:65-102.

Williams, P.H., and Carbonetti, N.H., 1984. The plasmid-specified aerobactin iron uptake system of *Escherichia coli*, In: "Plasmids in Bacteria", 741-757. (D.R. Helsinki, S.N. Cohen, D.B.Clewell, D.A. Jackson and A. Hollaender eds.), Plenum Publishing Corp., New York.

FERROUS COMPLEXES AND CHELATING COMPOUNDS IN SUPPRESSION OF FUNGAL DISEASES OF CEREALS

Averil E. Brown

Plant Pathology Research Division, Department of Agriculture for Northern Ireland and Faculty of Agriculture and Food Science, The Queen's University Belfast.

INTRODUCTION

The involvement of iron in fungal development has been studied from three main aspects; acquisition of iron in iron deficient environments, suppression of pathogenic fungi by iron deprivation and the displacement of inhibitory iron within conidia to activate the germination mechanism. Many fungi produce hydroxamate siderophores in iron-deficient conditions, e.g. compounds of the ferrichrome-type are produced by *Ustilago sphaerogena* Emery, 1974), *Aspergillus* sp. (Wiebe and Winkelmann, 1975) and *Epicoccum purpurascens* (Frederick et al., 1981). Several ectomycorrhizal fungi also produce hydroxamate siderophores and may provide plants with complexed iron (Szaniszlo et al., 1981). Hydroxamate siderophores specifically form ferric complexes which are actively transported across fungal cell walls (Emery, 1974) and ferric iron is utilized in fungal enzyme systems (Neilands, 1974). Iron deprivation has been shown to suppress fungal growth (Scher and Baker, 1982).

Anthranilic acid leached from banana fruits was rapidly metabolized by *Colletrotrichum musae,* the causal organism of banana anthracnose, to 2,3-dihydroxybenzoic acid (DHBA), a stimulant of spore germination and appressorium formation (Swinburne, 1981). Host leachates also stimulated germination and appressorium formation in pycnospores of the apple fruit pathogen *Diaporthe perniciosa,* and the compounds responsible were identified as p-coumarylquinic, chlorogenic and caffeic acids (Brown and Swinburne, 1978). The action of these compounds and DHBA is believed to depend on their iron-chelating properties. Stimulation of germination of conidia of *C. musae* was also achieved with a number of synthetic chelating compounds (Harper et al., 1980) and siderophores produced by bacteria (McCracken and Swinburne, 1979; McCracken and Swinburne, 1980). From subsequent investigations it has been suggested that chelating compounds remove iron from some site within the conidia, possibly on the ribosomes, so activating the germination mechanism (Graham and Harper, 1983). Stimulation of germination and the formation of mature appressoria in *C. musae* with iron-chelating compounds at the surface of the host resulted in increased aggressiveness of the pathogen (Brown and Swinburne, 1981; Swinburne and Brown, 1983). Similar observations of increased aggressiveness were made with *Botrytis cinerea* in the presence of iron-chelating compounds on leaves of *Vicia faba* (Brown and Swinburne, 1982). DHBA complexed with iron however suppressed spore germination and maturation of appressoria in *C. musae* (Swinburne and Brown, 1983) and inhibited spore germination in *B. cinerea* (Brown and Swinburne, 1982). As a consequence both pathogenic fungi failed to develop progressive lesions on their respective hosts. DHBA and other phenolic chelating compounds specifically chelate ferrous iron (Brown and Sharma, 1985) and this review will present evidence for a suppressive role for ferrous iron in the development of fungi pathogenic on ariel parts of cereal plants.

FERROUS COMPLEXES AND SUPPRESSION OF SPORE GERMINATION AND HYPHAL GROWTH IN FUNGI

The optimum concentration for most iron chelating compounds as stimulants of germination of *C. musae* was in the range 10^{-3} to 10^{-4} M (Swinburne, 1981). No such stimulation of germination by iron chelating compounds was however apparent in three cereal pathogens, *Septoria nodorum, Pyrenophora teres* and *Cochliobolus sativus* the spores of which all germinated well in water. Germination in the presence of these compounds was not significantly different from that in water. Ferric 1:1 complexes of ethylenediaiminetetraacetic acid (EDTA), Ethylenediamine -di (o-hydroxyphenyl -acetic acid) (EDDHA) and citric acid also showed no inhibitory activity. Germination of spores of these fungi was, however, almost totally suppressed in the presence of compounds complexed with ferrous iron both on glass slides and on the surface of leaves of their respective hosts (Table 1). Glycine and1:10-phenanthroline are known to form ferrous 2:1 and 1:1 complexes respectively (Albert, 1960) and the divalent state of iron complexed with DHBA and gallic acid (GA) was confirmed by polarography (Brown and Sharma, 1985). Both phenolic compounds form ferrous 1:1 complexes. Several other ferrous complexes including those of 2,2^1-dipyridyl, histidine, chlorogenic and caffeic acids also inhibited fungal spore germination while phenolic acids incapable of reducing and complexing dissolved iron were not inhibitory. Germination of spores of many other necrotrophic fungi including *Fusariun* spp., *Nectria galligena, B. cinerea* and *Colletotrichum* spp. was similarly inhibited by ferrous complexes.

Table 1. Germination of spores of *S. nodorum, P. teres* and *C. Sativus* in solutions (5 x 10^{-4} M, pH 5) of iron complexed with various chelating compounds (*S. nodorum* and *P. teres* data from (Brown and Sharma, 1985).

Chelating agent-iron complex	Germination (%)		
	S. nodorum	*P. teres*	*C. sativus*
Water	96	95	79
Fe^{II}-DHBA	0	2	0
Fe^{II}-gallic acid	0	0	0
Fe^{II}-glycine	0	12	6
Fe^{II}-1:10-phenanthroline	0	4	0
Fe^{III}-EDTA	96	97	82
Fe^{III}-EDDHA	91	86	75
Fe^{III}-citric acid	98	97	80
SE of means	3.2	3.9	1.9

Table 2. Mean elongation of germ-tubes in germinated spores of *S. nodorum* and *P. teres* placed in solutions of complexed iron (data from Brown and Sharma, 1985).

Chelating agent-iron complex	Germ-tube growth (μ)	
	S. nodorum	*P. teres*
H_2O	9.3 ± 3.6*	5.1 ± 3.1
Fe^{II}-DHBA	2.0 ± 1.5	3.2 ± 2.1
Fe^{II}-gallic acid	1.7 ± 1.2	3.6 ± 2.7
Fe^{III}-EDTA	9.9 ± 3.7	5.8 ± 3.3
Fe^{III}-citric acid	10.2 ± 4.0	6.3 ± 3.0

* Standard deviation of means from three replicate experiments

There is evidence from earlier studies on the role of iron in germination of conidia of *C. musae* that iron complexed with DHBA was transported across the walls of the conidia (Graham, 1981). Germination of spores of *S. nodorum* and *P. teres* was greatly suppressed even when washed and suspended in water following 24 h incubation in Fe^{II}-DHBA(5×10^{-4}M, ph5). DHBA (5×10^{-4}M, pH5) in the inoculum drop did not release the germination mechanism in Fe^{II}-DHBA-pretreated spores of either fungus. Even EDDHA, which has a very high affinity for ferric ions, only slightly revived germination in spores of *S. nodurum* pretreated with Fe^{II}-DHBA but not those of *P. teres*. In retrospect, it is possible that a compound such as 1:10-phenanthroline with a much higher affinity for ferrous ions than DHBA may have revived germination from which it could have been concluded that iron had remained in the divalent state in the spore.

The elongation of germ-tubes of germinated spores of *S. nodorum* and *P. teres* was suppressed but not totally inhibited by ferrous complexes of DHBA and GA (Table 2). The effect was much greater in *S. nodorum* than *P. teres*. Ferric ions complexed with EDTA and citric acid slightly enhanced germ-tube growth in both fungi perhaps reflecting the requirement for this essential ion as already demonstrated by Scher and Baker (1982).

The fate of ferrous ion and the chelating compounds in fungal spores and hyphae is unknown but it might be speculated that the excess iron remains in the reduced state. Complexes formed with purines or natural pteridines e.g. folic acid or riboflavines which have high affinities for ferrous ions (Albert, 1960) alone might be detrimental to fungal development. Ferrous ions are also known to bind to bacterial enzyme proteins, often activating the enzyme (Sagers, 1974). However, in some yeasts, e.g. *Candida gulliermondii* glycolysis was suppressed by high iron concentrations due to low amounts active aldolase (Light and Clegg, 1974). The formation of such ferrous-protein complexes in fungi might also result in inhibition of germination and/or suppression of germ-tube growth.

FERROUS COMPLEXES AND SUPPRESSION OF DISEASE DEVELOPMENT

Lesion development on detached leaves

Inhibition of spore germination of *S. nodorum*, *P. teres* and *C. sativus* in the presence of ferrous complexes was reflected in the virtual absence of lesion development at sites of inoculation on detached seedling leaves of their respective hosts. Lesions which developed in the presence of DHBA and GA were similar to those produced in water. Much larger lesions were, however, produced by *S. nodorum* and *C. sativus* in the presence of EDTA and EDDHA. Since little difference could be detected in percentage or rates of germination between treatments containing chelating compounds in the des-ferri form and water, EDTA and EDDHA may have increased the susceptibility of host leaves to the pathogens.

Germination, appressorium formation and lesion development by the three pathogens were also reduced by 50 to 60% on detached seedling leaves of host plants cut up to 5 days after a spray-application of Fe^{II}-DHBA (5×10^{-4} M, pH5) compared with leaves treated with DHBA at the same concentration and untreated leaves.

Foliar application and disease development on whole plants

The greatest losses of yield caused by *S. nodorum* and *P. teres* in wheat and barley often occur following infection of the flag leaves. Disease symptoms on the flag leaves of wheat cv Maris Ranger and barley cv Akka (cultivars highly susceptible to *S. nodorum* and *P. teres*, respectively) treated with Fe^{II}-DHBA 24 h prior to inoculation with the pathogen were reduced to approximately 40% of symptoms on inoculated, untreated leaves (Table 3). The lower disease levels on the Fe^{II}-DHBA treated flag leaves resulted in less spread of disease to the ears than on untreated plants and significantly smaller 1000 grain weight loss occurred.

Symptoms of *C. sativus*, the causal organism of leaf spot disease of barley seedlings, were significantly reduced on the susceptible cv Golf treated with Fe^{II}-GA, Fe^{II}-DHBA and also GA 24 h prior to inoculation (Table 4). Disease symptoms were significantly($P<0.01$) increased on leaves of seedlings treated with EDTA, Fe^{III}-EDTA and EDDHA. Barley seedlings cv Sundance were similarly treated with chelating compounds and iron complexes then inoculated with powdery mildew. Preliminary

Table 3. The effect of Fe^{II}-DHBA on disease development by *S. nodorum* on wheat (cv. Maris Ranger) and by *P. teres* on barley (cv. Akka) (Data from Brown and Sharma, 1985).

Pathogen/ host	% flag leaf area diseased		1000 grain weight*		
	Control	Fe^{II}-DHBA	Disease free	Inoculated Control	Inoculated Fe^{II}-DHBA
S. nodorum/ wheat	47	19	21.7	11.9	16.7
P. teres/ barley	60	29	24.2	16.9	20.3
S.E. of means	2.3			1.21	

* corrected to 85% dry matter

Table 4. Disease development by *C. sativus* on barley seedlings (cv. Golf) following foliar sprays of chelating compounds or iron complexes (10^{-3}M, pH 5)

Treatment	Leaf area infected (%)
Control	20.4
DHBA	20.6
Fe^{II}-DHBA	12.7
GA	14.2
Fe^{II}-GA	6.9
EDTA	30.0
Fe^{III}-EDTA	34.3
EDDHA	26.3
Fe^{III}-EDDHA	18.4
Fe SO_4	17.8
S.E. of means	1.41

results would suggest that Fe^{II}-GA and Fe^{II}-DHBA reduced disease development by approximately 50%. Fe^{III}-EDTA and Fe^{III}-EDDHA, however, also suppressed mildew development by up to 30% but chelating compounds in the desferri form and unchelated $FeSO_4$ neither hindered nor enhanced this disease.

Foliar application and leaf surface microflora

In the present climate of interest in siderophore producing bacteria and their influence on pathogenic organisms it was necessary to investigate any effect chelating compounds and complexed iron had on leaf surface microflora. This investigation has not been made in depth but it quickly became apparent that these treatments had a marked effect on the populations of fluorescent pseudomonads on leaves of barley seedlings. The largest populations of these bacteria were isolated from leaves treated with water and unchelated $FeSO_4$. Much smaller populations were isolated from leaves treated with Fe^{II}-DHBA, Fe^{II}-GA, EDTA or EDDHA and fluorescent pseudomonads were generally not isolated from leaves treated with DHBA, GA, Fe^{III}-EDTA or Fe^{III}-EDDHA. It was also noted that the mycoparasite *Trichoderma viride* was isolated from leaves which had received any of the iron treatments. Suspensions of fluorescent pseudomonads isolated from untreated barley leaves and mixed cultures including some fluorescent pseudomonads isolated from leaves treated with Fe^{II}-DHBA and EDTA caused some reduction in germination and term-tube growth in *C. sativus* on glass slides and slightly suppressed lesion development on detached leaves. However, since disease development on whole plants was suppressed by Fe^{II}-DHBA and enhanced by EDTA compared with untreated plants it would seem unlikely that these bacteria played any role in determining the effect of either chelating compounds or complexed iron on infection of barley by *C. sativus*. *T. Viride* germinated more slowly than *C. sativus* and had little effect on infection in the detached leaf assay.

Table 5. Effect of application of chelating compounds and iron complexes (10^{-3} M, pH 5) as soil drenches on disease development by *C, sativus* and Fe^{II} status of the leaves

Treatment	Leaf area infected (%)	Fe^{II} (% of total iron)
C	25.0	44
DHBA	14.4	65
Fe^{II}-DHBA	12.1	83
GA	9.4	89
Fe^{II}-GA	15.0	70
EDTA	24.8	56
Fe^{III}-EDTA	17.5	59
EDDHA	22.5	40
Fe^{III}-EDDHA	15.6	54
$Fe SO_4$	20.7	50
S.E. of mean	1.62	

FERROUS STATUS OF BARLEY LEAVES AND DISEASE DEVELOPMENT

Chelating compounds and iron complexes (10^{-3} M, pH 5) were applied to barley seedlings as soil drenches and two days after treatment the seedlings were inoculated with *C. sativus*. The John Innes No. 2 compost used to grow the seedlings was approximately pH 6. The lowest level of leaf infection developed on seedlings treated with GA and significantly ($P<0.01$) fewer lesions also developed on seedlings treated with Fe^{II}-DHBA than on any other treatment (Table 5). Disease development was significantly ($P<0.01$) less on seedlings treated with DHBA, Fe^{II}-GA, Fe^{III}-EDTA and Fe^{III}-EDDHA than on untreated seedlings.

The ferrous iron content of the leaves was determined by extraction with 1:10-phencenthroline and calculated as a percentage of total iron determined after wet digestion. Seedlings treated with GA and Fe^{II}-DHBA which showed the lowest levels of disease contained the highest percentage ferrous iron in the leaves (Fig. 1). A good negative correlation ($r=-0.87$) was obtained when disease development and ferrous status of the leaves were compared. The ultimate effect of these drench treatments on the growth of barley has not yet been determined but no obvious deficiency symptoms or retardation of growth were observed in young plants.

CONCLUSION

Ferrous ions have been used to control a number of fungal pathogens. Ferrous sulphate has been used to reduce the incidence of stem rot in *Capsularis* jute (Ji et al., 1976), and as a drench to increase resistance of apple trees to *Cephalosporium* wilt (Tsakadze et al., 1977) and apricot trees to Verticillium wilt (Kibishauri and Tsiklauri, 1980) in Russia. Evidence presented in this review gives some insight into the processes

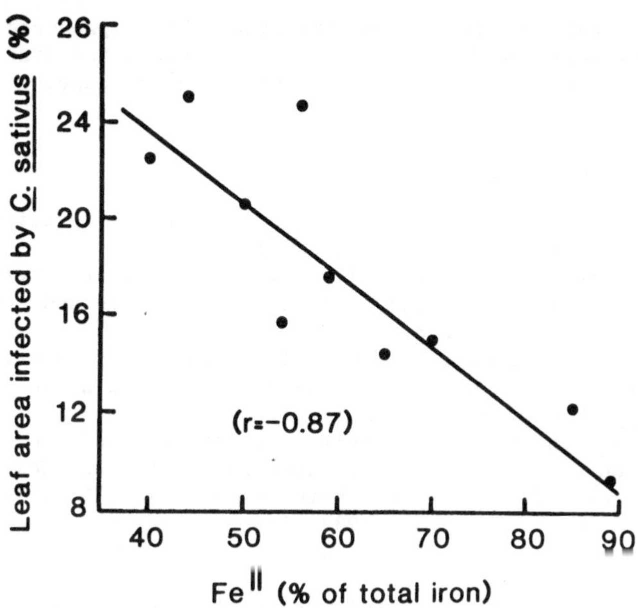

Fig. 1. Correlation of ferrous iron status of barley leaves and infection by *C. sativus*.

involved in fungal disease control with ferrous iron. Although the mechanism has not been elucidated it has been adequately demonstrated that ferrous (pH <7) and not ferric complexes suppress germination and appressorium formation in fungal pathogens and produce significant reductions in infection of host plants. The fate of ferrous iron from foliar sprays at the host surface has not yet been determined but that its absorption through leaf surfaces might directly influence host resistance mechanisms should not be overlooked.

A role for ferrous iron in the enhancement of disease resistance in plants was suggested from data obtained following the application of chelating compounds and complexes as soil drenches. Barley plants which contained a high percentage of iron in the ferrous, or "active" state, especially following drenches of GA and Fe^{II}-DHBA, proved to be much more resistant to *C. sativus* than untreated plants. Much of the GA will have reduced ferric compounds in the soil and formed Fe^{II}-GA before being utilized by the barley. Gallic acid in combination with coumarin and caffeic acid has, in fact, been reported as a control measure for root (wilt) disease of coconut (Dwivedi et al., 1980). These phenolic chelating compounds are similar to the ferric reductants produced by some plant roots under stress conditions (Hether et al., 1984; Olsen et al., 1982). Although barley reductants are reported to be glucose containing compounds (Hether et al., 1984) the ferrous iron from the phenolic complexes is apparently taken up by the roots with a high percentage remaining unoxidized as it is transported from the protoxylem to the metaxylem (Brown, 1978). A reduction in *C. sativus* disease development was also observed in barley plants drenched with ferric complexes. A higher percentage of ferrous iron was also detected in these plants compared with untreated plants, as has been observed in leaves of Fe^{III}-EDTA -treated spinach (El-Sherif et al., 1984).

Oat plants have been reported to utilize iron from ferrichrome much more readily than from synthetic chelates (Reid et al., 1984). I would like to suggest that iron complexed with siderophores produced by microorganisms in the soil could also increase the ferrous ion status of plants with a consequent increase in resistance to fungal disease.

REFERENCES

Albert, A., 1960. Metal-binding agents. In: "Selective Toxicity", pp. 145-168. Methuen and Co. Ltd., London.

Brown, A.E., and Sharma, H.S.S., 1985. A possible role for ferrous complexes in fungal disease suppression: glume blotch and net blotch or cereals. Phytopathol. Z., 113.

Brown, A.E., and Swinburne, T.R., 1978. Stimulants of germination and appressorial formation by *Diaporthe perniciosa* in apple leachate. Trans. Brit. Mycol. Soc., 71:405-411.

Brown, A.E., and Swinburne, T.R., 1981. The influence of iron and iron chelators on the formation of progressive lesions by *Colletotrichum musae* on banana fruits. Trans. Brit. Mycol. Soc., 77:119-124.

Brown, A.E., and Swinburne, T.R., 1982. Iron-chelating agents and lesion development by *Botrytis cinerea* on leaves of *Vicia faba*. Physiol. Plant Path., 21:13-21.

Brown, J.C., 1978. Mechanism of iron uptake by plants. Plant, Cell. Environ., 1:249-257.

Dwivedi, R.S., Amma, B.S.K., Mathew, S.C., and Ray, P.K., 1980. Control of root (wilt) disease of coconut (*Cocos mucifera*) with micronutrients, phenolic compounds and ascorbic acid. Plant Dis., 64:843-844.

El-Sherif, A.F., Osman, A.Z., Sadik, M.K., and Shata, S.M., 1984. Determination of ferrous and ferric iron ratio in spinach plants and their relation to iron application. J. Plant Nutr., 7:767-776.

Emery, T., 1974. Biosynthesis and mechanism of action of hydroxamate-type siderochromes. In: "Microbial Iron Metabolism," pp.107-122. (J.B. Neilands, Ed.), Academic Press, New York and London.

Frederick, C.B., Szaniszlo, P.J., Vickery, P.E., Bentley, M.D., and Shine, W., 1981. Production and isolation of siderophores from the soil fungus *Epicoccum purpurascens*. Biochemistry, 20, 2432-2436.

Graham, A.H., 1981. Studies on the role of iron in germination of conidia of *Colletotrichum musae*. Ph.D. thesis, The Queen's University of Belfast.

Graham, A.H., and Harper, D.B., 1983. Distribution and transport of iron in conidia of *Colletrotrichum musae* in relation to the mode of action of germination stimulants. J. Gen. Microbiol., 129:1025-1034.

Harper, D.B., Swinburne, T.R., Moore, S.K., Brown, A.E., and Graham, H., 1980. A role for iron in germination of conidia of *Colletotrichum musae*. J. Gen. Microbiol., 121:169-174.

Hether, N.H., Olsen, R.A., and Jackson, L.L., 1984. Chemical identification of iron reductants exuded by plant roots. J. Plant Nutr., 7:667-676.

Ji, T., Lal, H., and Singh, S.P., 1976. Influence of micronutrients on incidence of stem rot of *Capsularis* jute. Indian J. Mycol. Plant Pathol., 6:96-97.

Kibishauri, V.D., and Tsiklauri, M.S., 1980. The effect of chemical nutrition on the resistance of apricot to Verticillium Wilt. Rev. Plant Pathol., 60:6023.

Light, P.A., and Clegg, R.A., 1974. Metabolism in iron-limited growth. In: "Microbial Iron Metabolism," pp. 35-61. (J.B. Neilands, Ed.), Academic Press, New York and London.

McCracken, A.R., and Swinburne, T.R., 1979.Siderophores produced by saprophytic bacteria as stimulants of germination of conidia of *Colletotrichum musae*. Physiol. Plant Path., 15:331-340.

McCracken, A.R., and Swinburne, T.R., 1980. Effects of bacteria isolated from the surface of banana fruits on the germination of conidia of *Colletotrichum usae* (Berk & Curt.) Arx. Trans. Brit. Mycol. Soc. 74: 18-20.

Neilands, J.B., 1974. "Microbial Iron Metabolism,", 597 pp. (J.B. Neilands Ed.), Academic Press, New York and London.

Olsen, R.A., Brown, J.C., Bennett, J.H. and Blume, D., 1982. Reduction of Fe^{3+} as it relates to Fe chlorosis. J. Plant Nutr., 5:433-445.

Reid, C.P.P., Crowley, D.E., Kim, H.J., Powell, P.E., and Szaniszlo, P.J. 1984. Utilization of iron by oat when supplied as ferrated synthetic chelate or as ferrated hydroxamate siderophore. J. Plant Nutr., 7:437-447.

Sagers, R.D., 1974. Other iron-containing or iron-activated enzymes: enzymes acting on certain amino acids, amines and acetyl phosphate. In: "Microbial Iron Metabolism," pp. 446-452. (J.B. Neilands, Ed.) Academic Press, New York and London.

Scher, F.M., and Baker, R., 1982. Effect of *Pseudomonas putida* and a synthetic iron chelator on induction of soil suppressiveness to Fusarium wilt pathogens. Phytopathology, 72:1567-1573.

Swinburne, T.R., 1981. Iron and iron chelating agents as factors in germination, infection and aggression of fungal pathogens. In: Microbial Ecology of the Phylloplane, pp. 227-243. (J.P. Blakeman, Ed.), Academic Press, New York and London.

Swinburne, T.R., and Brown, A.E., 1983. Appressoria development and quiescent infections of banana fruit by *Colletrotrichum musae*. Trans. Brit. Mycol. Soc., 80:176-178.

Szaniszlo, P.J., Powell, P.E., Reid, C.P.P., and Cline, G.R., 1981. Production of hydroxamate siderophore iron chelators by ecto-mycorrhizal fungi. Mycologia, 73:1158-1174.

Tsakadze, T.A., Kikvadez, M.A., Kibishauri, V.A., and Giorgadze, R.G., 1977. Study on the effect of metal chelates of natural compounds on the resistance of apple (Kekhura) to Vrphslodpotium eily. Rev. Plant Path., 56:3621.

Wiebe, C., and Winkelmann, G., 1975. Kinetic studies on the specificity of chelate-iron uptake in *Aspergillus*. J. Bacteriol.,123:837-842.

Neilands, J.B., 1974. "Microbial Iron Metabolism," 597 pp. (J. B. Neilands Ed.), Academic Press, New York and London.

Olsen, R.A., Brown, J.C., Bennett, J.H., and Blume, D., 1982. Reduction of Fe[illegible] as it relates to Fe chlorosis. J. Plant Nutr. [illegible]

[illegible]

THE EFFECT OF CHELATING AGENTS ON A RUST FUNGUS DEVELOPING ON INDUCED RESISTANT PLANTS

H. von Alten and F. Schönbeck

Institut fur Pflanzenkrankheiten und Pflanzenschutz
Universitat Hannover, Herrenhäuser Str. 2
3000 Hannover 21

Susceptibility to diseases is not stable for the entire life of an individual plant. It may change with stages in the plant's development, but it is also possible to influence susceptibility by one application of an agent which induces resistance against plant diseases. Culture filtrates of some bacteria and fungi have been used to induce resistance against obligate biotrophic fungal pathogens in many plant species (Schönbeck et al., 1980). This principle of induced resistance is unspecific, success depending on the mode of parasitism of the pathogens, the obligate biotrophy. In plants with induced resistance fungi are able to produce very few sporulating pustules or colonies. In addition to this reduction of disease frequency there is also a reduction in the quantity (Denhe et al., 1984) and quality (Dehne et al., 1984; Von Alten and Schönbeck, 1985) of the spores produced.

With the bean rust caused by *Uromyces phaseoli* (Pers. ex Pers.) Wint. the alteration of sporulation-quality seems to have an added dimension in that not only are the uredospores which develop on plants with induced resistance of impaired quality, but so are successive generations, resulting from passages through totally normal host plants, (Von Alten and Schönbeck, 1981). Reduced spore quality remains measurable for a number of passages, after which the spores return to normal. Normal spores will be referred to as N-spores, while the altered spores from induced resistant bean plants will be referred to as R-spores.

The lack of infectivity is transmitted from generation to generation by the induced spores. In the first generation which develops on induced-resistant plants there is the possibility that substances were absorbed from the plants which altered their infectivity. If this were so it might be expected that such substances would also inhibit N-spores germinating on induced-resistant plants, but this is not so (Schönbeck et al., 1980). Moreover, an infectivity-altering substance would be highly diluted by two or three passages over new host plants. So such direct and simple modes of action have to be excluded and one has to search for those abilities or properties of the induced R-spores which are responsible for the altered infectivity.

Some properties of R-spores are listed in Table 1. The first significant differences are found in the quantity and the quality of germ tube respiration. The germ tubes of R-spores require much more oxygen, but show

Table 1. Comparison of some properties of N- and R-spores.

Property	R-spores different	Ref.
Form and size	0	(2)
Germination	faster start	(1)
Oil content	0	(2)
Fatty acids	0	(2)
Germ tube reaction with Janus-Green	100% more	(1)
Intensity of germ tube respiration	80% more	(1)
Antimycin-independent respiration	190% more	(4)

no visible advantage from this increase in their metabolism - except in the initial stage of germination in which they develop faster. This difference however, soon disappears. The normal infectivity of spores is restored with new passages over susceptible host plants - and related with this, increased germ tube respiration also returns to normal. Greater oxygen consumption may be related to a reduction in the efficiency of metabolism. Oxygen reduction via the normal respiratory chain allows three molecules of ATP to be formed. A side chain, which is Antimycin A independent, produces only one ATP without involvement of any cytochromes. Germ tubes of R-spores obviously use this less efficient way of respiration and increase the turnover to achieve normal development. This disturbance of the respiratory metabolism might be due to a limited supply of iron for the rust pustules on induced-resistant bean plants. Iron is the key ion of the cytochrome molecule and the lack of iron may hinder cytochrome biosynthesis.

The infectivity of rust spores is measured in terms of pustule density which results from inoculation, using spore suspensions of equal density. The success of infection is not the same with N- and R-spores. Germ tubes of R-spores must compensate for the poor efficiency of their energy metabolism as germination rate and germ tube elongation are normal. The next sequence in bean rust pathogenesis is germ tube orientation towards stomates, development of an appressorium and penetration into the leaf. Germ tubes of R-spores are capable of reducing a higher amount of oxygen, resulting in the production of more CO_2. Germ tube orientation and location of stomates are initiated by stimuli from the bean leaves: in addition to the form of leaf surface and shape of stomates (Wynn, 1976) there is natural gas exchange through the stomates (Von Alten, 1983) which serves as a signal of orientation. The germ tubes of R-spores are capable of producing a higher amount of CO_2 which could endanger their orientation towards the stomates, provided that CO_2 serves as signal for germ tube orientation.

Appressoria develop over stomates and consequently a measure of the orientation of germ tubes is the frequency of appressoria. A high correlation was found between the infectivity of uredospores and the frequency of appressoria (Fig. 1) which must be ascribed to lack of orientation of germtubes emerging from R-spores. Similar effects on orientation have been found with high germination temperatures (Von Alten, 1983).

In the experiments described here the bean plants became resistant following application with a culture filtrate of an isolate of *Bacillus subtilis*. The culture filtrate was sprayed onto the leaves. In addition to the resistance inducing agent the filtrate also contained all the material which the bacterium excreted onto the growth medium. If there is a limited supply of iron for the rust fungus in induced resistant leaves, it is probably caused by substances in the culture filtrate, either by

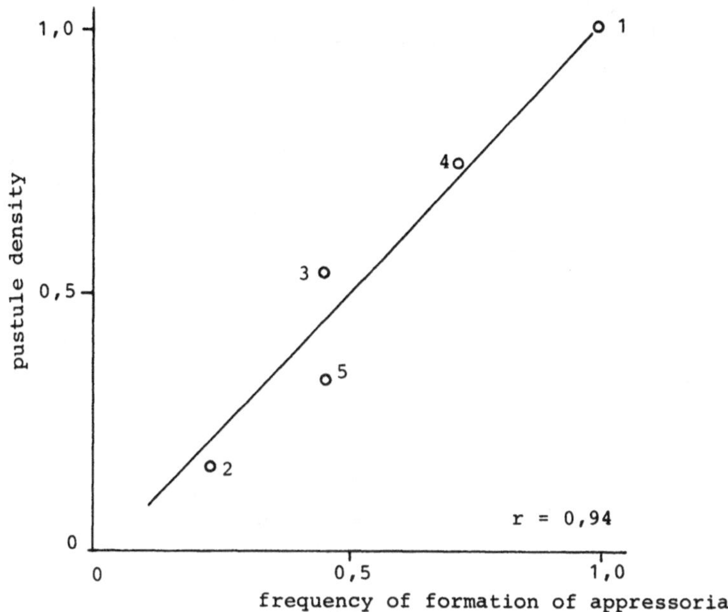

Fig. 1. Infectivity of *U. phaseoli* uredospores (postule density) correlated to the efficiency of the orientation of their germ tubes (frequency of formation of appressoria), relative amounts (Von Alten, 1982).

1 = N-spores inoculated on young leaves
2 = N-spores inoculated on adult leaves
3 = N-spores inoculated on young leaves, with constant light during leaf penetration phase
4 = R-spores inoculated on young leaves
5 = R-spores inoculated on adult leaves.

influencing the iron metabolism of the plant or binding the iron ions directly. This direct binding becomes visible when potassium hexacyanoferrate is competing with the culture filtrate to bind iron *in vitro* (Von Alten and Schönbeck, 1985). In the absence of culture filtrate potassium hexacyanoferrate and iron react at once (see A in Fig. 2). When potassium hexacyanoferrate is dissolved in the culture filtrate and iron is added subsequently the reaction is slower and less intense (see B in Fig. 2) - one part of the iron is bound irreversibly to the culture filtrate. When iron is added to the culture filtrate alone there is an additional reversible binding. When potassium hexacyanoferrate is added to the culture filtrate-iron mixture there is a delay in the reaction of the mixture (t_1 in Fig. 2) before binding by potassium hexacyanoferrate occurs. This compound has a stronger affinity for Fe^{3+} ions than the reversible iron-binding sites in the culture filtrate. The binding potential in the culture filtrate can be saturated by adding more iron thereby reducing t_1 (Fig. 2). When free iron is available in the culture filtrate the reaction with potassium hexacyanoferrate starts at once (Von Alten, 1982). The resistance inducer found in the *B. subtilis* culture filtrate can be inactivated by boiling (Schönbeck et al., 1980). Also the components in the culture filtrate which influence the infectivity of the bean rust spores are destroyed by heat. The capacity for iron binding in this culture filtrate can also be reduced, but not totally (Von Alten, 1982). Therefore only one part of the iron binding substances in the culture filtrate may be responsible for altering the sporulation quality on induced resistant plants.

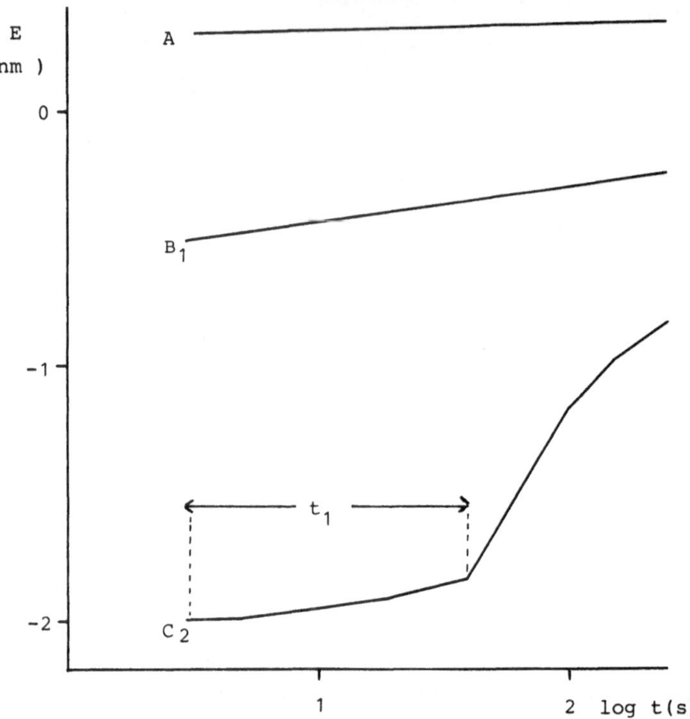

Fig. 2. Formation of $Fe_4(Fe(CN)_6)_3$ in water or culture filtrate of *B. subtilis* (Von Alten and Schönbeck, 1985). Components were mixed in different sequences:

A = water + $FeCl_3$+ potassium hexacyanoferrate
B = culture filtrate + potassium hexacyanoferrate + $FeCl_3$
C = culture filtrate + $FeCl_3$+ potassium hexacyanoferrate

As shown above, the culture filtrate binds iron *in vitro* and is presumed to do the same in the bean leaves, causing the production of less infective uredospores due to a limited supply of iron for the rust pustules. To substantiate this fact, iron was added to the culture filtrate before it was applied to the bean leaves. The iron binding sites in the culture filtrate were thus saturated so that the fungal iron nutrition could not have been limited. In fact, such a culture filtrate with saturated iron binding sites had no effect on the infectivity of uredospores (see C in Table 2): uredospores which developed on plants, treated in this way, showed a normal germ tube orientation and resulted in a normal pustule density (Von Alten and Schonbeck, 1985; Von Alten, 1982), when they were inoculated on new bean plants.

Table 2. Effect of $FeCl_3$, added to the culture filtrate of *Bacillus subtilis*, on the infectivity of R-spores.

Inoculation	Spores harvested from bean plants which were susceptible A(N-spores)	treated with culture filtrate B(R-spores)	treated with culture filtrate + $FeCl_3$(200 umol) C
To young leaves (4)	normal infection density	less pustules	no difference to A
to adult leaves (4)	reduced infection density	more pustules	no difference to A
to genetic resistant plants (2)	necrotic spots	less spots	no difference to A
high temp. during germination + penetration (2)	reduced infection density	more pustules	no difference to A

REFERENCES

Von Alten, H., and Schönbeck, F., 1981. Zum Einfluß induzierter Wirtzresistenz auf Vitalität und Infektiosität von *Uromyces phaseoli* (Pers.) Wint. Phytopathol. Z. 101:271-274.

Von Alten, H., 1982. Zur Pathogenität von Uredosporen des *Uromyces phaseoli* (Pers.) Wint. von Pflanzen mit induzierter Resistenz. Dissertation Universitat Hannover, Fachbereich Gartenbau.

Von Alten, H., 1983. The effect of temperature, light and leaf age on the frequency of appressoria formation and infection with *Uromyces phaseoli* (Pers.) Wint. Phytopathol. Z. 107:327-335.

Von Alten, H., and Schönbeck, F., 1985. Zur Infektiosität von *Uromyces phaseoli* -Uredosporen von induziert resistenten Wirtspflanzen. Z. Pflanzenkr. Pflanzenschutz, in press.

Dehne, H.-W., Stenzel, K., and Schönbeck, F., 1984. Zur Wirksamkeit induzierter Resistenz unter praktischen Anbaubedingungen. III. Reproduktion Echter Mehltaupilze auf induziert resistenten Pflanzen. Z. Pflanzenkr. Pflanzenschutz. 91:258-265.

Schönbeck, F., Dehne, H.-W., and Beicht, W., 1980. Untersuchungen zur Aktivierung unspezifischer Resistenzmechanismen in Pflanzen. Z. Pflanzenkr. Pflanzenschutz, 87:654-666.

Wynn, W.K., 1976. Appressorium formation over stomates by the bean rust fungus: Response to a surface contact stimulus. Phytopathology, 66:136-146.

HIGH AFFINITY IRON TRANSPORT IN *USTILAGO MAYDIS*

S.A. Leong

USDA-ARS, Department of Plant Pathology
University of Wisconsin
Madison, U.S.A.

The mechanism by which pathogenic organisms sequester and regulate uptake of iron is central to an understanding of the etiology of infectious disease. The ability of microbial pathogens to acquire this essential element from mammalian hosts is an important component of virulence. In contrast, the importance of iron acquisition to plant infection remains a virtually unexplored field.

The corn smut fungus *Ustilago maydis* offers an ideal system to assess the impact of high affinity iron transport on plant infection: this fungus has well-defined genetics, sexual crosses are feasible, the organism produces haploid, yeast-like spores, auxotrophic mutants are readily isolated, the chemical structure of it's siderophores, ferrichrome and ferrichrome A, are known, and some information about the biosynthesis and transport of these compounds is available for the closely related species *Ustilago sphaerogena*. Final formation of the smut gall requires the arid growth of the fungus in host tissue. This growth is predominantly intracellular with the smut gall being composed of hypertrophied plant cells filled with fungal spores and mycelium. Thus specialized mechanisms for fungal iron assimilation may be required for gall formation.

An understanding of the biochemistry and genetics of siderophore biogenesis and uptake in *U. maydis* is pre-requisite to assessing the role of siderophores in the development of this disease. Little is currently known about the biosynthetic pathway that leads to ferrichrome and ferrichrome A in *Ustilago*. Based on isotope labelling experiments in *U. sphaerogena*, Emery (1966) deduced that L-ornithine is first hydroxylated to yield δ-N-OH-L-ornithine. This compound is then acetylated via acetyl CoA to give δ-N-acetyl-N-OH-L-ornithine, a precursor of ferrichrome. The acetylase has been partially purified from this organism (Ong and Emery, 1972). The latter steps in ferrichrome synthesis are unknown but may proceed, as in the biosynthesis of the cyclic peptide antibiotic gramicidin, via covalently bound thioester intermediates on a multifunctional polypeptide. This mechanism has been implicated in both rhodotorulic acid and ferrichrome biosynthesis since an δ-N-acetyl-δ-N-OH-L-ornithine-ATP-PPi exchange activity has been observed in extracts prepared from *Rhodotorula pilimanae* (Akers and Neilands, 1978) and *Aspergillus quadricinctus* (Hummel and Diekman, 1981). Ferrichrome and ferrichrome A most likely share the first step in their biosynthesis since both contains δ-N-OH-L-ornithine. The pathways must diverge after this reaction since ferrichrome A contains trans-β-methyl glutaconic acid as the acyl group of the hydroxamate. Also the amino acid composition of the peptides differs: ferrichrome contains three glycines while ferrichrome A contains one glycine and two serines.

We have initiated a molecular genetic analysis of siderophore biosynthesis in *U. maydis*. Four classes of siderophore mutants were isolated after nitrosoguanidine mutagenesis. Among these were ferrichrome-minus, ferrichrome A-minus, ferrichrome/ferrichrome A-minus, and constitutive. A mutant that was defective in the production of both siderophores was analyzed genetically. Examination of the random basidiospores obtained from sexual crosses between the wild-type and mutant suggested that this mutant phenotype is determined by a single gene. Moreover, the mutant could be biochemically complemented by supplementation of low iron media with δ-N-OH-L-ornithine. These data are consistent with a mutation in an L-ornithine-δ-N-oxygenase. This is the first genetical evidence for a common biosynthesis route to ferrichrome and ferrichrome A in *Ustilago* spp. We are currently developing a plant model for assessing phytopathogenicity of these siderophore-less mutants.

REFERENCES

Akers, H.A., and Neilands, J.B., 1978. Biosynthesis of rhodotorulic acid and other hydroxamate type siderophores. In: "Biological Oxidation of Nitrogen". 429-436. (J.W. Gorrod, Ed.) Elsevier/North Holland Biomedical Press.

Emery, T., 1966. Initial steps in the biosynthesis of ferrichrome. Incorporation of δ-N-hydroxyornithine and δ-N-acetyle-δ-N-hydroxyornithine. Biochemistry, 5:3694-3701.

Hummel, W., and Diekman, H., 1981. Preliminary characterization of ferrichrome synthetase from *Aspergillus quadricinctus*. Biochim. Biophys. Acta, 657:313-320.

Ong, D.E., and Emery, T.F., 1972. Ferrichrome biosynthesis: enzyme catalyzed formation of the hydroxamic acid group. Arch. Biochem. Biophys., 148: 77-83.

PRODUCTION OF FUSARININE AND IRON ASSIMILATION BY PATHOGENIC AND NON-PATHOGENIC *FUSARIUM*

P. Lemanceau[1], C. Alabouvette[2] and J.M. Meyer[3]

E.N.I.T.H., rue Le Nôtre, Angers, France[1]

I.N.R.A., 17 rue Sully, Dijon, France[2]

Université Louis Pasteur, 4 rue Blaise Pascal
Strasbourg, France[3]

INTRODUCTION

The soil from Chateaurenard (France) is naturally suppressive to fusarium-wilts (Louvet et al., 1976). Nutritional competition particularly for carbon between pathogenic and non-pathogenic *Fusarium*, seems to be an important mechanism for explaining the suppressiveness of this soil (Alabouvette et al., 1985). In contrast, *Fusarium* suppressiveness in the soil of Salinas Valley (California) is linked to the competition for iron associated with the presence of fluorescent *Pseudomonas* (Scher and Baker, 1982). Thus Kloepper et al., (1980) managed to render a fusarium-wilt conducive soil suppressive by addition of either a fluorescent *Pseudomonas* strain isolated from a suppressive soil or its siderophore. Finally, Vandenbergh et al., (1983) reported a decrease in the tomato plant mortality inoculated with both a *Fusarium oxysporum* f. sp. *lycopersici* and a *Pseudomonas putida* strain in comparison with control inoculated with just the *Fusarium* strain. The results presented in this workshop by Alabouvette show that competition for iron in the soil of Chateaurenard also seemed to play a part in the mechanism of suppressiveness of this soil.

Following the recognition of the importance of the population of non-pathogenic *Fusarium* spp. (Rouxel et al., 1979), and of their participation in competition for carbon sources we also studied the competition for iron between non-pathogenic and pathogenic *Fusarium* strains *in vitro* and *in vivo*.

Like most aerobic and facultatively aerobic micro-organisms *Fusarium* spp. produce siderophores when cultured on media containing low iron concentrations. These siderophores, called fusarinines, are a series of basic ninhydrin-positive hydroxamates (Neilands, 1973). Their *in vitro* biological activity can be characterised by their affinity for iron and by the relationship between the quantities produced and the iron content of the medium. The latter forms the subject of this paper.

MATERIALS AND METHODS

Microorganisms

The non-pathogenic *Fusarium* spp. were isolated from soil from Chateaurenard and were identified as strains of *F. oxysporum* (F.o.20, F.o.47) and *F. solani* (F.s.2, F.s.3). Pathogenic strains were isolated from plants with wilt and included *F. oxysporum* f.sp. *lini* (F.o.ln.3) *lycopersici* (F.o.l.8) and *meloni* (F.o.m.15). The bacteria, *Pseudomonas fluorescens* (CFBP 2392, identified as biovar 2) was isolated from the rhizosphere of *Phaseolus vulgaris* and selected for intensity of siderophore production, antagonism *in vitro* and *in vivo* to pathogenic fungi and PGPR effects on *P. vulgaris* (Lemanceau and Samson, 1983).

Media and growth conditions

Two media were used, one described by Meyer and Abdallah (1978) (Succinic acid, $(NH_4)_2SO_4$, K_2HPO_4,KH_2PO_4, $MgSO_4.{}_7H_2O$) and one described by Scher and Baker (1982) (Sucrose, L-asparagine K_2HPO_4, $MgSO_4.{}_7H_2O$). Traces of iron were removed from the standard media by complexing with 8-hydroxyquinolin (Waring and Werkman, 1942) followed by chloroform extraction for the iron complex. In some experiments, the sucrose medium was supplemented with $FeCl_3$.

Cultures were grown, 48 hours for the bacteria and 5 days for the fungi, at 25°C in Erlenmeyer flasks subject to mechanical agitation.

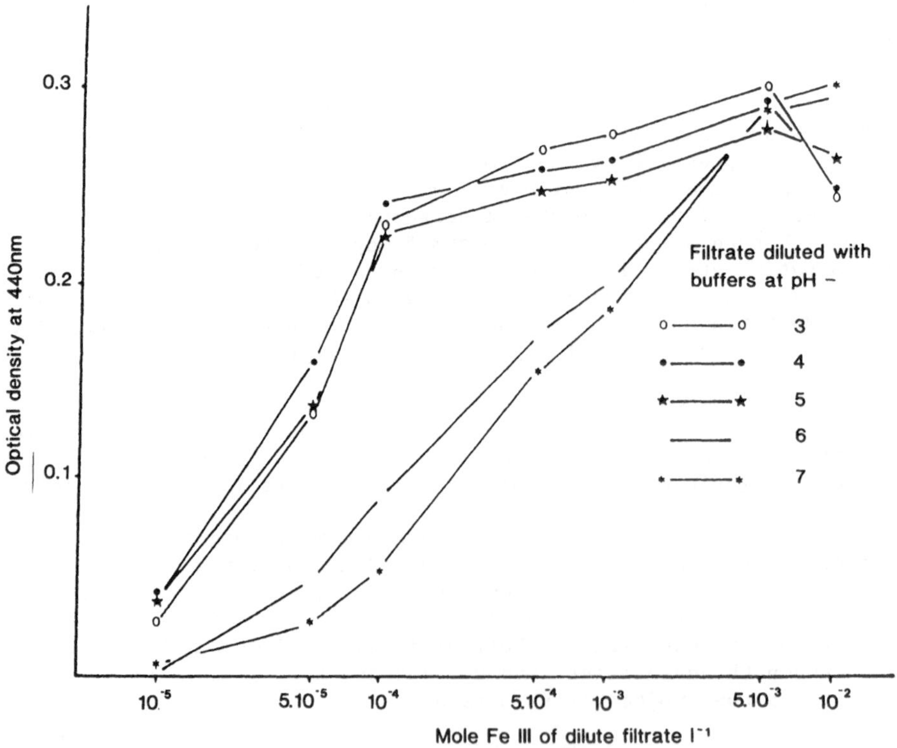

Fig. 1. Absorbance of a culture filtrate of *Fusarium* (strain F.o.l.8, growth medium of Scher and Baker) after addition of increasing amounts of FeIII.

Measurements of microorganisms growth and siderophore synthesis

Growth was estimated turbimetrically at 600 nm for the bacteria and either by the absorbance at this wavelength after having been ground in a Waring blender or by the weight of dry matter for the fungi.

Bacteria and siderophore were removed by centrifugation and filtration through polycarbonate membrane filters (0.2 µm and 0.45 µm respectively). As already described (Meyer and Abdallah, 1978), the *Pseudomonas* siderophore production was assessed by measuring the absorbance of the filtrate (adjusted at pH 7) at 400 nm.

Siderophore production could only be assessed after addition of FeIII which induced the formation of a fusarinine-FeIII complex. The presence of this complex was indicated by the absorbance of the filtrate at 440 nm (Emery, 1965), under precise conditions determined in preliminary experiments using strain F.o.1.8. In Fig. 1, the absorbance at 440 nm increased sharply from 10^{-5} to 10^{-4} Mol. As detailed in Figs. 1 and 2 FeIII/1 of diluted filtrate was optimal at 5.10^{-3} Mol FeIII/1. The addition of $FeCl_3$ to filtrate diluted with pH 6 or 7 buffers caused a precipitation which was removed by filtration. From a consideration of the effect of pH and the interval of time between addition of $FeCl_3$ and measurement of absorbance, the standard procedure adopted was to adjust the culture filtrate to pH 3 (citrate, chlorhydric acid) and to 5.1×10^{-3} M Fe^{+++} and to measure absorbance at 440 nm 60 min after addition of the iron.

Tests for suppressive properties of non-pathogenic *Fusarium* spp

Chateaurenard soil was rendered conducive by steaming (Alabouvette et al., 1985). The pathogenic strain F.o.ln. 3 was introduced to the soil either

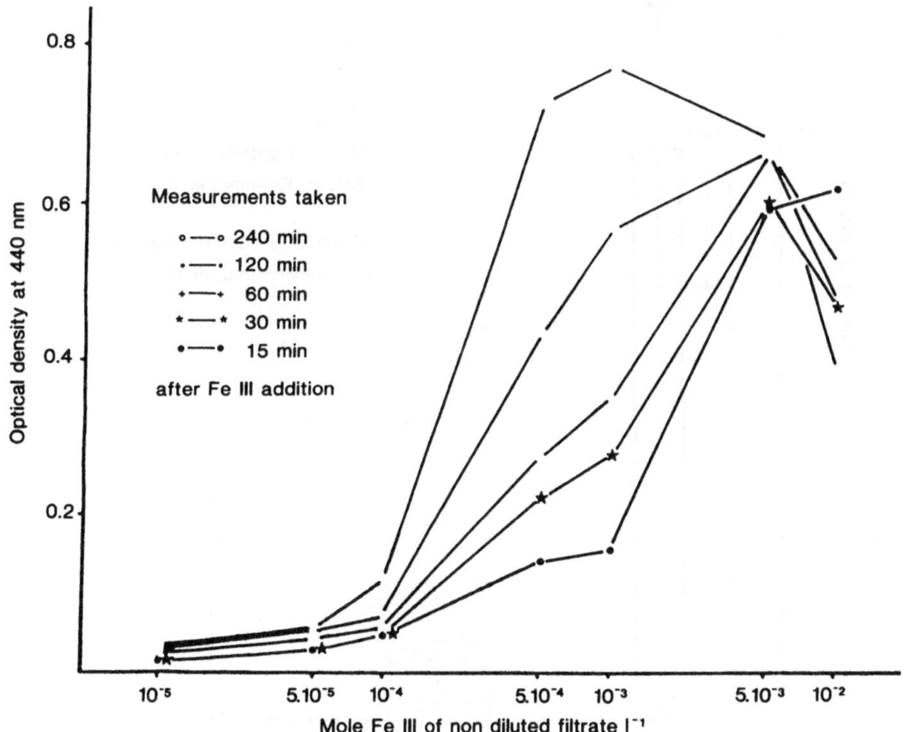

Fig. 2. Absorbance of a culture filtrate of *Fusarium* (strain F.o.1.8, growth medium of Meyer and Abdallah) after addition of increasing amounts of FeIII.

alone or paired with one of the non-pathogenic strains F.o.20, F.o.47, F.s.2 and F.s.3. The percentage of healthy flax plants surviving was recorded between 4 and 9 weeks following inoculation.

RESULTS

Effect of growth conditions on fusarinine production

Following work already carried out on siderophore production (Meyer and Abdallah, 1978; Scher and Baker, 1982), we compared two growth media. The results expressed in Fig. 3 show that the specific production of fusarinine (ratio of siderophore synthesis (O.D. at 440 nm) to growth (O.D. at 600 nm)), of the strain F.o.1.8 was a lot higher in the medium of Scher and Baker. Indeed for a similar biomass, the siderophore production was greater when grown in this medium. With both growth media, the difference of specific production of fusarinine in 250 ml and 500 ml Erlenmeyer flasks was not noticeable. For all further experiments, the *Fusarium* was then grown in 500 ml Erlenmeyer flasks containing 50 ml of the Scher and Baker growth medium.

The *Fusarium* strain tested reacted in a different way from *Pseudomonas* to iron content of growth medium.

From iron as low as 10^{-2} mg FeIII/l, the fusarinine synthesis decreased while the fungus growth was still very low (Fig. 4). The addition of 10^{-2} mg FeIII/l completely repressed this production and induced a growth increase. The absorbance at 600 nm of the growth suspension continued

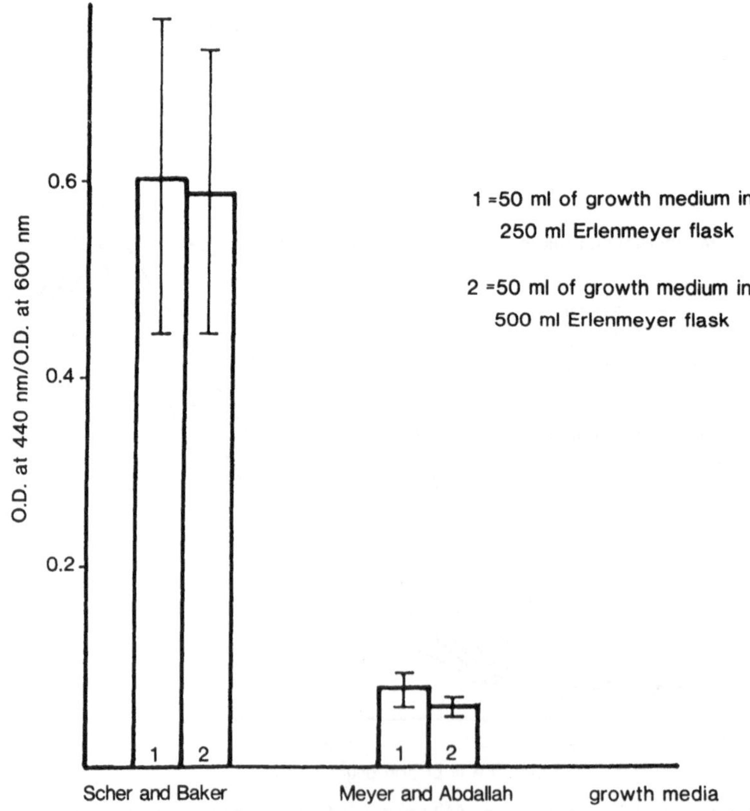

Fig. 3. Effect of two growth media on the siderophore production by *Fusarium* (strain F.o.1.8).

increasing with the iron content of the medium.

The results obtained with the *Pseudomonas* strain CFBP 2392 grown in the Scher and Baker medium were similar to those of Meyer and Abdallah (1978) (Fig. 5). At the lowest FeIII contents (10^{-3} and 10^{-2} mg FeIII/l), the limited growth of the bacteria enabled only a low pyoverdine production. At 10^{-2} mg FeIII/l, the bacterial growth increased noticeably, any siderophore production reached its maximum level. For higher FeIII concentrations, the siderophore production decreased, then was repressed while the bacterial growth continued increasing to finally enter its stationary phase.

Fusarinine production by several *Fusarium* strains

Significant differences occurred in fusarinine production between the strains (Fig. 6). The highest siderophore synthesis was obtained with the pathogenic *Fusarium oxysporum formae speciales meloni* (F.o.m.15) and *lycopersici* (F.o.l.8). But the pathogenic strain of flax (F.o.ln.3) produced a lot less siderophore than the previous ones. Among the pathogenic *F. oxysporum* tested, the level of this production was then not homogenous. The level of siderophore synthesis of the two non-pathogenic *F. oxysporum* strains was lower than the one of F.o.m.15 and F.o.l.8 but higher than the one of F.o.ln.3. The lowest production occurred with the two non-pathogenic *F. solani* strains (F.s.2, F.s.3).

The Fig. 7 shows that the four non-pathogenic strains (F.o.20, F.o.47, F.s.2, F.s.3) tested decreased the incidence of fusarium-wilt caused by the pathogenic strain (F.o.ln.3). But the effectiveness of the *Fusarium oxysporum* strains was much higher than the *Fusarium solani* strains.

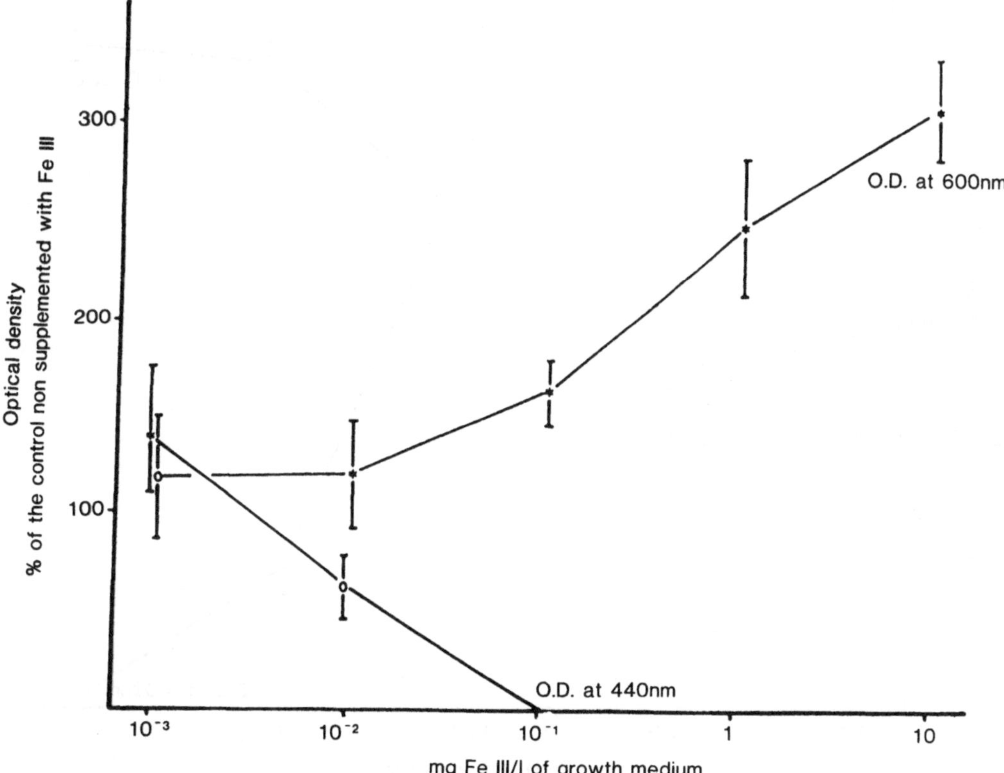

Fig. 4. Effect of growth medium FeIII concentration on the growth and siderophore production of *Fusarium* (strain F.o.l.8, medium of Scher and Baker).

The non-pathogenic strains (F.o.20, F.o.47) which re-established the suppressiveness of the steamed suppressive soil were then those which synthetised more fusarinine than the pathogenic one (F.o.ln.3).

DISCUSSION, CONCLUSION

Some preliminary experiments have enabled us to specify the best conditions to assess fusarinine production. Whilst in our conditions, the maximum absorption spectrum peak wavelength (440 nm), of the growth filtrate after addition of iron occurred when diluted with pH 3 buffer, Emery (1965) described that, for a pH of about this level, protons compete with iron for the hydroxamates group and 2:1 chelate is favoured instead of a 3:1 chelate at higher pH values. This modification is indicated by a shift of the absorption spectrum to higher wavelength and a decrease in absorbance.

Contrary to *Pseudomonas fluorescens* for which only the growth iron medium content regulates pyoverdine synthesis (Meyer and Abdallah, 1978), the intensity of the fusarinine production seems to be linked with the composition and the iron content of the growth medium. Thus, for a similar growth, the fusarinine synthesis was much higher with the medium of Scher and Baker than with that of Meyer and Abdallah. This difference could be explained by the carbon source (sucrose, succinic acid) and/or nitrogen source (L-asparagine, $(NH_4)_2SO_4$). Moreover, we have noticed, during our experiments, that for a similar biomass the amount of siderophore synthetised varied a lot with the oxygenation level of the Meyer and Abdallah medium. This observation might suggest that *Fusarium* could grow and then

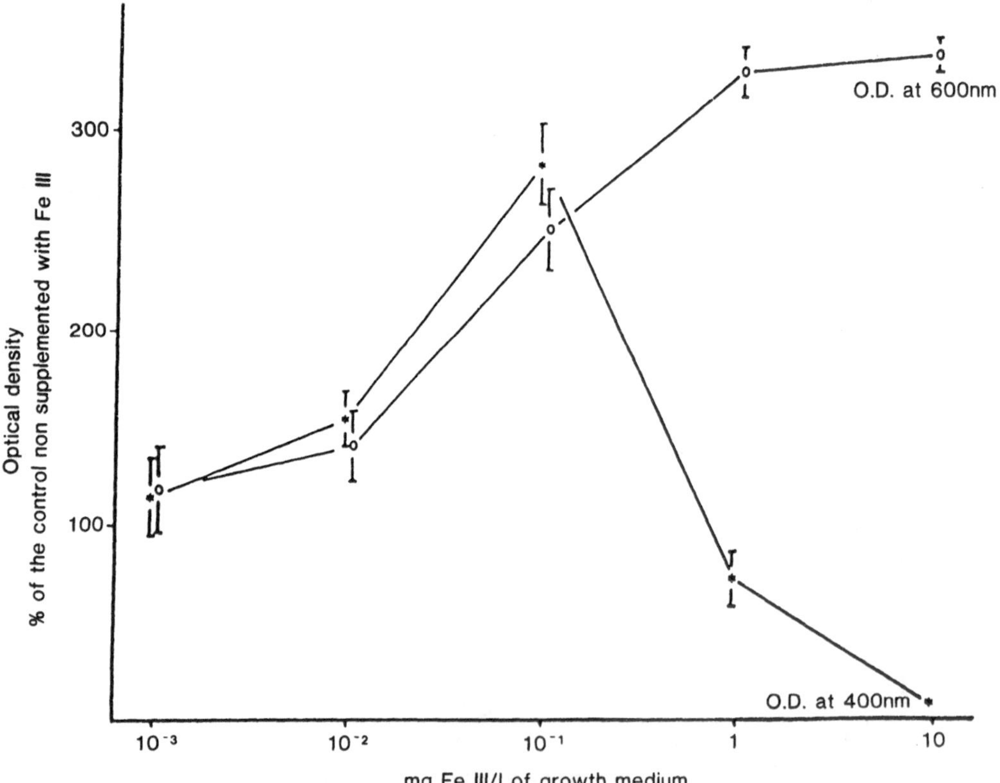

Fig. 5. Effect of growth medium FeIII concentration on the growth siderophore production of *Pseudomonas fluorescens* (strain CFBP 2392, growth medium of Scher and Baker).

use iron from the medium by some other mechanism than fusarinine. This was not true with Scher and Baker medium.

As already described by Emery (1965) and Neilands (1973), our results show the great sensitivity of the fusarinine synthesis to the growth medium iron content. The level of the latter which repressed the siderophore synthesis was 10 mg l^{-1} for the *Pseudomonas* and only 10 mg l^{-1} for the *Fusarium*. The sensitivity of the, *in vitro*, fusarinine production to these different factors (composition, oxygenation, iron content of the medium) leads us to wonder about the intensity of this synthesis, *in vivo*.

However, among the four non-pathogenic *Fusarium* strains (F.o.20, F.o.47, F.s.2, F.S.3) tested in biological control, the most effective of these synthetised more fusarinine, *in vitro*, than the pathogenic strain (F.o.ln.3) and than the two other non-pathogenic strains (F.s.2, F.s.3). As expressed in the introduction, the suppressiveness to fusarium-wilt of the Chateaurenard soil is explained by the hypothesis of nutritional competition between pathogenic and non-pathogenic *Fusarium*. Our first results would then suggest that the non-pathogenic strains effective in biological control, synthetize more siderophore and are then better equipped for the iron competition than the pathogenic strain with which they are confronted.

The results obtained are in accordance with the hypothesis that non-pathogenic *Fusarium* spp. are able to suppress pathogenic *Fusarium* spp. in the wilt suppressive soils of Chateaurenard through their ability to compete more effectively for iron. However, further work is required on a wider range of isolates and with a better understanding of the structures of fusarinines and their affinity for iron before definite conclusions can be drawn.

ACKNOWLEDGEMENTS

We wish to thank Dr. R. Samson, I.N.R.A. Station de Phytobacteriologie Angers, in whose laboratory all the work on fusarinine was performed, for her invaluable advice and discussion. We also gratefully acknowledge the interest shown and the help given by Dr. L.M. Riviere, E.N.I.T.H. department de Sciences du Sol, Angers.

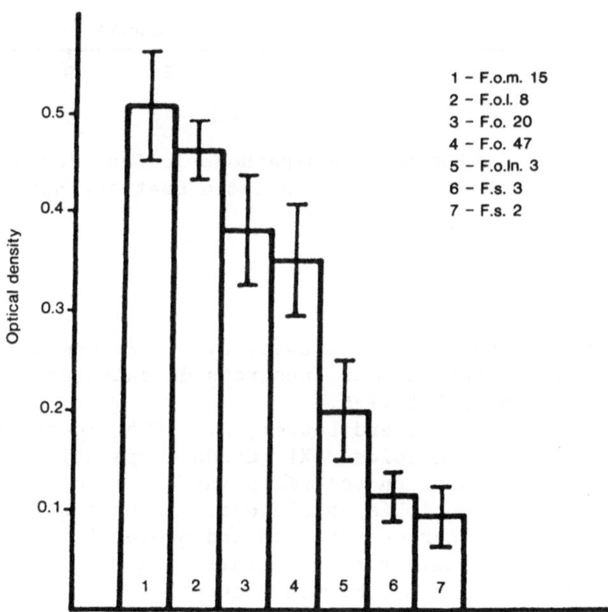

Fig. 6. Siderophore production, *in vitro* by the pathogenic and non-pathogenic *Fusarium* spp.

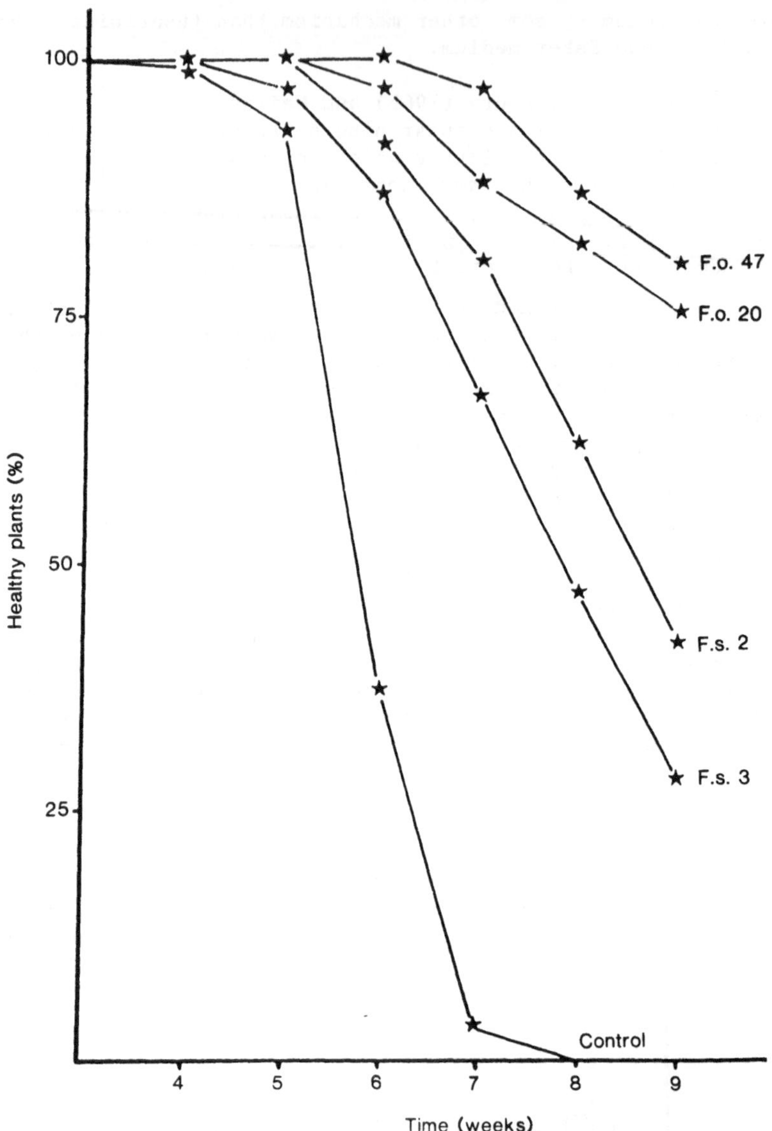

Fig. 7. Effect of inoculation with non-pathogenic *Fusarium* strains on fusarium-wilt incidence in the steamed Chateaurenard soil.

REFERENCES

Alabouvette, C., Couteaudier, Y., and Louvet. J., 1982. Comparaison de la réceptivité de differents sols et substrats de cultures aux fusarioses vasculaires. Agronomie, 2(1):1-6.

Alabouvette, C., Couteaudier, Y., and Louvet, J., 1985. Recherches sur la résistance des sols aux maladies. XI. Etude comparative du comportement des *Fusarium* spp. dans un sol résistant et un sol sensible aux fusarioses vasculaires enrichis en glucose. Agronomie, 5(1):63-68.

Emery, T., 1965. Isolation, characterization and properties of fusarinine, a δ-hydroxamic acid derivative of ornithine. Biochemistry, 7:1410-1417.

Kloepper, J.W., Leong, J., Teintze, M., and Schroth, M.N., 1980. *Pseudomonas* siderophores: a mechanism explaining disease suppressive soils. Curr. Microbiol., 4:317-320.

Lemanceau, P., and Samson, R., 1983. Relation entre quelques caractèristiques *in vitro* de 10 *pseudomonas* fluorescents et leur effet sur la croissance du haricot (*Phaseolus vulgaris*). In: "Les Antagonismes Microbeins. 327-328. (Ed. INRA Publ.) 24ème colloque SFP, Bordeaux.
Louvet, J., Rouxel, F., and Alabouvette, C., 1976. Recherches sur la résistance des sols aux maladies. I. Mise en évidence de la nature microbiologique de la résistance d'un sol au developpement de la Fusariose vasculaire du Melon. Ann. Phytopathol., 3(2), 171-188.
Meyer, J.M., and Abdallah, M.A., 1978. The fluorescent pigment of *Pseudomonas fluorescens*: biosynthesis, purification and physicochemical properties. J. Gen. Microbiol., 107:319-328.
Neilands, J.B., 1973. Microbial iron transport compounds (siderochromes). In: "Inorganic Biochemistry". Vol. 1. 167-202. (G.L. Eichorn, Ed.) Elsevier, Amsterdam.
Rouxel, F., Alabouvette, C., and Louvet, J., 1979. Recherches sur la résistance des sol aux maladies. IV. Mise en évidence du rôle des *Fusarium* autochtones dans la résistance d'un sol à la Fusariose vasculaire du Melon. Ann. Phytopathol., 11(2):199-207.
Scher, F.M. and Baker, R., 1982. Effect of *Pseudomonas putida* and a synthetic iron chelator on induction of soil suppressiveness to *Fusarium* wilt pathogens. Phytopathology, 72:1567-1573.
Vandenbergh, P.A., Gonzalez, C.F., Wright, A.M., and Kunka, B.S., 1983. Iron-chelating compounds produced by soil Pseudomonads: correlation with fungal growth inhibition. Applied Environ. Microbiol., 42:737-739.
Waring, W.S., and Werkman, C.H., 1942. Growth of bacteria in an iron-free medium. Arch. Biochem. 1:303-310.

CELL SURFACE MUTANTS OF *ERWINIA CHRYSANTHEMI*:

POSSIBLE INVOLVEMENT OF IRON ACQUISITION IN PHYTOPATHOGENICITY

D. Expert[1], E. Schoonejans[2] and A. Toussaint[2]

Laboratoire de Pathologie Végétale, INA-PG[1]
16 rue C. Bernard, 75231, Paris cedex 05, France

Laboratoire de Génétique, ULB, 67 rue des Chevaux[2]
1640 Rhode St Genèse, Belgium

INTRODUCTION

Compatible association between plant and phytopathogenic *Erwinia chrysanthemi* results in the development of soft rot symptoms in which bacterial depolymerizing activities are involved. Indeed it is now well established that pectate lyase isoenzymes, which degrade the pectic components of the plant cell walls play an essential role in plant tissue maceration and thus serve as virulence factors (Chatterjee and Starr, 1980; Collmer et al., 1982). In addition, *E. chrysanthemi* strains present a high degree of host specificity (Dickey, 1981). Infection of an incompatible host may also trigger a defense reaction in the plant which can correspond to the so-called hypersensitive response (Starr and Chatterjee, 1972; M. Lemattre, personal communication). These facts suggest the existence of a recognition event(s) which could depend on bacterial cell surface components.

During the last decade, a lot of work has focused on the basic mechanisms which lead to successful infection. For several other phytopathogenic bacteria, there is some evidence for a role for bacterial outer membrane components in the establishment of a compatible relation (Garrett et al., 1974; Sequeira and Graham, 1977; Whatley et al., 1978). For a better understanding of *E. chrysanthemi* virulence, we undertook a study of mutants altered in the outer membrane (O.M.) structure, isolated from the wild type strain 3937J which is pathogenic on *Saintpaulia ionantha*.

RESULTS

Spontaneous Bacteriocin-insensitive mutants and Phage-resistant mutants

Selection of the mutants. One of the classical procedures for selecting cell surface mutants is to use bacteriophages or bacteriocins which are known to bind specifically to O.M. components. Nine different bacteriocins (Table 1) isolated from several *Erwinia* species, a temperate bacteriophage of *E. chrysanthemi*- phage PhiEC-2- (Resibois et al., 1984) and the coliphage Mu (Toussaint and Schoonejans, 1982) were used for selection. About ten independant mutants obtained with each bacteriocin and with phage

Table 1. Characterization of the mutants based on their cross resistance patterns and sensitivity to active agents

Agent used for selection[a]	Designation of prototype mutant	Resistance to bacteriocins[b]	Resistance to PhiEC-2[c]	Resistance to Mu[c]	Sensitivity to active agents[d] DOC	SDS	EDTA
Bacteriocins							
1277	P1277	P	R	S	I	I	I
1455	P1455	P	R	S	S	I	I
1871	R1871	R	R	S	S	I	I
3912	R3912	R	R	S	S	I	I
1277	R1277	R	R	S	S	I	S
1455	R1455	R	R	S	S	I	S
1500	R1500	R	R	S	S	I	S
1521	R1521	R	R	S	S	I	S
1884	R1884	R	R	S	S	I	S
1456	R1456	R	R	R	S	S	I
20D3	R20D3	T	R	S	I	I	I
Phages							
PhiEC-2	1	S	R	S	I	I	I
	3	I	R	S	S	I	I
	18	P	R	S	I	I	I
	19	I	R	I	I	I	I
	20	I	R	I	I	I	I
	15	I	R	R	S	S	I
Mu	2.3a	S	S	I	I	I	I

(a) Designation adopted for bacteriocins refer to the name of the strains producing them. (b) Mutants were tested for their cross-resistance to all bacteriocins. R. Complete resistance to all tested bacteriocins; P. complete resistance to only some bacteriocins and reduced sensivivity to the others; T. tolerance i.e. alteration in a step subsequent to binding to receptor (for certain bacteriocins): I. insensitivity (resistance and tolerance not determined). (c) S. Sensitive; R. resistant and unable to adsorb phage particles. (d) DOC, 0.1%; EDTA, 0.1%; SDS, 0.1%; I, insensitive; S, sensitive. The efficiency of plating (e.o.p) of sensitive mutants was $<10^{-3}$ for DOC, $<10^{-6}$ for EDTA and $<10^{-7}$ for SDS.

PhiEC-2, and one Mu insensitive isolate were further analyzed.

Characterization of the mutants. Most of the mutants were affected in the receptor function for the agent used for selection. All mutants selected with PhiEC-2 failed to adsorb phage particles (>95% unadsorbed p.f.u.) and mutants selected with bacteriocins were unable to adsorb the bacteriocin used for selection. Evidence was obtained that the Mu insensitive derivative was affected at a stage subsequent to binding to the receptor but early in the phage development.

Most of the mutants were shown to be insensitive to infectious agents other than the one used for selection, (Table 1). Roughly, all mutants fall into the following classes based on their cross-resistance pattern: (i) mutants both resistant to bacteriocins (to all or several of them or tolerant i.e. affected in a step subsequent to binding to receptor) and to phage PhiEc-2, (ii) mutants only resistant to phage PhiEc-2, (iii) mutants both resistant to all bacteriocins and to phages PhiEC-2 and Mu and finally, (iv) one derivative only insensitive to phage Mu.

A great part of the mutants were shown to be sensitive to cell surface active agents (Table 1). Indeed, mutation to bacteriocin or phage resistance sometimes resulted in an increased sensitivity to detergents or chelating agents (Davies and Reeves, 1975). Most of the mutants selected with bacteriocins showed increased sensitivity to sodium deoxycholate (DOC) and in some cases to EDTA, whereas mutants selected with PhiEC-2 remained in most cases DOC insensitive and in all cases EDTA insensitive. The two mutants sensitive to SDS (and also DOC sensitive) were those previously characterized as resistant to all the agents. None of the mutants exhibited increased sensitivity to Triton X100.

Table 2. Pathogenicity of the bacteriocin resistant mutants on Saintpaulia plants.

Inoculated Strain[a]	No. of Tests [b]	Total No. of inoculated plants	No. of susceptible responses
3937J	2	6	6
R1277	2	6	3
R1455	2	6	0
R1456	2	6	2
R1500	2	6	1
R1521	2	6	0
R1871	1	3	1
R1884	2	6	0
R3912	2	6	1
R20D3	2	6	1
P1277	1	3	1
P1455	2	6	1
control	2	6	0

(a) Designations of mutants strains correspond to those given in Table 1
(b) Tests were performed by infiltrating the leaf parenchyna of Saintpaulia plants with a bacterial suspension (100 µl) in 0.15 M NaCl, containing about 3.10^8 cfu/ml. Pathogenicity was scored 4 to 8 weeks after inoculation. Non-pathogenic mutants were shown to trigger a hypersensitive response.

Pathogenicity

When tested by the cup plate technique (detection of enzyme activities) and by electrofocusing (detection of isoenzymes) (Bertheau et al., 1984) all mutants appeared to produce normal amounts of extracellular polygalacturonate lyases and endoglucanase. All mutants also retained the ability to macerate leaves isolated from Saintpaulia plants.

On the other hand, when tested on living plants, the mutants selected with bacteriocins and the Mu insensitive isolate appeared defective in the ability to induce soft rot symptoms; they caused a hypersensitivity reaction (HR) resulting in the appearance of a brown patch at the inoculation site. This patch, due to leaf tissue maceration, causes the death of the infected area (Table 2). Furthermore, a preliminary experiment done with some of these mutants indicated that they can protect the plant against further infection with wild-type strain. Pathogenicity of the mutants selected as resistant to phage PhiEC-2 is under current analysis. One preliminary test showed that all but two of them were able to induce soft rot symptoms (data not shown). The two non-pathogenic derivatives (R3 and R15 as noted in Tables 1 and 3) failed to trigger HR as described above: 4-5 days after inoculation only, a yellow patch could be detected at the inoculation site and eventually chlorosis of the infiltrated area.

Outer Membrane analysis

O.M. proteins. Since O.M. proteins often function as bacteriocin or phage receptors (Inouye, 1979), mutation to bacteriocin or phage resistance might be expected to alter the O.M. proteins pattern of the mutants. Therefore, profiles of O.M. proteins from parental strain and resistant mutants were compared by SDS-PAGE. When strains were grown under standard conditions, in L-broth or minimal medium M63 with glucose or glycerol as carbon source, no significant difference could be detected. Thus the mutations responsible for bacteriocin or phage resistance might affect a minor protein not easily detectable under the conditions used. One possibility was that the mutation affected iron regulated proteins, analogous to some colicin or phage resistant mutants of *Escherichia coli* K12 (Neilands, 1982). Therefore, we looked for additional O.M. proteins induced under iron starvation in both parental and mutant strains. When strain 3937J was grown in iron depleted medium, three bands corresponding to polypeptides with apparent molecular weights of 78, 82 and 88 kD appeared as major bands, whereas under high iron conditions, such polypeptides were either minor bands or could not be detected (Fig.1). Most of the bacteriocin resistant mutations caused the complete or partial disappearance of the 88 kD polypeptide, both the 88 and 82 kD polypeptides or all three of them. Mutants selected with bacteriocin from strain 20D3 produced large amounts of the three polypeptides even under high iron conditions (Fig. 1, Table 3). The Mu insensitive derivative also lacked the 82 and 88 kD polypeptides. The changes in O.M. proteins profile were not correlated with bacteriocin-resistance patterns.

On the other hand, all the resistant mutants selected with PhiEC-2 retained the three low iron inducible proteins and one of them appeared to express these proteins constitutively (analogous to R20D3: Fig. 1, Table 3).

Lipopolysaccharide (LPS). Since no significant alteration could be detected in the O.M. proteins profile of most of the mutants selected as resistant to PhiEC-2, the possibility remained that these mutants were affected in the LPS (which also often function as phage receptor (Braun, 1981).This was supported by the fact that at least one of them is Mu

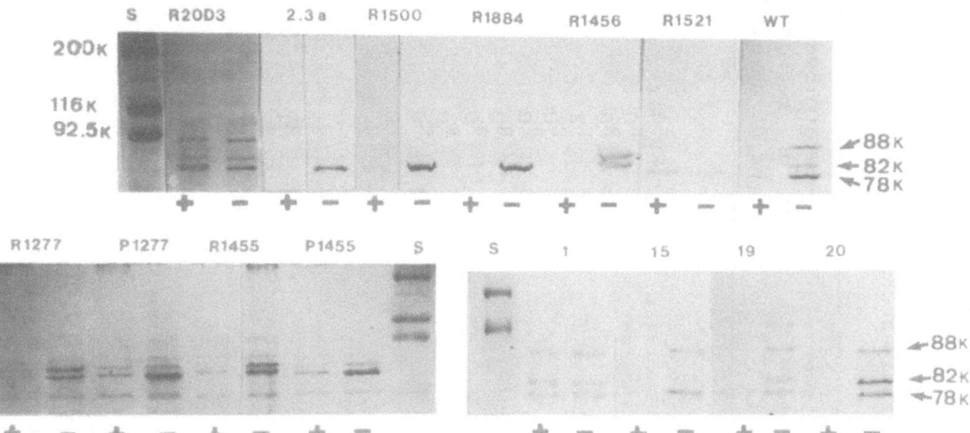

Fig. 1. Protein composition of the Triton insoluble cell envelope fraction (O.M.) prepared from wild type strain (W.T.) and, bacteriocin-resistant and PhiEC-2 resistant mutants (strains are labeled above each lane). SDS-PAGE in 11% acrylamide was carried out on 20 to 40 µg samples of O.M. proteins prepared from 5 ml cultures in M63. Only proteins (revealed by Coomassie blue staining) with high molecular weight are shown. The apparent molecular sizes of standard proteins (lanes labeled S) are indicated in kilodaltons. The numbers with an arrow refer to the molecular sizes of proteins detected as major bands under low iron conditions.

resistant, this phage being known to bind to the LPS molecule (Kamp and Sandulache, 1983). This possibility was tested by comparing LPS profiles from parental strain and PhiEC-2 resistant mutants by SDS-PAGE followed by silver staining (Fig. 2) or PAS (specific for carbohydrates) staining.

In the wild type strain, the LPS profile shows two banding regions: a few (2 to 4) lower-molecular-weight bands and a higher-molecular-weight heterogeneous region containing many bands. The same profile was seen after PAS staining (data not shown). A great part of the higher-molecular-weight bands are separated from one another by a small and apparently constant interval. This heterogeneous region, which possibly reflects the presence of O-antigenic side chains contains also in its higher-molecular-weight part, a continuous stained region indicating a high level of heterogeneity. This pattern presents similarities to that obtained for the LPSs from *E. coli* and *Salmonella typhimurium* smooth strains (Goldman and Leive, 1980; Hitchcock and Brown, 1983; Pavla and Makela, 1980) but more closely resembles in its complexity that of *Rhizobium* species (Carlson, 1984).

Most of the PhiEC-2 resistant mutants lack at least the upper heterogeneous region and have alterations in the lower-molecular-weight bands but without change to their electrophoretic mobilities (Fig. 2). The two mutants resistant to all infectious agents (R1456 and R15) and the PhiEC-2 resistant mutant R3 have additional changes in the lower-molecular-weight bands. Mu receptor activity has been shown to be associated with the core part of LPS (Sandulache et al., 1985) and it is probable the Mu resistant mutants have alterations to core LPS. Such changes in LPS pattern are reminiscent of the different chemotypes identified in rough mutants of *Salmonella* species.

Since a large proportion of mutants selected as bacteriocin-

Table 3. Outer membrane alterations of the mutants

Agent used for selection[a]	Designation of prototype mutant[a]	Low iron inducible proteins[b]: Size of proteins still present[d] (kD)			Lipopolysaccharide[c]: Presence of higher mol. wt. region[e] (O-antigen ?)	Lipopolysaccharide[c]: Type of lower mol. wt. region[e] (Type of core?)
Bacteriocins						
1277	R1277	78	82		–	R_0
1455	R1455	78	82		–	R_0
1871	R1871	78	82		–	R_0
1456	R1456	78	82		–	R_2
1277	P1277	78			–	R_0
1455	P1455	78			–	R_0
1500	R1500	78			–	R_0
1884	R1884	78			–	R_0
3912	R3912	78			–	R_0
1521	R1521				–	R_0
20D3	R20D3	78	82	88	–	R_0
Phages						
PhiEC-2	15	78	82	88	–	R_3
	3	78	82	88	–	R_1
	18	78	82	88	–	R_0
	19	78	82	88	–	R_0
	20	78	82	88	–	R_0
	1	78	82	88	+	W.T.
Mu	2.3a	78			+	W.T.
Wild Type Phenotype		78	82	88	+	W.T.

(a) see Table 1. (b) Low iron inducible proteins were detected as described in Fig. 1. (c) LPS was detected as described in Fig. 2. (d) Sizes in kilodaltons (kD). For mutants R20D3 and PhiEC-2^R 1, production of the three proteins was also high under iron conditions. (e) Higher- and lower-molecule-weight regions as described in the text.

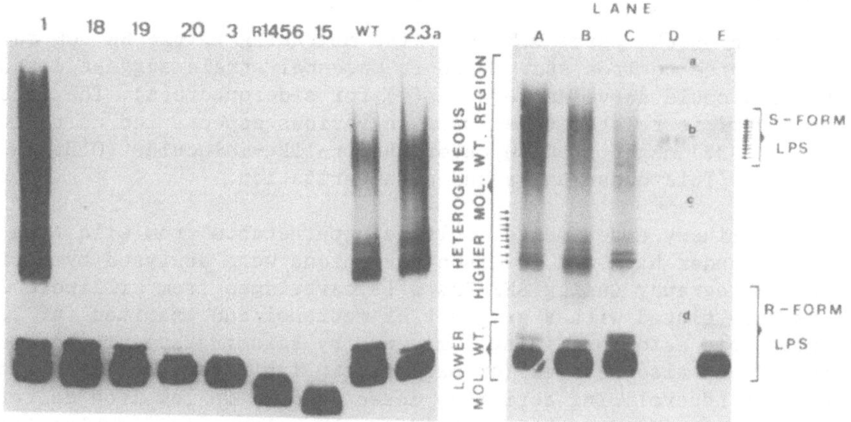

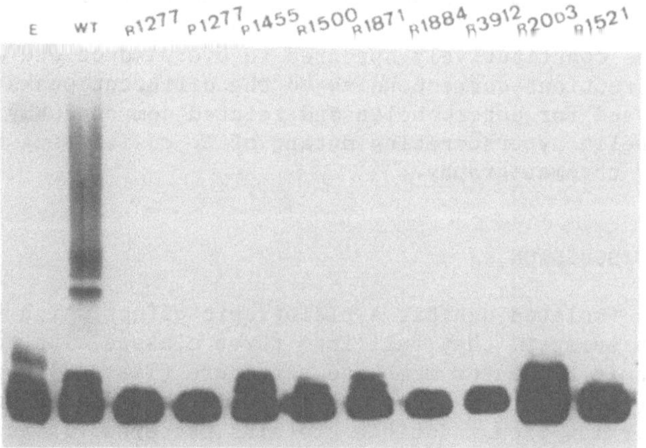

Fig. 2. Analysis of LPS from wild type strain (W.T.) and mutant derivatives (strains are labeled above each lane). SDS-PAGE (13.5% acrylamide) was carried out on 5 μg of LPS (Triton insoluble cell envelope fraction, digested with pronase). The gels were stained by the silver method as described by Hitchcock and Brown (1983). The same patterns appeared after periodic acid-Schiff staining (data not shown). No material could be stained with Coomassie blue. Wild type LPS extracts purified according to the procedure of Westphal and Jann (1965) were analyzed before (lane A) and after (lane B) digestion with lysozyme and were compared to the Triton insoluble mini-scale extraction (lane C). Arrows indicate bands separated from one another by a small and constant interval corresponding to the smooth-type LPS of *Escherichia coli* 026:B6 (lane E). The apparent molecular sizes of standard proteins (lane D) are: 92.5 kD (a), 45 kD (b), 31kD (c) and 14.4 kD (d).

resistant appeared resistant to phage PhiEC-2, it seemed possible that these mutants were also altered in the LPS. All the mutants selected with bacteriocins appeared to lack at least the higher-molecular-weight heterogeneous region (Fig. 2).

Search for Siderophores in the wild type strain 3937J

Induction of the three O.M. proteins migrating in the 80,000 dalton range in response to iron starvation in parental strain suggested that such proteins could serve as receptor(s) for siderophore(s). The fact that *E. chrysanthemi* is related to *E. coli* in various aspects led us to suppose that strain 3937J might produce enterochelin-like-molecules (O'Brien and Gibson, 1970). This question is under investigation.

In preliminary experiments, culture supernatants from wild type strain grown under high and low iron conditions were analysed by affinity column chromatography (using SEP-PAK C 18 cartridges from Millipore Co.). Fractions were eluted with a gradient of methanol and examined for 2,3-dihydroxybenzoic acid containing compounds by scanning spectrophotometry (300-350 nm) and also assayed for bacteriocin inhibitory activity; indeed, a bacteriocin neutralizing activity, detected only in low iron culture supernatants, was found to be retained to the column. The results (Fig. 3) indicate that strain 3937J grown under low iron conditions secrete several hydrophobic compounds, able to neutralize the bacteriocin activity and not detected under high iron conditions. The mutant R20D3 which expresses the low iron proteins constitutively appeared to overproduce one of these compounds. The fractions corresponding to the different peaks detected will be further analysed for enterochelin and related compounds by comparison with an enterochelin hypersecreting mutant of *E. coli*, using two-dimensional ultra-thin layer chromatography.

CONCLUSION AND DISCUSSION

The mutants isolated exhibit a pleiotropic effect which remains to be explained. In summary, they fall into three classes with respect to the detected changes in the outer membrane structure (Table 3) and their pathogenicity: (i) Mutants selected as bacteriocin-insensitive are all affected in their low iron inducible proteins patterns. In addition, all these mutants are altered in the LPS, lacking the O-antigen-like-structure. One of them (R1456) is also affected in the core part of the molecule. They are non-pathogenic and trigger the hypersensitivity reaction. (ii) Mutants selected as phage PhiEC-2 resistant appeared to be only affected in the LPS, lacking also the O-antigenic side chain; two of them (R3 and R15) are, in addition, affected in the core part. They are not significantly changed with regard to the low iron inducible proteins. They were able to elicit soft rot symptoms on Saintpaulia plants, except the latter showing a rough phenotype: these two mutants became non-pathogenic and unable to trigger the HR induced by mutants affected in iron proteins. (iii) Two mutants remained apparently unaffected in the LPS but are affected in the low iron inducible proteins; one PhiEC-2 resistant derivative is derepressed for these proteins and is pathogenic whereas the Mu insensitive isolate lacks the two 82-88 kD proteins and induces HR.

None of the mutants appeared to be affected in their pectinolytic and cellulolytic activities.

These results represent the first evidence for a role of certain O.M. components in the expression of phytopathogenicity in *E. chrysanthemi*. We do not yet have genetic evidence that changes in outer membrane structure, and absence of phytopathogenicity are caused by a single mutation, but this seems highly likely in view of the large number of independent mutants tested. The results obtained so far do not yet allow a definitive interpretation concerning the role of each of the O.M. components i.e. low iron inducible proteins and LPS, in pathogenicity. However, our data show that induction of HR is correlated with the disappearance of one to three

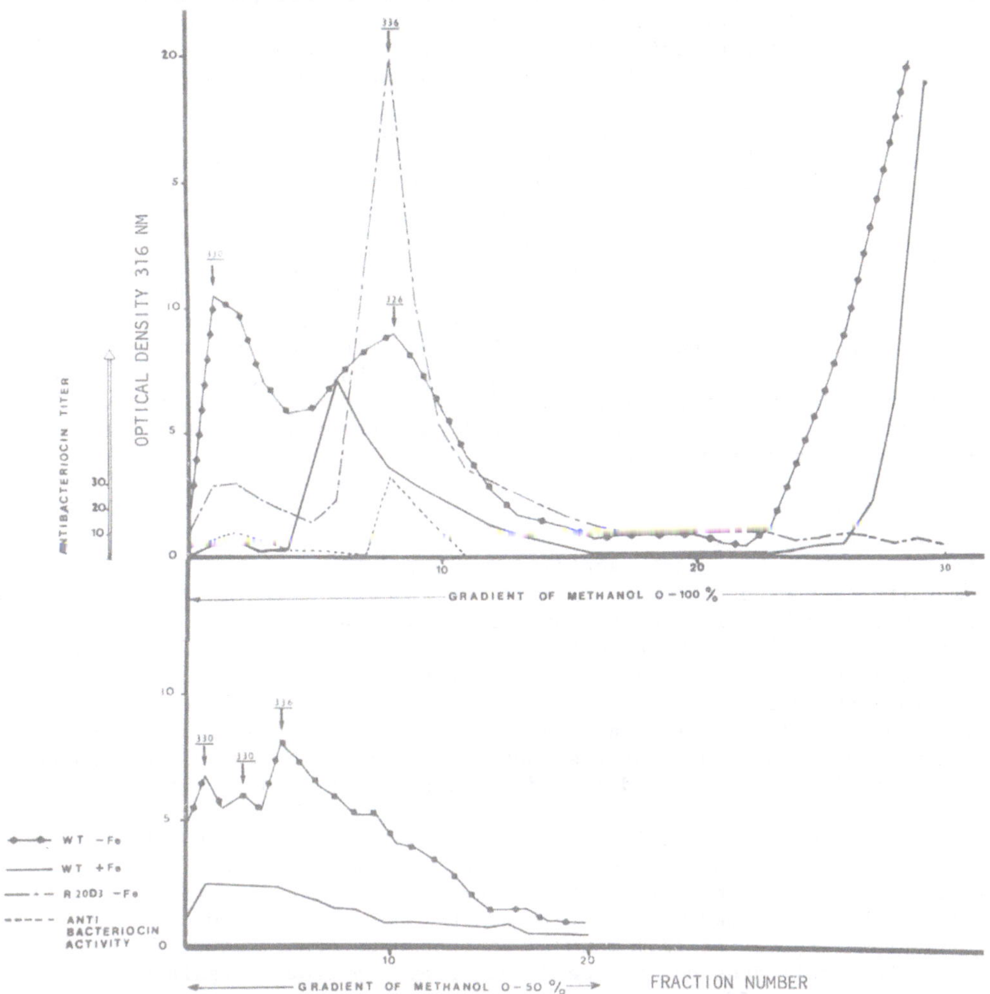

Fig. 3. Fractionation of culture supernatants of wild type strain grown under high iron (solid line) and low iron (solid line with closed squares) conditions by affinity column chromatography. Fractions were eluted with a 0-100% methanol gradient (upper graph, 30 fractions) or with a 0-50% methanol gradient (lower graph, 20 fractions). They were assayed for the presence of 2,3 dihydroxybenzoic acid compounds by measuring absorbance at 316 nm and for antibacteriocin activity by scoring the last dilution at which inhibition of the bacteriocin zone could be detected (dotted line). Fractions corresponding to each peak were scanned between 300 and 350 nm: the wavelength of maximum absorption (max) is noted with an arrow above these fractions. The elution profile of the low iron culture supernatant of the mutant R20D3 (constitutive for the iron proteins) is shown (upper graph, broken line).

of these low iron inducible proteins. The most plausable interpretation of the role these proteins play is that the low iron inducible O.M. proteins are receptors for iron siderophores and, as such, are essential for development of the bacteria *in planta*. Therefore, the development of an efficient iron capture system might be considered as a possible virulence factor. In this respect, any mutant severely affected in iron transport might become non pathogenic. Such an assumption could account for the following observations: (i) a correlation exists between the loss of pathogenicity and the loss of low iron regulated O.M. proteins expected to be involved in iron uptake. Additional evidence suggesting that such protein(s) might serve as receptor(s) for possible siderophore(s) includes the fact that growth of mutants, but not of the parental strain, was strongly reduced in iron depleted medium (data not shown). Furthermore, the present study shows that the wild type strain grown in iron depleted medium produces several hydrophobic compounds able to neutralize the bacteriocin activity, which could not be detected under high iron culture conditions. (ii) The loss of pathogenicity in mutants derepressed for all three proteins might result from a mutation affecting iron transport at one of the translocation steps of a ferric complex subsequent to O.M. binding. This possibility can be supported by the fact that such mutants, rather than receptor mutants, were shown to be tolerant for several bacteriocins. One possibility would be that these mutants carry a mutation in a locus analogous to *tonB* of *E. coli* K-12 (Pugsley and Reeves, 1976).

Such an interpretation implies that infecting bacteria would be under iron restricted conditions *in planta,* as was shown for animal host infection (Weinberg, 1978). Thus, pathogenicity would involve interaction between the ability of the host to deny iron to the invading pathogen and the ability of the infecting bacteria to recover this vital ion. The role of iron in the infection of plants by bacterial pathogens remains unknown, but, Leong and Neilands (1982) have looked for production of siderophores in various phytopathogenic bacteria. They showed that biosynthesis of the siderophore agrobactin by *A. tumefaciens* was associated with the production of several low iron-induced envelope proteins, but no correlation between siderophore activity and infectivity could be made (Leong and Neilands, 1981).

Finally the loss of ability to induce HR as detected with mutants altered in iron proteins can be correlated with the presence of LPS severely affected in the core part of the molecule. These data are reminiscent to those obtained with certain LPS mutants of *Pseudomonas solanacearum* (Whatley et al., 1980). It should be mentioned that in our system, LPS itself does not induce HR but seems to protect against HR induced by mutants affected in iron proteins and against pathogenic response (data not shown).

REFERENCES

Bertheau, Y., Madjidi-Hervan, E., Kotoujansky, A., Nguyen-The, C., Andro, T., and Coleno, A. 1984. Detection of depolymerase isoenzymes after electrophoresis or electrofocusing, or in titration curves. Anal. Biochem., 139:383-389.

Braun, V., and Handke, K., 1981. Bacterial Cell Surface Receptors. In: "Organization of Prokaryotes Cell Membranes", Vol.II. (B.K. Ghosh, Ed.) CRC Press, Boca Raton.

Carlson, R.W., 1984. Heterogeneity of *Rhizobium* lipopolysaccharides. J. Bacteriol., 158:1012-1017.

Chatterjee, A.K., and Starr, M.P., 1980. Genetics of *Erwinia* species. Annu. Rev. Microbiol., 34:645-676.

Collmer, A.P., Berman, P., and Mount, M.S., 1982. Pectate lyase regulation and bacterial soft-rot pathogenesis. In: "Phytopathogenic Prokaryotes" Vol.I. (M. Mount and G. Lacy, Eds.) Academic Press, New York.

Davies, J.K., and Reeves, P., 1975. Genetics of resistance to colicins in *Escherichia coli* K-12 : cross resistance among colisins of group A. J. Bacteriol., 123:102-117.

Dickey, R.S., 1981. *Erwinia chrysanthemi*: reaction of eight plant species to strains from several hosts and to strains of other *Erwinia* species. Phytopathology, 71:23-29.

Garrett, C.M.E., Crosse, J.E., and Slettern, A., 1974. Relations between phage sensitivity and virulence in *Pseudomonas morsprunorum*. J. Gen. Microbiol., 80:457-483.

Goldman, R.C. and Leive, L., 1980. Heterogeneity of antigenic-side-chain length in lipopolysaccharide from *Escherichia coli* O111 and *Salmonella typhimumurium* LT2. Eur. J. Biochem., 107:145-153.

Hitchcock, P.J., and Brown, T., 1983. Morphological heterogeneity among *Salmonella* lipopolysaccharide chemotypes in silver-stained polyacrylamide gels. J. Bacteriol., 154:269-277.

Inouye, M., 1979. "Bacterial Outer Membranes". Wiley-Interscience Publication, New York.

Kamp, D., and Sandulache, R., 1983. Recognition of cell surface receptors is controlled by invertible DNA of phage Mu. FEMS Microbial. Lett., 16:131-135.

Leong, S.A., and Neilands, J.B., 1981. Relationship of siderophore mediated iron assimilation to virulence in crown gall disease. J. Bacteriol., 147:482-491.

Leong, S.A., and Neilands, J.B., 1982. Siderophore production by phytopathogenic microbial species. Arch. Biochem. Biophys., 218:351-359.

Neilands, J.B., 1982. Microbial envelope proteins related to iron. Annu. Rev. Microbiol., 36:285-309.

O'Brien, I.G., and Gibson, F., 1970. The structure of enterochelin and related 2,3-dihydroxy-N-benzoylserine conjugates from *Escherichia coli*. Biochim. Biophys. Acta, 215:393-402.

Pavla, T., and Makela, P.H., 1980. Lipopolysaccharide heterogeneity in *Salmonella typhimurium* analyzed by sodium dodecyl polyacrylamide gel electrophoresis. Eur. J. Biochem., 107:137-143.

Pugsley, A.P., and Reeves, P., 1976. Characterization of group B colicin resistant mutants of *Escherichia coli* K-12: colicin resistance and the role of enterochelin. J. Bacteriol., 127:218-228.

Resibois, A., Colet, M., Faelen, M., Schoonejans, E., and Toussaint, A., 1984. PhiEC2, a new generalized transducing phage of *Erwinia chrysanthemi*. Virology, 137:102-112.

Sandulache, R., Prehms, P, and Kamp, D., 1985. Cell wall receptor for bacteriophage Mu G(-) in *Erwinia chrysanthemi* and *E. coli* C. J. Bacteriol. (in press).

Sequeira, L., and Graham, T.L., 1977. Interaction of bacteria and host cell walls: its relation to mechanisms of induced resistance. Physiol. Plant Pathol., 11:43-54.

Starr, M.P., and Chatterjee, A.K., 1972. The genus *Erwinia*: enterobacteria pathogenic to plants and animals. Annu. Rev. Microbiol., 26: 389-426.

Toussaint, A., and Schoonejans, E., 1982. Production and modification of Mu (G-) phage particles in *E. coli* K-12 and *Erwinia*. Genet. Res. 41: 145-154.

Weinberg, E., 1978. Iron and infection. Microbiol. Rev. 42:45-66.

Westphal, O., and Jann, K., 1985. Bacterial lipopolysaccharide. Extraction with phenol-water and further application of the procedure. In: "Methods in Carbohydrate Chemistry", Vol.V, (R.L. Whistler, Ed.) Academic Press, New York.

Whatley, M.H., Hunter, N., Cantrell, M.A., Hendrick, C., Keegstra, K., and Sequeira, L., 1980. Lipopolysaccharide composition of the wilt pathogen *Pseudomonas solanacearum*. Plant Physiol., 65:557-559.

Whatley, M.H., Margot, J.B., Schell, J., Lippincott, B.B., and Lippincott, J.A., 1978. Plasmid and chromosomal determination of *Agrobacterium* adherence specificity. J. Gen. Microbiol., 107:393-396.

IRON AND PHYTOTOXINS AS EXEMPLIFIED BY STEMPHYLOXINS AND OTHER TOXINS

Isaac Barash and Shulamit Manulis

Department of Botany
The George S. Wise Faculty of Life Science
Tel Aviv University, Tel Aviv, Israel

INTRODUCTION

Microbial phytopathogens are known to produce, in culture, many toxins of diverse structures that affect plant cells. When such phytotoxins are involved in pathogenicity they become chemical determinants of disease (Schefer, 1983). Phytotoxins can be classified as either host-selective or non-selective (Yuder, 1980). The host-selective toxins exhibit exclusive and extreme toxicity to hosts of the producing pathogen and in most cases act as primary disease determinants (Scheffer, 1983; Yoder, 1980). The vast majority of the phytotoxins which have been identified in cultures of microbial pathogens are non-selective and in all cases which were investigated appear to act as secondary disease determinants (Rudolph, 1976; Scheffer, 1983; Stoessl, 1981; Yoder, 1980). Most of the non-selective toxins have not yet been adequately demonstrated to play a role in disease development (Scheffer, 1983). This could result, in part, from difficulties in obtaining suitable genetic systems through which unequivocal proof for their involvement in disease can be obtained (Yoder, 1980). Until their function in the infection process is substantiated they may be considered as potential chemical determinants of disease.

A few non-selective toxins, such as stemphyloxins, marasmins, naphthazarins, fusaric acid and ascochitine, exhibit chelating properties. Iron may either increase or decrease the activity of these toxins. Iron may also affect the biosynthesis of certain toxins, e.g. stemphyloxins and syringomycin. The present discussion will deal with the interaction of iron with these toxins.

STEMPHYLOXINS

Stemphylium botryosum Wallr. f. sp. *lycopersici* is the causal agent of leaf spot and foliage blight disease of tomato. Culture filtrates of this fungal pathogen contain two novel phytotoxic compounds designated stemphyloxin I and stemphyloxin II (Fig. 1), which are capable of producing the symptoms associated with this disease (Barash et al., 1982; Barash et. al., 1983; Manulis et al., 1984). Stemphyloxin I has been crystallized and its structure was fully determined by x-ray diffraction analysis (Barash et al., 1983). Stemphyloxin I is a highly functionalized β-ketoaldehyde trans decalone derivative. Stemphyloxin II is closely related to

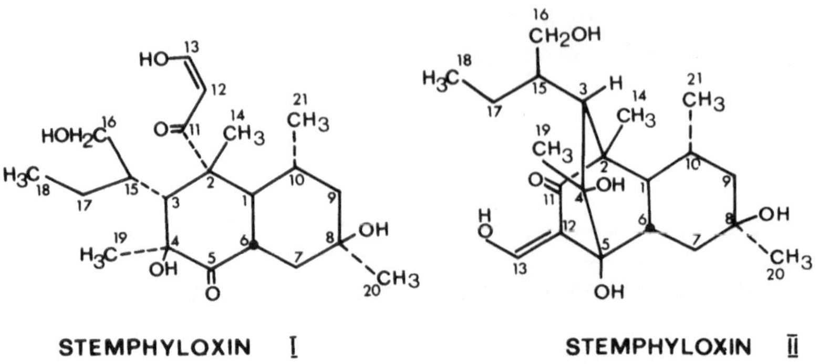

STEMPHYLOXIN I

STEMPHYLOXIN II

Fig. 1. Structural formula of stemphyloxins

stemphyloxin I and can be readily obtained from the latter by mild catalysis (Manulis et al., 1984). The two major changes in the structure of stemphyloxin II as compared to I are the disappearance of the saturated ketone at C-5 and the alteration in the position of the β-ketoaldehyde group. These changes are best explained by an aldol type condensation of the active methylene at C-12 with the carbonyl at C-5 leading to a tricyclododecane system as observed in stemphyloxin II. The phytotoxicity of stemphyloxin I is at least 200 times greater than that of stemphyloxin II, when measured by either symptom development on tomato leaves or inhibition of ^{14}C-amino acid incorporation into proteins of exponentially growing tomato cell suspension (Manulis et al., 1984). Stemphyloxins are nonselective toxins but show differential toxicity towards various plant species, tomato being the most sensitive (Barash et al., 1982).

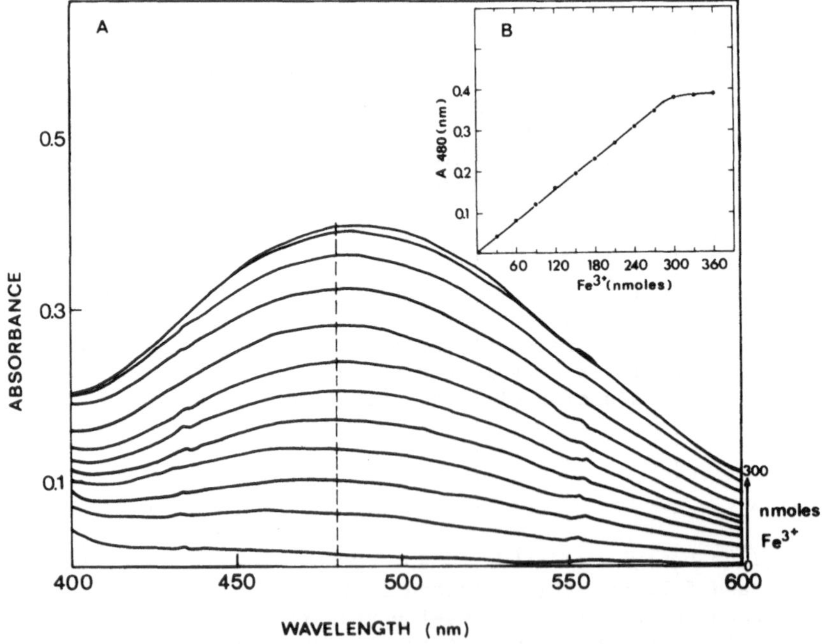

Fig. 2. Determination of the stoichiometry of the Fe (III)-stemphyloxin I complex. (A) changes in the absorption spectrum of stemphyloxin I as a function of the amount of ferric ion added; (B) increase in absorption at 480 nm (derived from Fig 2A) as a function of added Fe^{3+}. (Reproduced by permission from Manulis et al., 1984).

Stemphyloxins appear to act as chelates of ferric ion (Manulis et. al., 1984). Addition of aqueous solution of ferric ion to stemphyloxin I produces a reddish complex with a maximal absorption at 480 nm (Fig. 2A). The absorption spectrum of Fe (III)-stemphyloxin II complex has a maximum at 510 nm. Stemphyloxins I and II show high affinity for ferric but not ferrous ion. The stoichiometry of Fe (III)-stemphyloxin complexes for both toxins is 1 : 3 (Fig. 2B). The apparent stability constants for the Fe (III)-stemphyloxin complexes of stemphyloxin I and II at pH 7 were found to be 1.7×10^{24} and 1.6×10^{24} respectively (Manulis et. al., 1984). The reaction of stemphyloxins with ferric ion has been used for their quantitative estimation by spectrophotometry (Barash et. al., 1983.

Production of stemphyloxins in culture begins early and quickly becomes proportional to the growth rate of the fungus (Barash et. al., 1983). From various cations which were examined for an effect on toxin production, only iron appeared to induce stemphyloxins accumulation by approximately 5 fold. The iron effect reached an optimum at 2 mg/l but decreased significantly in the presence of higher iron concentration (Fig. 3B).

Production of non-toxic iron-chelating pigments, which were also present in cultures of *S. botryosum,* was regulated by iron in a similar manner to stemphyloxins (Fig. 3B). In contrast to the inducing effect of

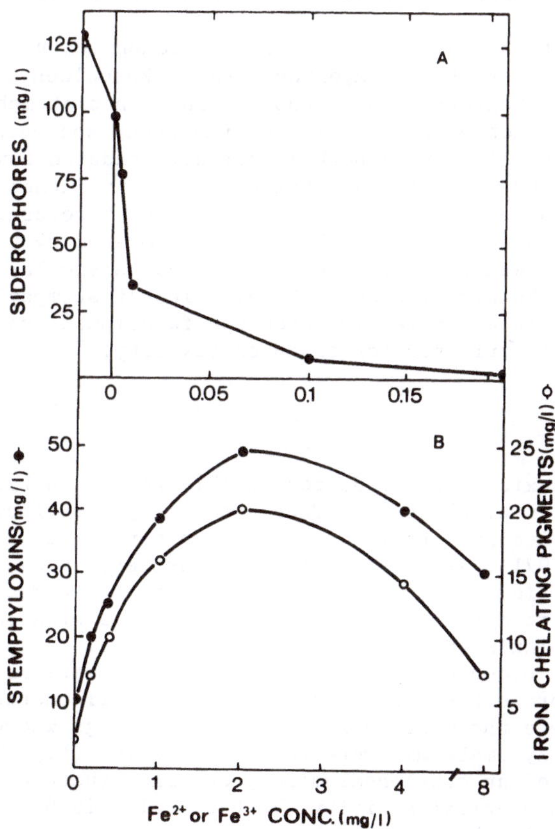

Fig. 3. Secretion of different iron chelating agents by *S. botryosum* as a function of iron concentration. The culture was grown on a sucrose-glutamate synthetic medium, according to Barash et. al., (1982) with various iron concentrations. Iron concentration below zero refers to medium in which traces of iron were removed by 8-hydroxyquinoline.

iron on stemphyloxin production, the presence of 0.1 mg/l iron in the medium completely repressed the secretion of three hydroxamate siderophores (Fig. 3A), which have been identified in cultures of *S. botrysum* (Manulis, unpublished results).

The role of hydroxamate siderophores in iron acquisition by fungi has been reviewed by Nielands (1981). Stemphyloxins share some properties with siderophores, namely, regulation by iron and preferential binding of ferric rather than ferrous ion. However, they differ from siderophore compounds in their dependence on low iron concentration for optimal biosynthesis and distinctive lower affinity for ferric ion ($K_2 = 10^{24}$ as compared to $K_2 = 10^{30-34}$ of hydroxamate siderophores). Recently, Manulis and Barash (1985) studied uptake of iron by *S. botryosum* under non-limiting conditions. They found that both Fe (II) and Fe (III) were readily adsorbed on the mycelial surface. Ferric ion was rapidly reduced, mainly by reducing compounds and it appeared that transport of iron into the cell was predominantly, if not exclusively, in the ferrous form. Stemphyloxin I induced iron adsorption on the mycelial surface by approximately 2 fold but did not seem to exert a significant direct effect on iron transport itself. It seems therefore, that stemphyloxins are part of a mechanism for solubilization and acquisition of iron under conditions of low iron concentration. In contrast, the high affinity transport of ferric ion mediated by siderophores occurred without marked adsorption on the mycelial surface and appears to be a system which functions only under conditions of extreme iron deficiency.

The mechanism of the phytotoxic action of stemphyloxin I is not yet known. Two features appear to be important, the β-ketoaldehyde group and the presence of iron. Conversion of stemphyloxin I into stemphyloxin II which alters the position of the β-ketoaldehyde group and methylation of this group (Barash et. al., 1982) both drastically reduced toxicity. Iron appeared to increase the toxicity of stemphyloxin I in intact plants (Manulis, unpublished results). Obstruction of iron therefore, does not seem to be a factor in toxicity and this is corroborated by the finding that stemphyloxin II, which is much less toxic, has a similar binding constant. Possibly, the association of 3 molecules of stemphyloxin I with a ferric ion results in a complex in which the functional groups of the toxin are particularly oriented for enhanced toxicity.

MARASMINS

Marasmins are toxic imino acids responsible for disturbance of water balance and wilting in plant leaves. They are widely distributed among fungi and have been isolated from *Fusaria* (Hardegger et. al., 1963; Trouvelot et. al., 1971), *Aspergillus flavus* (Barbier, 1972), *Colletotrichum gloeosporioides* (Ballio et. al., 1969) and *Pyrenophora teres* (Bach et. al., 1979). Marasmins exert differential toxicity towards a wide range of mono- and dicotyledonous plants (Ballio, 1981; Gaumann, 1951; Kern, 1972). Gaumann was first to propose that the toxic action of marasmins is related to their chelation properties (Gaumann, 1951). Aspergillomarasmin B and two synthetic analogs were the most toxic to tomato cuttings whereas aspergillomarasmin A and lycomarasmin were the least active (Fig. 4). (Ballio, 1981). The presence of an ethylenediamine group in structure appeared to be significant for biological activity (Bach et. al., 1979; Barbier, 1984). Wilt symptoms could also be produced by ethylenediamine tetra acetic acid but anhydroderivatives of marasmins were inactive (Fig. 4F). (Barbier, 1984).

The activity of marasmins is considerably enhanced by the presence of ferric ions (Ballio et. al., 1969; Gaumann, 1951). Gaumann et. al. 1955

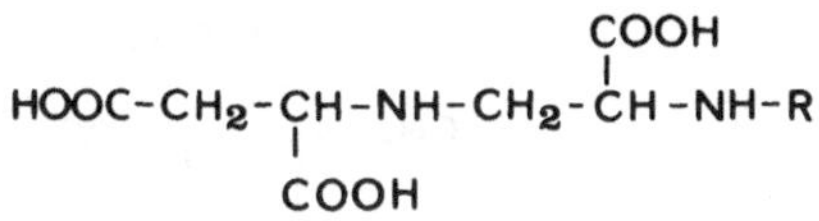

$$HOOC{-}CH_2{-}\underset{\displaystyle COOH}{CH}{-}NH{-}CH_2{-}\overset{\displaystyle COOH}{CH}{-}NH{-}R$$

a. $R = CH_2\,CH(NH_2)\,COOH$

b. $R = CH_2\,COOH$

c. $R = CH_2\,CONH_2$

d. $R = H$

e. $R = CH\,(COOH)\,CH_2\,COOH$

f.

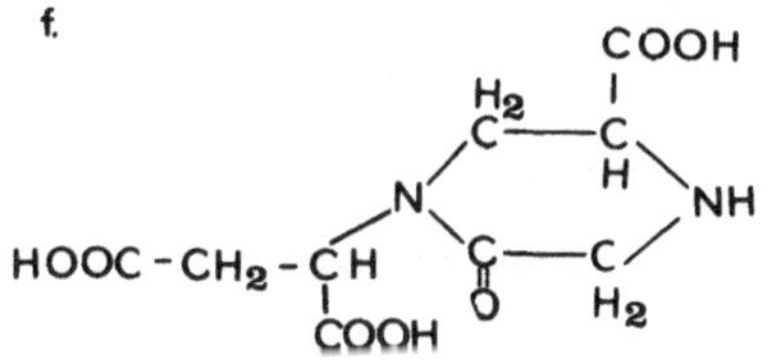

Fig. 4. Some natural and synthetic analogs of marasmins. (a) Aspergillomarasmin A. (b) Aspergillomarasmin B. (c) Lycomarasmin. (d) Synthetic analog I. (e) Synthetic analog II. (f) Anhydroaspergillomarasmin B.

have shown that after application of lycomarasmin to tomato cuttings it forms a water soluble, unstable chelate with iron, which is translocated into the leaves. During accumulation of the chelate in the leaves, iron is set free and causes symptoms typical of those produced by infection with *Fusarium*. Furthermore, Gaumann also observed that the biological effect of lycomarasmin was dependent on the photophase, darkness having a protective role (Gaumann, 1951). Gaumann and his associates (Gaumann, 1951; Kern, 1972) suggested that the modification of the osmotic pressure by iron and other metal chelates played a decisive role in wilt development through the disruption of leaf cell permeability.

Barbier (1984) has recently demonstrated that marasmin-Fe^{3+} chelates are capable of catalyzing photooxidations in plant leaves. Thus, the ferrous ion which is formed from ferric compounds by photoreduction, reduces H_2O_2 and generates the highly reactive hydroxyl radical through Fenton's reaction (Fridovich, 1978). The latter radical is responsible for subsequent oxidation of various substances in the plant leaves. Although the modification of osmotic pressure described previously (Gaumann, 1951) cannot be excluded as a mechanism of toxicity, the existence of photooxidations through a bio-Fenton general reaction is more likely to be the determinant factor. The latter mechanism provides an adequate explanation for the enhanced photo-phase effect and the reduced toxicity caused by copper (Gaumann, 1951; Kern, 1972). This metal ion forms a stabel chelate with marasmins and thus prevents the formation of the iron chelate.

OTHER TOXINS

The naphthazarins are red pigments which exhibit phytotoxic, antibacterial, antifungal and insecticidal properties (Kern, 1972). Six different naphthazarins have been isolated from phytopathogenic strains of *Fusarium*. The isomeric compounds marticin and isomarticin (Fig. 5) are structurally distinguished from the other compounds by an additional C_3 residue to the 5, 8 dihydroxynaphthoquinone (Stoessl, 1981). These two

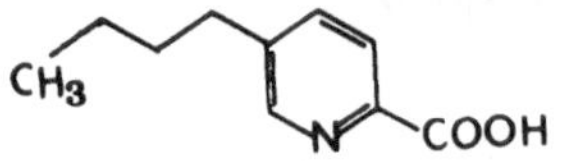

FUSARIC ACID

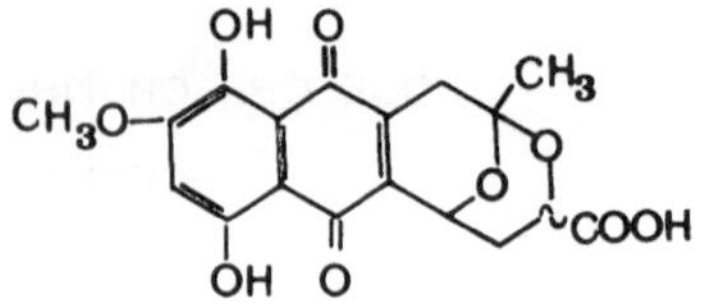

MARTICIN OR ISOMARTICIN

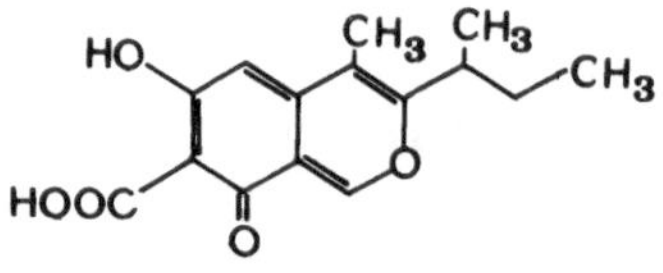

ASCOCHITINE

Fig. 5. Structures of various toxins.

compounds are essentially phytotoxic, whereas the rest are mostly antimicrobial (Kern, 1972; Kern, 1978). The marticins inhibit glutamine synthetase more efficiently than do other naphthazarins (Ballio, 1981). On the other hand, the four remaining naphthazarins react with thiamine pyrophosphate and inhibit decarboxylation reactions in which this compound is a factor. The degree of inhibition is highly correlated with their antimicrobial activity.

The naphthazarins can readily chelate multivalent metals (Kern, 1978). In the presence of copper, aluminum or iron, the marticins lose their phytotoxicity (Kern, 1972). The degree of detoxification varies according to the metal added, copper being the most effective. Isomarticin is readily transported from stems to leaves, where it affects membrane function (Roos, 1977). It has been suggested by Kern (1972) that copper and other metals which chelate the toxin molecules may interfere with their transport through vessels and protoplasmic membranes.

Fusaric acid (5-n butylpicolinic acid, Fig. 5) is toxic to higher plants, interfering with membrane function in various ways and possessing antimicrobial activity (Kern, 1972). It is produced by several *Fusarium* species pathogenic on a wide host range and its production is accompanied by other closely related compounds such as dehydrofusaric acid (Kern, 1972). Toxicity of fusaric acid is linked to two different molecular features, the α-carboxyl group and the alkyl side chain. The severity of damage to plant cells increases with the size of the alkyl substituent from methyl to butyl (Ballio, 1981). However, the unsubstituted compound α-picolinic acid, produced by *Pyrucilaria oryzae* is also toxic to rice plants (Tamari and Kaji, 1954.

Fusaric acid chelates metals, the binding occurring with the N-atom of the pyridine nucleus and the carboxyl group (Tamari and Kaji, 1954). Of the heavy metals, the ferric ion is most strongly chelated followed by the cupric ion (Tamari et. al., 1963). The toxicity of fusaric acid and other α-carboxy pyridines is markedly reduced by chelation. A direct correlation between chelation efficiency and growth inhibition of rice seedlings was observed (Tamari and Kaji, 1954). Destruction of the capacity of these compounds to chelate by methylation of the N-atom or esterification of the carboxyl group results in detoxification (Braun, 1960). This also occurs if fusaric acid is saturated with heavy metal ions. Addition of iron, and to a lesser degree cobalt, copper, nickel, zinc and manganese, causes inactivation, especially at higher pH (Bar, 1963). Fusaric acid and related derivatives inhibit metal containing enzymes such as Fe-porphyrin oxidase or catalase (Tamari and Kaji, 1954).

Ascochitine (Fig. 5) has been isolated from *Ascochyta fabae* and *A. pisi* and possesses phytotoxic and antibiotic properties (Oku and Nakanishi, 1963). Its biological activity is antagonized by iron.

Syringomycin, a non-selective phytotoxin, which also exhibits wide spectrum antibiotic activity, is produced by *Pseudomonas syringae* pv. *syringae*. The toxin is considered essential for pathogenicity (Gross and DeVay, 1977). Production of the toxin *in vitro* is enhanced by the presence of iron (Hemming, 1982). The precise structure of syringomycin is not yet known but results obtained by Gross *et al.*, (1977) suggested that it contained an hexapeptide structure consisting of serine (2 mole), phenylalanine (1 mole) arginine (1 mole) and 2 unidentified basic amino acids. The latter were recently identified by Hemming (1982) as δ-N-hydroxyornithine. Interestingly, this amino acid is a major component of hydroxamate siderophores (Neilands, 1981). Hemming and Strobel (1983) hypothesized that syringomycin and a number of other N-hydroxylated amino acid containing peptides produced by fluorescent pseudomonads "are derived from or are precursors of strains or species specific siderophores that differ in the peptide portion of the molecules". Such an assumption may provide some clues for the understanding of the relationship of iron to the biosynthesis of syringomycin.

CONCLUSIONS

Very few toxins are so far known to interact with iron. Their biological activity may be related to iron by the following mechanisms: (a) sequestration of iron by chelation; (b) provision of iron for catalysis of secondary toxic effects and (c) alteration of the spatial structure of the toxin's molecule.

Chelation of iron as a significant factor in toxicity is best demonstrated by fusaric acid which is effective both *in vivo* and *in vitro*. However, it might be misleading to assign its toxic effect exclusively to the obstruction of iron. A similar situation exists with the herbicide glyphosate, N-(phosphonomethyl) glycine, which exhibits metal chelating properties and forms a strong chelate with iron (Nilsson, 1985). Its mode of action covers a complex of effects including inhibition of aromatic acid synthesis. Iron reduces the toxic effect of glyphosate (Nilsson, 1985). Nevertheless, the interaction of glyphosate with iron and other metals might not be connected to the central mechanism of its action, but rather related to other structural features of the molecule. (Nilsson, 1985).

The example of marasmins suggests that toxins may also serve as vehicles for translocation and excessive accumulation of free iron in the leaves resulting in iron toxicity. Iron in the ferrous form may then reduce hydrogen peroxide, as it does in the well known Fenton's reaction [Fe (II) + H_2O_2 > Fe (III) + OH^- + OH•] forming the very potent hydroxyl radical (Fridovich, 1978). The latter reacts avidly with many substances causing severe toxic effects.

Binding of iron may modify the three dimentional structure of a compound and consequently alters its biological activity. For example, the deferri form of hydroxamate siderophores such as ferrichrome, are not transported inside the cell across cellular membrane, whereas their ferri forms are (Neilands, 1981). Similarly, the binding of iron to stemphyloxin might bring about changes in the configuration of the molecule which render it more toxic.

Iron is an essential nutrient for most living cells and ferric ion is the predominant oxidation state prevailing in aerobic environments (Neilands, 1981). Owing to the profound insolubility of the ferric ion, microbes release specific chelating compounds to ensure its availability. The production of such chelates appears to be iron regulated. Some of these compounds, as for instance stemphyloxins or syringomycin, may have a pleiotropic effect of phytotoxicity.

REFERENCES

Bach, E., Christensen, S., Dalgaard, L., Larsen, P.o., Olsen, C.E., and Smedegard-Petersen, V., 1979. Structures, properties and relationship to the aspergillomarasmines of toxins produced by *Pyrenophora teres*. Physiol. Plant Path., 14:41-46.

Ballio, A., Bottalico, A., Buonocore, V., Carilli, A., DiVittorio, V., and Graniti, A., 1969. Production and isolation of aspergillomarasmin B (lycomarasmic acid) from cultures of *Colletotrichum gloeosporioides* Penz. Phytopathol. Mediterr., 8:187-196.

Ballio, A. 1981. Structure-activity relationships. In: "Toxins in Plant Disease," p. 395. (R.D. Durbin, Ed.), Academic Press, New York.

Bar, H. 1963. Untersuchungen uber die wirkungsweize der fusarinsaure. Phytopathol. Z., 48:173-177.

Barash, I., Pupkin, G., Netzer, D., and Kashman, Y., 1982. A novel enolic β-ketoaldehyde phytotoxin produced by *Stemphylium botryosum* f. sp. *lycopersici*. Plant Physiol., 69:23-27.

Barash, I., Manulis, S., Kashman, Y., Springer, J.P., Chen, M.H.M., Clardy, J., and Strobel, G.A., 1983. Crystallization and x-ray analysis of stemphyloxin I, a phytotoxin from *Stemphylium botryosum*. Science, 220:1065-1066.

Barbier, M., 1972. The chemistry of some amino-acid derived phytotoxins. In: "Phytotoxins in Plant Diseases," P.91. (R.K.S. Wood, A. Ballio and A. Graniti, Eds.), Academic Press, New York.

Barbier, M., 1984. Marasmin-Fe^{3+} chelates catelyzed photo-oxidations in plant leaves as a mechanism for their biological activity through bio-Fenton reactions. Photochem. Photobiophys., 7:53-57.

Braun, R., 1960. Uber Wirkungsweise und umurandlungen der fusarinsaure. Phytopathol. Z., 39:197-241.

Fridovich, I., 1978. The biology of oxygen radicals. Science, 201:875-880.

Gaumann, E., 1951. Some problems of pathological wilting plants. Adv. Enzymol., 11:401-437.

Gaumann, E., Naef-Roth, S., and Kern, H., (1955). Uber die chelierende Wirkung einiger Welke toxine III Die Verschiebungen der toxizitat durch steigende absaftingung mit Eisenionen. Phytopathol. Z. 24:373-374.

Gross,D.C., and DeVay, J.E., 1977. Production and purification of syringomycin, a phytotoxin produced by *Pseudomonas syringae*. Physiol. Plant Path., 11:13-28.

Gross, D.C., DeVay, J.E., and Stadtman, F.H., 1977. Chemical properties of syringomycin and syringotoxin: toxigenic peptides produced by *Pseudomonas syringae*. J. Appl. Bacteriol., 43:453-463.

Hardegger, E., Liechti, P., Jackman, L.M., Boller, A., and Plattner, Pl.A., 1963. Die konstitution des lycomarasmins. Helv. Chim. Acta, 46:60-74.

Hemming, B.C., 1982. Plant-associated fluorescent Pseudomonas: their systematic analysis, microbial antagonism and iron interaction. Ph.D. dissertation, Montana State University, Bozeman, Montana.

Hemming, B.C., and Strobel, G.A., 1983. Bacterial disease - the iron status of plants. Int. Congr. Plant Pathol., Yugoslavia.

Kern, H., 1972. Phytotoxins produced by Fusaria. In: "Phytotoxins in Plant Diseases," P. 35., (R.K.S. Wood, A. Ballio and A. Graniti, Eds.), Academic Press, New York.

Kern, H., 1978. Les Napthazarines des *Fusarium*. Ann. Phytopath., 10:327-345.

Manulis, S., Barash, I., Kashman, Y., and Netzer, D., 1984. Phytotoxins from *Stemphylium botryosum*: structural determination of stemphyloxin II, production in culture and interaction with iron. Phytochemistry, 23:2193-2198.

Manulis , S. and Barash, I., 1985. Characterization of a ferrous ion transport system in *Stemphylium botryosum* (in preparation).

Neilands, J.B., 1981. Iron absorption and transport in mocroorganisms. Annu. Rev. Nutr., 1:27-46.

Nilsson, G., 1985. Interaction between glyphosate and metals essential for plant growth. In: "The herbicide glyphosate," p. 35, (A. Grosbard, Ed.), Butterworths, London.

Oku, H., and Nakanishi, T., 1963. A toxic metabolite from *Ascochyta fabae* having antibiotic activity. Phytopathology, 53:1321-1324.

Roos, A., 1977. Zur physiologie und pathologie von *Neocosmospora vasinfecta* E. F. Smith. Phytopathol. Z., 88:238-271.

Rudolph, K., 1976. Non-specific toxins. In: "Physiological Plant Pathology" p. 270. (R. Heitefuss and P.H. Williams, Eds.), Springer-Verlag, Herlin and New York.

Scheffer, R.P., 1983. Toxins as chemical determinants of plant disease. In: "Toxins and Plant Pathogenesis," p.1, (J.M. Daly and B.J. Deverall Eds.), Academic Press, New York.

Stoessl, A., 1981. Structure and Biogenetic relations: Fungal non host-specific. In: "Toxins in Plant Disease," P.109, (R.D. Durbin, Ed.), Academic Press, New York.

Tamari, K., and Kaji, J., 1954. The mechanism of the growth inhibitory action of fusaric acid on plants. J. Biochem., 41:143-165.

Tamari, K., Ogaswara, N., and Kaji, J., 1963. In: "The Rice Blast Disease," pp. 63-68. The Johns Hopkins Press, Baltimore, Maryland.

Trouvelot, A., Camporota, P., Barbier, M., and Pouteau-Thouvenot, M., 1971. Isolement de l'aspergillomarasmin A du milieu de culture de *Fusarium oxysporum* f. sp. *melonis*. Comptes Rendus Hebdomadaires des S'eances de l'Academie des Sciences, Serie D: Sciences Naturelles, 272:754-756.

Yoder, O.C., 1980. Toxins in pathogenesis. Annu. Rev. Phytopath. 18: 103-129.

CHARACTERIZATION OF CELL ENVELOPE PROTEIN PATTERNS OF CROP YIELD INCREASING ROOT-COLONIZING *PSEUDOMONAS* SPP.

L. de Weger[1], R. van Boxtel[2], B. van der Burg[2]
R. Gruters[2], P. Geels[3], B. Schippers[3], B. Lugtenberg[1]

Department of Plant Molecular Biology[1]
State University of Liden, Botanical Laboratory,
Nonnensteeg 3, 2311 VJ Leiden, The Netherlands

Department of Molecular Cell Biology[2]
State University of Utrecht, Transitorium 3
Padualaan 8, 3584 CH Utrecht, The Netherlands

Phytopathological Laboratory 'Willie Commelin Scholten' [3]
Javalaan 20, 3742 CP Baarn, The Netherlands

INTRODUCTION

Yield reduction in high frequency cropping soil is often caused by unknown deleterious micro-organisms. Such yield depressions can be reduced by seed inoculation with fluorescent *Pseudomonas* spp., which display *in vitro* antagonism to other micro-organisms (Geels and Schippers, 1983). To explain the beneficial effect of these *Pseudomonas* spp. on crop yield a mechanism has been suggested in which, at the root-surface a competition for Fe^{3+} takes place between *Pseudomonas* spp. and deleterious micro-organisms (Kloepper et al., 1980). Under the limiting Fe^{3+} conditions in soil *Pseudomonas* spp. are thought to produce a high affinity Fe^{3+}-uptake system, consisting of (i) siderophores, which are Fe^{3+}-chelating agents, and (ii) outer membrane receptor proteins, which have a high affinity and specificity for the matching Fe^{3+}-siderophore complex (Neilands, 1982). These siderophores bind Fe^{3+}, thereby making this essential element unavailable to other micro-organisms, including deleterious micro-organisms. This results in a reduction of the number of deleterious micro-organisms, thus creating a more favourable environment for the plant.

As part of a study on the cell surface of these antagonistic *Pseudomonas* spp., a structure which can reasonably be expected to be involved in the interaction with the plant, we describe here the characterization of thirty of these fluorescent root-colonizing *Pseudomonas* spp. (Table 1) by analysis of their cell envelope protein pattern, using sodium dodecylsulphate (= SDS) polyacrylamide gel electrophoresis (Lugtenberg et al., 1975). The value of these cell envelope protein patterns for strain identification is discussed. As the presence of the high affinity Fe^{3+}-uptake system in these *Pseudomonas* spp. is thought to be very important for their plant growth stimulating ability, special attention was paid to cell envelope proteins regulated by Fe^{3+}-ions.

Table 1. Properties of the fluorescent root-colonizing *Pseudomonas* strains

strain[a]	host[b] plant	degree of[c] antagonism	cell envelope protein class	Fe^{3+}-limitation inducible proteins	
				minimal number	apparent M_R
WCS 345	P	2,0	A	2	90K, 92K
WCS 348	P	2,0	A	2	90K, 92K
WCS 357	P	2,0	A	2	90K, 92K
WCS 358	P	2,0	A	2	90K, 92K
WCS 359	P	2,0	A	2	90K, 92K
WCS 360	P	2,0	A	2	90K, 92K
WCS 364	P	2,0	A	2	90K, 92K
A1	P	ND	A	2	90K, 92K
WCS 141	W	1,5	B	2	92K, 88K
WCS 312	P	2,0	B	2	92K, 88K
WCS 317	P	0,5	B	2	92K, 88K
WCS 321	P	1,0	B	2	92K, 89K
WCS 327	P	0,5	B	2	92K, 88K
WCS 374	P	3,0	B	2	92K, 88K, 76K
WCS 375	P	3,0	B	3	92K, 88K, 76K
WCS 007	W	1,5	-	2	81K, 86K
WCS 085	W	2,5	-	1	86K
WCS 134	W	1,5	-	2	81K, 85K
WCS 307	P	2,0	-	2	78K, 83K
WCS 314	P	1,5	-	1	85K
WCS 315	P	1,0*	-	2	80K, 86K
WCS 324	P	1,5	-	2	82K, 83K
WCS 326	P	1,5*	-	2	80K, 86K
WCS 361	P	3,0*	-	2	68K, 81K
WCS 365	P	2,5	-	4	74K, 80K, 83K, 91K
WCS 366	P	2,0	-	3	70K, 81K, 83K
WCS 374	P	1,5*	-	2	76K, 79K
WCS 429	W	ND	-	1	79K
B10	P	ND	-	2	73K, 76K
E6	C	ND	-	4	70K, 72K, 74K, 78K

CELL ENVELOPE PROTEINS

The cell envelope proteins patterns of the *Pseudomonas* spp., obtained by SDS-polyacrylamide gel electrophoresis, are dominated by a number of prominent bands (Fig. 1). These bands represent outer membrane proteins, as was shown by physical separation of outer- and cytoplasmic membranes for three *Pseudomonas* strains.

The cell envelope protein patterns of the thirty *Pseudomonas* strains revealed a great diversity among these rhizobacteria. Fifteen strains (Table 1) showed unique cell envelope protein patterns (Fig. 1A), while the remaining strains could be divided into two classes, A (Fig. 1B) and B (Fig. 1C).

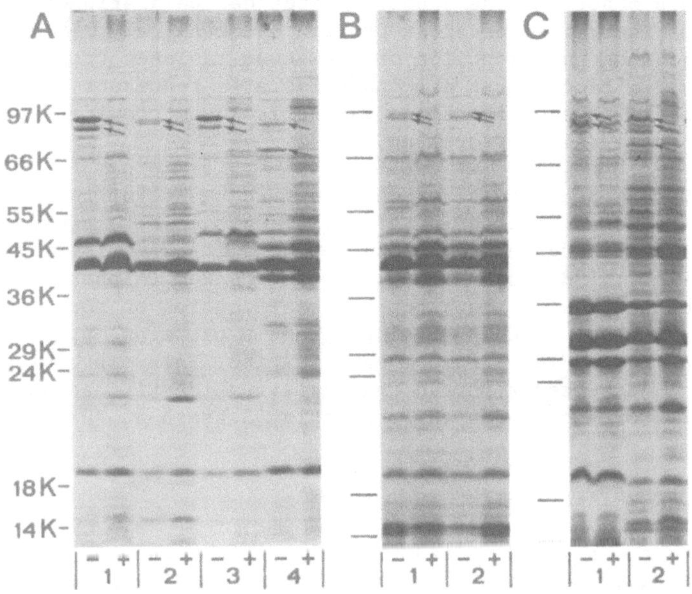

Fig. 1. (a) Strains with prefix WCS were isolated at the Phytopathological Laboratory 'Willie Commelin Scholten' in Baarn, the Netherlands. Strains A1, B10, E6 were obtained from J.W. Kloepper and M.N. Schroth, Berkeley, U.S.A. (Kloepper and Schroth, 1981). (b) Strains were isolated from roots of wheat (W), potato (P) or celery (C). (c) The *in vitro* antagonistic activity was calculated from a test versus 13 test organisms (fungi, bacteria) and rated in a 0-5 scale, where 0 indicates no inhibition and 5 indicates total inhibition of the test organisms. (Geels and Schippers, 1983). Values coded by asterisk indicate that the antagonism is not inhibited by the presence of 100 μM Fe^{3+} in the medium. (d) ND = not determined.

FE^{3+} LIMITATION INDUCIBLE PROTEINS

In all thirty fluorescent *Pseudomonas* strains, growth under Fe^{3+}-limiting conditions resulted in the appearance or increase of certain cell envelope proteins (Fig. 1A, B, C). A large variety in number and apparent molecular weight of Fe^{3+}-limitation inducible proteins was observed (Table 1).

IMPLICATIONS FOR THE NUMBER OF ANTAGONISTIC CLONES

Of the eight class A strains, one (strain A1) is an American potato isolate and seven were isolated from one potato plant. Since the latter isolates were also indistinguishable with respect to antagonistic behaviour and Fe^{3+}-limitation inducible proteins (Table 1), they probably are descendants from one clone. The cell envelope protein pattern of class B (Fig 1C) was shared by seven strains (Table 1), one wheat isolate and six isolates from different potato fields. As the protein patterns are not entirely identical (Fig. 1C) and as considerable differences exist among these isolates with respect to the degree of antagonism and, to a less extent, the number and electrophoretic mobilities of Fe^{3+}-limitation inducible proteins, it is likely that most group B isolates differ from each other. As the other isolates can be distinguished easily on the basis of their cell envelope protein patterns only, it must be concluded that, with the exception of the seven isolates from the same plant (class A), all or almost all other isolates represent distinct strains. Therefore these strains and future isolates provide a valuable pool of genes for strain improvement.

Outer membrane proteins have been shown to be useful for classifying clones within bacterial species, e.g. *Escherichia coli* (Achtman et al., 1983; Overbeeke and Lugtenberg, 1980), *Bordetella bronchiseptica* (Lugtenberg et al., 1984) and *Haemophilus influenzae* (Alphen et al., 1983). Our present results show that a classification of root-colonizing *Pseudomonas* spp. based on cell envelope protein patterns, provides a useful and fast method for strain identification. This especially is important, as it has been reported that existing identification methods are not satisfactory for these rhizobacteria (Schroth and Hancock, 1981).

FE^{3+}-SIDEROPHORE UPTAKE SYSTEMS

Although most analyzed strains clearly differ from each other, the possibility still exists that they share the same Fe^{3+}-uptake system, e.g. they could all harbour the same plasmid. However, as most strains carry neither small nor megaplasmids (unpublished results) and as the numbers and electrophoretic mobilities of the Fe^{3+}-limitation inducible proteins, which are known to be absolutely specific with respect to the matching Fe^{3+}-siderophore complex (Neilands, 1982), show large differences, it seems likely that a large variety exists among the isolates with respect to the siderophores that they can take up.

ACKNOWLEDGEMENTS

These investigations were supported by the Netherlands Technology Foundation (STW).

REFERENCES

Achtman, M., Mercer, A., Kusecek, B., Pohl, A., Heuzenroeder, M., Aaronson, W., Sutton, A. and Silver, R.P., 1983. Six widespread bacterial clones among *Escherichia coli* K1 isolates. Infect. Immun., 39:315-335.

Alphen, L., van Riemens, T., Poolman, J., Hopman, C. and Zanen, H.C., 1983. Homogeneity of cell envelope protein subtypes, lipopolysaccharide serotypes, and biotypes among *Haemophilus influenzae* type b from patients with meningitis in the Netherlands. J. Infect. Dis., 148:75-81.

Geels, F.P. and Schippers, B., 1983. Selection of antagonistic fluorescent *Pseudomonas* spp. and their root colonization and persistence following treatment of seed potatoes. Phytopathol. Z., 108:193-206.

Geels, F.P. and Schippers, B., 1983. Reduction of yield depressions in high frequency cropping soil after seed tuber treatments with antagonistic fluorescent *Pseudomonas* spp. Phytopathol. Z., 108: 207-214.

Kloepper, J.W., Leong, J., Teintze, M. and Schroth, M.N. 1980. Enhanced plant growth by siderophores produced by plant growth-promoting rhizobacteria. Nature, 286:885-886.

Kloepper, J.W. and Schroth, M.N., 1981. Relationship of *in vitro* antibiosis of plant growth-promoting rhizobacteria to plant growth and the displacement of root microflora. Phytopathology, 71:1020-1024.

Lugtenberg, B., Meijers, J., Peters, R., van der Hoek, P. and van Alphen, L., 1975. Electrophoretic resolution of the 'major outer membrane protein' of *Escherichia coli* K12 into four bands. FEBS Lett., 58:254-258.

Lugtenberg, B., van Boxtel, R., van den Bosch, R., de Jong, M. and Storm, P., 1984. Biochemical and immunological analyses of the cell surface of *Bordetella bronchiseptica* isolates with special reference to atropic rhinitis in swine. Vaccine, 2:265-273.

Neilands, J.B., 1982. Microbial envelope proteins related to iron. Ann. Rev. Microbiol., 36:285-309.

Overbeeke, N. and Lugtenberg, B., 1980. Major outer membrane proteins of *Escherichia coli* of human origin. J. Gen. Microbiol., 121:373-380.

Schroth, M.N., and Hancock, J.G., 1981. Selected topics in biological control. Ann. Rev. Microbiol., 35:453-476.

A SAGA OF SIDEROPHORES

J.B. Neilands

Biochemistry Department
University of California
Berkeley, CA 94720

INTRODUCTION

This is the first international symposium the title of which contains the term "siderophore". Accordingly, I take this unique opportunity in a "review type" paper to survey some of the data emanating from our laboratory over the past three or four decades. The account will be a more-or-less chronological one and is organized mainly under the names of the microorganisms studied. Only very recent references and those particularly pertinent to plants will be cited. The subject has been treated *in extenso* in reviews which have been published at regular intervals since 1957 (Hider, 1984; Neilands, 1957; Neilands, 1984).

USTILAGO SPHAEROGENA

In the early 1950's the late Paul Allen of the Botany Department, University of Wisconsin, was engaged in a study of the nutrition and physiology of the smut fungus, *Ustilago sphaerogena.* He observed that under certain conditions of growth the cells, which in their sporidial stage grow in simple laboratory media, assumed a reddish coloration. He suspected the colored substance to be cytochrome, but was unable to isolate the pigment. Having just had experience in this line, I was called in as an expert. At that time the organism was thought to have some special nutrient requirement for pigment synthesis, but since it gave the red compound on synthetic amino and vitamin assay media, very commonly available at that time in the Biochemistry Department at Madison, it was a simple matter to delete substances one-by-one and trace the causative agent. Allen found it to be zinc, but to this day the role of this metal ion in cytochrome synthesis in *Ustilago* remains unknown.

When the cytochrome was isolated -- it proved to be cytochrome c -- a substantial level of color remained in the saturated ammonium sulfate solution. The reddish-orange material could be taken into benzyl alcohol and crystallized from methanol. The new compound $C_{27}H_{42}N_9O_{12}Fe$, was named ferrichrome. Like the word television, this is an unfortunate mix of Latin and Greek. However, ferrichrome is a hybrid, half organic and half inorganic. The properties of the compound, viz., an exclusive affinity for the higher oxidation state of iron, indicated that it could not function as a redox catalyst and given that the source was a cell capable of making outrageous levels of a small protein with 0.4% iron, a role for ferrichrome in iron assimilation was suspected.

Since ferrichrome was studied *de novo* as an iron carrier, it is generally regarded as the prototypical siderophore although it is not the first ferric hydroxamate detected in nature. The year 1952 also saw the isolation of coprogen, a growth factor for coprophilic fungi, and the description of certain bacterial growth factors synthesized by soil bacteria. One of the latter was eventually characterized as a ferric hydroxamate (Terregens factor = arthrobactin). In the preceding decade Dutcher (1947) described an hydroxamic acid from *Aspergilli* and in the USSR a group initiated studies on an iron-containing antibiotic, albomycin. A satisfactory structure for this compound has only recently been forthcoming and, as could be deduced from its biology, it is very closely related to ferrichrome. Finally, mention should be made of the elegant investigations of Snow (1970) on a growth factor for mycobacteria, mycobactin. Although the complete structure, steriochemistry and even the metal binding activities were deduced, mycobactin was not immediately realized to be an iron carrier. However, the natural product chemistry worked out for mycobactin paved the way for subsequent characterization of ferrichrome.

In the era before NMR and MS the determination of structure by "Emil Fischer style" degradation consumed prodigious amounts of material. Variations in the Grimm-Allen medium for the culture of *Ustilago* were sought, and it was then discovered that the biosynthesis of the ferrichrome ligand is regulated by iron (Garibaldi and Neilands, 1955). This seemed to confirm the nascent hypothesis that the compounds functioned biologically as scouts for iron.

DIVERSE FUNGI AND BACTERIA SYNTHESIZE IRON BINDING COMPOUNDS AT LOW IRON NUTRITION

To probe the generality of the observation just described, various fungi and bacteria were cultured at different levels of iron and the supernatants were tested for reactivity with ferric chloride. The results indicated that the phenomenon was a general one and the findings were recorded in a short paper in Nature titled "Formation of Iron-Binding Compounds by Microorganisms" (Garibaldi and Neilands, 1956). The survey included *Aspergillus niger* and two spore-forming Gram positive bacteria, *Bacillus subtilis* and *Bacillus megaterium.* All three responded to iron starvation in a manner comparable to *Ustilago*, except that the ferri chromophore generated by *B. subtilis* was deep purple in color and hence obviously different from the other compounds. Ferrichrome could be crystallized from *A. niger* and the *B. megaterium* compound; although not isolated at that time, was judged by its spectrum to be related to the ferrichrome series.

BACILLIS SUBTILIS

The low iron induced iron-binding compound produced by *B. subtilis* was crystallized and characterized by chemical synthesis as 2,3-dihydroxybenzolglycine. At this stage the existence of bacteria and fungi requiring chelated iron was recognized. Accordingly, it was proposed that organisms be classed as autosequesteric or anautosequesteric as regards their mode of acquisition of iron: "in the former category iron sequestering agents can be synthesized from simple precursor molecules in the nutrient medium, while in the latter category the pre-formed organic sequestering agent must be added to the diet. Since the autosequesteric species are able to combat iron deficiency through the synthesis of abnormally large amounts of sequestering agents, such forms of life enjoy a distinct advantage in the competition for the essentials, insoluble mineral elements (of which iron is the prime example (Ito and Neilands, 1958). Although this classification has not been adopted, the isolation of the catechol from *B. subtilis* implanted the idea that there would be at least two lines of ferric-specific ligands encountered in the microbial world, namely, hydroxamate and catechols.

RHODOTORULA PILIMANAE

The red yeast, *Rhodotorula pilimanae,* appeared as a contaminant in a batch of ferrichrome production medium. It proved to synthesize a novel hydroxamic acid, rhodotorulic acid. The most remarkable feature of the species was the extremely high levels of the compound produced. For the first time we had available a biologically active iron-binding compound in virtually unlimited quantities and it was possible to cast about in biology for additional effects of this series of compounds, which were collectively referred to at that time as siderochromes.

It was noted that extracts of *Ustilago nuda* forms Green Islands when applied to detached grain leaves. A test of representative siderochromes including ferrichromes, rhodotorulic acid and ferrioxamine, showed that all are potent inducers of the Green Island response (Atkin and Neilands, 1972). Use of the *Arthrobacter* JG-9 assay for siderochrome activity in four plant pathogens, *Heterosporium iridis, Septoria lycopersici, Helminthosporium* sp. and *Septoria* sp. showed that all stimulated the growth of the test organism while only the first two gave Green Islands. Should the latter two only form siderochromes in laboratory media and not *in planta,* a perfect correlation would be observed between production of siderophore and the delay of senescence.

Recent advances in our knowledge of aging and degenerative disease may provide an explanation for the effect of siderophores just described. Siderophores may block the iron catalyzed generation of active oxygen species by preferential binding of the higher oxidation state of the metal ion as the latter occurs loosely complexed to cell constituents. From the Nernst equation relating equilibrium constants to the redox potential, it is apparent that any such ligand will have an E_0^1 too low to be reducible by, for example, superoxide. Since siderophores display an exceptionally high ratio of affinity for Fe(III/Fe(II), their potentials must be very low and outside of the range of reduction by superoxide.

Of the four hydroxamate type siderophores examined, two (ferrichrome, ferrichrome A) were applied to the excised bean leaves while the remaining two (rhodotorulic acid, deferriferrioxamine B) were used in the iron-free form. All four were active in combating senescence. We assume, tentatively, that the enzymatic systems in the plant reduce the ferri form of the siderophore and that the free Fe^{3+} in the aerobic environment is then rapidly reoxidised. This cycle will gradually precipitate iron as the oxyhydroxide polymer even in the presence of the ligand. The free ligand may then escape by diffusion into a zone peripheral to the site of application where it acquires and neutralizes the loosely bound iron responsible for potentiation of the toxicity of O_2, thus preserving green tissue in the shape of a ring. Whether or not this is the true explanation for the Green Island effect must, of course, await experimental verification.

SALMONELLA TYPHIMURIUM

In the course of mapping the chromosome of the Gram negative enteric bacterium, *Salmonella typhimurium*, B.N. Ames isolated a series of "iron mutants". These mutants required the application of iron salts for growth on minimal agar medium containing citrate (Pollack et al., 1970). Some of the mutants produced very large amounts of a catechol, which proved to be identical to the catechol previously isolated from *B. subtilis*, namely, 2,3-dihydroxybenzoic acid. A second class of mutants formed little catechol and was stimulated by an ether extract of the supernatant fluid of wild type cells. The siderophore was characterized as the cyclic trimer of 2,3-dihydroxybenzoylserine, enterobactin. At the same time an Australian group isolated the identical compound in the course of studying the biosynthesis of aromatic compounds in *Escherichia coli* and gave it the trivial name enterochelin.

The enb mutants of *S. typhimurium* were found to ulilize a series of hydroxamate type siderophores derived from other bacteria and fungi. From the enb mutants a second line of mutants, termed sid, was generated by selecting for resistance to albomycin, a close structural analog of ferrichrome. Most of the sid mutants were found to map at panC on the *S. typhimurium* chromosome.

ESCHERICHIA COLI

Early in the study of ferrichrome transport in *Escherichia coli*, a bacterium even more genetically accessible than *S. typhimurium*, it immediately became apparent that a classical genetic function could be involved in the uptake of this siderophore. The transport of vitamin B12 had been shown to require an outer membrane receptor which served also as binding site for the E group colicins, and enterobactin had been found to compete with colicin B for a common receptor in the cell envelope. As the map position of PanC in *E. coli* coincided with that of tonA, it was first speculated and then demonstrated that this function is in reality the ferrichrome receptor. We now know that outer membrane receptors for ferrichrome, enterobactin and aerobactin may exist singly or side-by-side in *E. coli* and that all of these ca. 80K outer membrane proteins act as binding sites for specific lethal agents, namely, antibiotics, phages or bacteriocins.

E. coli, normally an innocuous endosymbiont of man, is nonetheless the most common Gram negative bacterium isolated in clinical laboratories. It may establish infections in practically every organ and tissue in the body. In 1979 Peter H. Williams at the University of Leicester, discovered a novel iron assimilation system coded on a large plasmid long known to be correlated with virulence in *E. coli* and hence designated plasmid V. As the plasmid encoded synthesis of a bacteriocin, also designated V, it is generally referred to as pColV. A collaborative study disclosed the siderophore system on pColV to be aerobactin, previously isolated from the Gram enteric bacterium *Aerobacter aerogenes* 62-I. Williams showed that the colicin V is irrelevant to virulence in an animal model and that separate genes for aerobactin synthesis and utilization could be demonstrated. These were designated iuc ("iron uptake chelate") and iut ("iron uptake transport"), respectively.

Work in several laboratories then identified the outer membrane receptor for ferric aerobactin as a 74K protein which also acts as a binding site for cloacin, a bacteriocin from *Enterobacter cloacae*. This observation greatly facilitated the subsequent cloning of the aerobactin biosynthesis and transport determinants since mutants defective in the cloacin-aerobactin receptor were completely resistant to the bacteriocin. Susceptibility to cloacin signaled the presence of the 74K outer membrane receptor protein for ferric aerobactin as a product of any particular recombinant plasmid used to transform the host mutant cells.

By this selection method the aerobactin gene complex was cloned on a 16kb *Hind*III section of plasmid V-K30. The resulting plasmid, pABN1, was shown to contain all of the biosynthetic and transport functions for aerobactin while an 8 kb sub-clone, pABN5, was observed to have only the genes for aerobactin biosynthesis. A series of deletion plasmids were constructed and placed in minicells charged with ^{35}S-methionine. Five labelled polypeptides were detected which, in SDS-PAGE gels were estimated to have molecular weights of 74, 63, 53, 32 and 33K. The first of these is obviously the ferric aerobactin receptor while the 32K protein may be derived from the vector. In theory, a minimum of three proteins should be required for aerobactin synthesis. The biogenesis of aerobactin must involve first oxidation of the N^{ε} atom of lysine, acetylation of the hydroxylamino group thereby formed, and, finally, condensation of the N^{ε}-acetyl-N^{ε}-hydroxy function with citrate. The 53K protein may be required for uptake of ferric aerobactin; the 74K protein alone is sufficient for sensitivity to cloacin. The aerobactin gene complex is bounded by insertion elements, defined by fine structure mapping as ISI elements. On the right side of the gene cluster one copy of ISI follows immediately after the gene for the 74K protein while at the upstream end of the gene complex the second ISI sequence occurs about 5 kb to the left of the transcription start site. It hence has the structural features expected of a transposable element.

The promoter of the aerobactin gene complex was localized by *in vitro* run-off transcription to a 0.7 kb *Hind*III-*Sal*I fragment (Bindereif and Neilands, 1985). The standard S1 nuclease protection assay was used, in conjunction with *in vitro* and *in vivo* transcribed RNA, to map the transcriptional start site. One major start site was identified at 30 bp and a minor start site was found at 80 bp upstream from the initiation codon of the first structural gene. The DNA sequence around the promoter region was shown to be characterized by a very large number of potential secondary structures. The sequence cannot be compared with any other involved in iron assimilation since no others are available at the present time from *E. coli* or from any other organism, microbe, plant or animal.

A quantitative S1 nuclease protection mapping assay was used to probe the level of specific RNA present as a function of iron nutrition of the cell. *This proved directly that the iron is acting at the transcriptional level*. Further confirmation that this is so was obtained by construction of a *iuc*A-*lac*Z protein fusion in which the *lac*Z gene, minus its promoter, was attached directly to the aerobactin promoter-operator. In this construction the expression of β-galactosidase was again shown to be regulated by iron. The level of induction at low iron was many orders of magnitude below that found for an operon fusion but this may reflect inefficient translation of the message as a consequence of diminished affinity for the ribosome. The entire iron-regulatory sequences are contained within a 152 bp region spanning the major and minor promoters.

The next step was to define genetically the nature of the system interacting at the upstream control region of the aerobactin gene complex. In 1978 Ernst et al. had described a mutant of *S. typhimurium* which was derepressed in all iron-regulated functions and a similar mutation was subsequently identified in *E. coli* by Hantke (1981). The isolation of such a mutant was facilitated by having in hand the cloned aerobactin system. In particular, the two closely spaced *Bam*H1 sites in *iuc*C of pABN5 proved to be a convenient site at which to introduce *lac*Z. Two derepressed mutants were found which overproduced all iron components of the iron assimilation system of *E. coli*. The mutants were mapped at 15.7 min and the gene, designated by Ernst et al., (1978) as fur (ferric uptake regulation), cloned on a derivative of pBR322 (Bagg and Neilands, 1985).

AGROBACTERIUM TUMEFACIENS

With iron firmly established as a virulence factor in infections of man and animals, it was natural to inquire if the analogy could be extended to plants. Bacterial rather than fungal phytopathogens were chosen for first examination since genetic manipulations can be most readily performed in the prokaryotic world. *Agrobacterium tumefaciens*, the causative agent of tumor-like galls on dicotyledonous plants, was chosen for investigation. The bacterium proved to produce a linear catechol strikingly similar to the Compound III described some few years earlier by Tait (1975) from *Micrococcus* (now *Paracoccus*) *denitrificans*. The siderophore from *A. tumefaciens*, unlike Compound III, contained no salicylic acid. There was an additional difference. At low pH the deep ultraviolet absorption spectrum of agrobactin indicated the protonation of an oxazoline ring. Based on the duplicated NMR spectra seen at temperatures below 80°, it was assumed that the oxazoline must be substituted into the secondary amide N of the spermidine backbone of the molecule. The structure of agrobactin (Ong et al., 1979) has been confirmed by cystallography and by chemical synthesis.

Mutants of *A. tumefaciens* defective in the synthesis of agrobactin were found capable of generating galls in sunflower. The mutants could be re-isolated from the infected plant and had apparently not reverted *in planta* (Leong and Neilands, 1982). We hence assume that agrobactin is not a virulence factor. However, it may play an important role in the iron nutrition of the bacterium *ex planta*.

PARACOCCUS DENITRIFICANS

The discovery of the oxazoline ring in agrobactin prompted the thought that this structural detail might have been overlooked in Compound III. An investigation of the NMR spectra and electronic absorption spectra vs pH confirmed the presence of this ring in Compound III, now re-named parabactin. The pioneering work of Snow (1970) established the presence of this structural feature in all members of the mycobactin family of siderophores. It affords a unique coordination site for metal ions in spite of the very low affinity of the oxazolium N atoms for protons. In agrobactin the configuration of the ring hydrogen atoms is trans and it is likely that a similar configuration occurs in parabactin.

Tait (1975) has laid much of the groundwork necessary for an investigation of the molecular genetics and regulation of the parabactin biosynthesis and transport.

BACILLUS MEGATERIUM

We recall that in the original survey of microorganisms producing iron-binding compounds when starved for this element, *B. megaterium* excreted a product which gave ferric chloride tests resembling those given by the ligand of the ferrichromes, i.e. reminiscent of hydroxamates. Subsequently a compound was described from *B. megaterium* which reduced the division lag of the bacterium and it was accordingly named schizokinen.

In our studies of this siderophore we initially isolated an organism from a fish tank which gave in low iron media a strong ferric chloride test. The isolate was classified taxinomically and found to be a typical *B. megaterium*. We than acquired a copy of strain ATCC 19213 and, using the isolation procedure devised for the strain obtained from the fish tank, we obtained and characterized a pure compound. The structural work was facilitated by the finding, two years earlier in Australia, that a closely related siderophore aerobactin, is citric acid symmetrically substituted by N^{ϵ}-acetyl-N^{ϵ}-hydroxylysin side chains. Schizokinen is unusual in that the side chains are exceptionally short, only 3 carbon atoms in length. Yet the siderophore is able to clutch iron in an embrace which involves the α-hydroxycarboxylate functionality as well as the two hydroxamic acid groups in the side arms.

CYANOBACTERIA

The cyanobacteria or blue-green algae represent a most fundamental prokaryotic life form, some members of which can fix both N_2 and CO_2. They were thus prime candidates for production of siderophores. Culture of an *Anabaena* sp. at low iron led to the unexpected finding that schizokinen, first detected and characterized structurally from *B. megaterium*, is also present in the former bacterium. Schizokinen has been isolated from a soil sample collected from a dried rice paddy. One can surmise that a pond of blue-greens once flourished at the site and that the siderophore survived in the sediment after evaporation of the water phase. A number of other cyanobacteria have been investigated and found to produce schizokinen while no detectable hydroxamate or catechol could be detected when other strains of the species were cultured in low iron laboratory media.

RHIZOBIUM MELILOTI

This laboratory has long been interested in the siderophore profile of members of the *Rhizobiacae* and the interest itensified when nitrogenase was demonstrated to be an iron containing enzyme. Various strains of *Rhizobia* when cultured at low iron gave the Arnow test for catechols but no definitive structural assignments could be made. Finally, when reversal of EDDA induced growth inhibition was substituted for the chemical assays, a siderophore was isolated and characterized from *Rhizobium meliloti* DM4 (Smith et al., 1985).

Rhizobactin (I) is best described as a complexone, a family of synthetic chelating agents related to EDTA and characterized by the presence of the ethylenediamine moiety. The latter is the most classic of all coordination compounds, but its presence has not hitherto been reported in any siderophore. The compound is named as a derivative of lysine, since this amino acid comprises the major bulk of the molecule, and is designated N^2-[2-[(1-carboxyethyl)amino]ethyl]-N^6-(3-carboxy-3hydroxy-1-oxopropyl) lysine. The chirality at the α-amino acid carbons has not been established but the malic acid residue is of the L-configuration. We are greatly in need of a chemical test for detection of rhizobactin type compounds and much information remains to be deduced regarding its biology and metal binding properties. Although rhizobactin activity can be detected in nodules, the siderophore has not been isolated from that source. It is not a simple matter to screen *R. meliloti* strains for rhizobactin synthesis but having the pure compound in hand one can readily test its power to alleviate iron stress. Initially we were surprised to find the siderophore to stimulate only half of a collection of *R. meliloti* isolates. However, both production and utilization of rhizobactin may be highly strain specific.

$$\begin{array}{l} \quad\ \ CO_2H \qquad\qquad\quad\ \ CO_2H \\ CH_3\text{-}\overset{|}{C}H\text{-}NH\text{-}(CH_2)_2\text{-}NH\text{-}\overset{|}{C}H\text{-}(CH_2)_4\text{-}NH\text{-}CO\text{-}CH_2\text{-}CH(OH)\text{-}CO_2H \end{array}$$

Rhizobactin, I.

CONCLUSIONS

The technique of low-iron induced expression of siderophore systems has enabled the conclusion that this mode of iron assimilation is widespread, although not universal, in the microbial world. A few aerobic and facultative anaerobic species of bacteria seem devoid of this system of iron absorption, as judged by the failure of low-iron grown supernatant media to reverse the inhibitory action of EDDA. However, we cannot be sure that organisms growing in other environments, such as in the tissues of a host may well call forth the siderophore pathway of iron uptake.

A correlation between siderophores and virulence in animal pathogens has been demonstrated beyond any reasonable doubt and it is logical that this relationship should be extended to phytopathogenic species. The general presence of siderophores in bacterial and fungal pathogens of plants seems assured but, prior to this symposium, the literature contained no reference to iron uptake as a determinant of virulence.

The fact that both the promoter of the aerobactin gene complex and the fur gene of *E. coli* have been cloned means that we will soon be in possession of detailed information, at the molecular level, on the mechanism of regulation of siderophore expression by iron. The fur protein of *E. coli* also regulates expression of receptors for ferrichrome, ferric enterobactin and colicin Ia as well as certain other outer membrane proteins. It will be interesting to compare the sequence of the ferric aerobactin promoter with those for these other iron-regulated genes in *E. coli*. A logical extention of this work would be to clone siderophore systems from both Gram positive bacteria and fungi. The regulatory mechanisms in these species can then be compared with those established for *E. coli*.

The characterization of rhizobactin proves that high affinity ligands for Fe(III) are not restricted to catechol and hydroxamate functions. The ethylenediamine moiety standing alone would exhibit a preference for Fe(II), but in the case of rhizobactin the excess carboxyl and α-hydroxycarboxyl groups provide the excess oxygen atoms needed to tip the stability preference in favor of Fe(III).

Regarding the correlation of siderophore systems with virulence, it is clear that the elaboration of the systems *in planta* needs to be established, following which mutants should be selected which are defective specifically in expression of this means of iron uptake.

Siderophores are currently under investigation as therapeutic agents for the treatment of certain forms of arthritis. No such practical application has been made of the observation that they may delay senescence in plant as well as in animal tissue.

By examination of a full spectrum of microbial species it has been possible to conclude that siderophores play a very significant role in microbial iron assimilation. It now remains for those with the proper motivation to exploit this reservoir of basic research information.

REFERENCES

Atkin C.L., and Neilands, J.B., 1972. Leaf infections: Siderochromes (natural polyhydroxamates) mimic the "Green Island" effect. Science, 176:300-302.

Bagg, A. and Neilands, J.B., 1985. Mapping of a mutation affecting iron uptake systems in *Escherichia coli* K-12. J. Bacteriol., 161:450-453.

Bindereif, A. and Neilands, J.B., 1985. Promoter mapping and transcriptional regulation of the iron assimilation system of plasmid ColV-K30 in *Escherichia coli* K-12. J. Bacteriol., 162:1039-1046.

Dutcher, J.D., 1947. Aspergillic acid, an antibiotic substance produced by *Aspergillus flavus*. I. General properties. J. Biol. Chem. 171: 321-340.

Ernst, J.F., Bennett, R.L. and Rothfield, L.I. 1978. Constitutive expression of the iron-enterochelin and ferrichrome uptake systems in a mutant strain of *Salmonella typhimurium*. J. Bacteriol., 135: 928-934.

Garibaldi, J.A., and Neilands, J.B., 1955. Isolation and properties of ferrichrome A. J. Am. Chem. Soc., 77:2429-2430.

Garibaldi, J.A., and Neilands, J.B., 1956. Formation of iron-binding compounds by microorganisms. Nature, 177:526-527.

Hantke, K., 1981. Regulation of ferric iron transport in *Escherichia coli* K-12. Isolation of a constitutive mutant. Molec. Gen. Genet. 182:288-292.

Hider, R.C., 1984. Siderophore mediated absorption of iron. Struct. Bond., 58:25-87.

Ito, T., and Neilands, J.B., 1958. Products with "low iron fermentation" with *Bacillus subtilis:* Isolation, characterization and synthesis of 2,3-dihydroxybenzoylglycine. J. Am. Chem. Soc., 80:4645-4647.

Leong, S.A., and Neilands, J.B., 1982. Siderophore production by phytopathogenic microbial species. Arch. Biochem. Biophys., 218:351-359.

Neilands, J.B., 1957. Some aspects of microbial iron metabolism. Bacteriol. Revs., 21:101-111.

Neilands, J.B., 1984. Siderophores of bacteria and fungi. Microbiol. Sci. 1:9-14.

Ong, S.A., Peterson, T., and Neilands, J.B., 1979. Agrobactin, a siderophore from *Agrobacterium tumefaciens*. J. Biol. Chem., 254:1860-1865.

Pollack, J.R., Ames, B.N., and Neilands, J.B., 1970. Iron transport in *Salmonella typhimurium:* Mutants blocked in the biosynthesis of enterobactin. J. Bacteriol., 104:635-639.

Smith, M.J. Schoolery, J.N., Schwyn, B., Holden, I. and Neilands, J.B., 1985. Rhizobactin, a structurally novel siderophore from *Rhizobium meliloti*. J. Am. Chem. Soc. 107:1739-1743.

Snow, R.A., 1970. Mycobactins, iron chelating growth factors from mycobacteria. Bacteriol. Revs., 34:99-125.

Tait, G.H., 1975. The identification and biosynthesis of siderochromes formed by *Micrococcus denitrificans*. Biochem. J. 146:191-204.

GENETIC ANALYSIS OF THE IRON-UPTAKE SYSTEM OF TWO PLANT GROUPS PROMOTING *PSEUDOMONAS* STRAINS

P.J. Weisbeek, G.A.J.M. van der Hofstad, B. Schippers and J.D. Marugg

Department of Molecular Cell Biology, Institut of Molecular Biology and Department of Phytopathology University of Utrecht, Utrecht, The Netherlands

INTRODUCTION

Pseudomonas species can be found in a wide variety of environments; in association with animals and plants and in water. This is a consequence of their ability to use for their metabolism a great number of different compounds and to synthesize for different purposes many secondary metabolites like complex aromatic molecules and unusual amino acids and peptides. These properties together with the great stability of the organism are the reason why *Pseudomonas* spp. have become one of the major groups of colonizers of the plant-root. Many of the rhizosphere-*Pseudomonas* species are known plant-pathogens but more recently a number of root-associated fluorescent *Pseudomonas* strains have received increasing attention because of their ability to benefit plant growth. (Kloepper et al., 1980: Geels and Schippers, 1983a, Geels and Schippers, 1983b).

All plant-specific fluorescing *Pseudomonas* species produce under iron-limiting conditions a fluorescing compound of low molecular weight (approx. 1500 dalton) which is excreted from the cell with the function of binding iron(III) ions present in low concentrations in the soil and to deliver them to the cell. The structures of the siderophores produced by *Pseudomonas* strains tested have close similarities but differ from all other siderophores found in nature. The main differences are the presence of a fluorescing group, the presence of several different aminoacids and the combination of hydroxamate, catechol and α-dihydroxy acid groups for optimal binding of iron(III) ions (Hider, 1984). In addition to these pseudobactin-like siderophores, other iron-binding compounds are found among different Pseudomonas species like pyochelin and ferribactin (Cox et al., 1981; Philson and Llinas, 1982).

Of the rhizosphere *Pseudomonas* species tested so far all of those with plant growth promoting ability belong to the *P. putida-fluorescens* group.

The standard laboratory *Pseudomonas* strains (mainly *Pseudomonas aeruginosa* and *Pseudomonas putida*) can be analyzed with the complete set of genetic techniques developed with *E. coli*, like transformation, conjugation, transduction and transposition-mutagenesis (Haas, 1983). The analysis of new isolates however can be hampered by the fact that new strains often resist the standard genetic techniques for introduction of DNA into the cells. Adaptation of these methods to each particular strain can be very difficult and time-consuming.

To avoid these problems we have screened our collection of plant growth-stimulating *Pseudomonas* isolates for their susceptibility to conjugation (with RP4 derived plasmids) and to transformation. On this basis we have selected two strains (WCS358 and WCS374) that appeared to be both growth-stimulating and amenable to molecular-genetic analysis.

CHARACTERISTICS OF THE STRAINS WCS358 AND WCS374

Both strains were isolated from the roots of potato-plants grown in continuously cropped soil. They were selected on the basis of their ability to inhibit, under laboratory conditions, the growth of a number of plant-pathogenic microorganisms like *Erwinia carotovora* and *Gaeumannomyces graminis*. This interference was only effective when the media used were limiting in available iron. Above 20 μM $FeCl_3$ no effect was observed on the growth of the pathogenic and other microorganisms tested (Geels and Schippers, 1983b). *E. coli* was inhibited in a manner to the plant pathogens and for convenience was used as the test organism in most experiments.

Strain WCS358 was shown by standard microbiological techniques to belong to the *P. putida* group and WCS374 to be a *P. fluorescens* strain. Both strains grew at temperatures between 4° and 30° with an optimum at 25°. At this temperature the cells divided in approximately 30 min. Under conditions of iron-limitation the cells produced and excreted large quantities of a fluorescing Fe^{3+}-complexing compound (Marugg et al., 1985); this product was not formed at temperatures above 35° although WCS358 still grew at 41°. As is found with most *Pseudomonas* strains, WCS358 and WCS374 were resistant to several antibiotics, as shown in Table 1.

In most genetic analysis antibiotic resistance is used as a selectable marker, therefore the applicability of many available markers, plasmids and transposons is dependent on the innate resistance of the *Pseudomonas* strain in question. Plasmids which confer amipicillin-resistance on their host therefore cannot be used in most *Pseudomonas* strains

Table 1. Antibiotic resistance of *Pseudomonas* strains WCS358 and WCS374

	WCS 358	WCS 374
Ampicillin	300	⩾ 400
Cycloheximide	⩾ 100	⩾ 100
Chloramphenicol	50	25
Kanamycin	< 10	< 10
Streptomycin	50	50
Tetracyclin	10	< 5
Nalidixic acid	⩾ 300	25
Erythromycin	⩾ 100	⩾ 100

The resistance is expressed as maximum antibiotic concentration (μg/ml) which permitted coloby formation. Growth was on solid KB-medium at 30°.

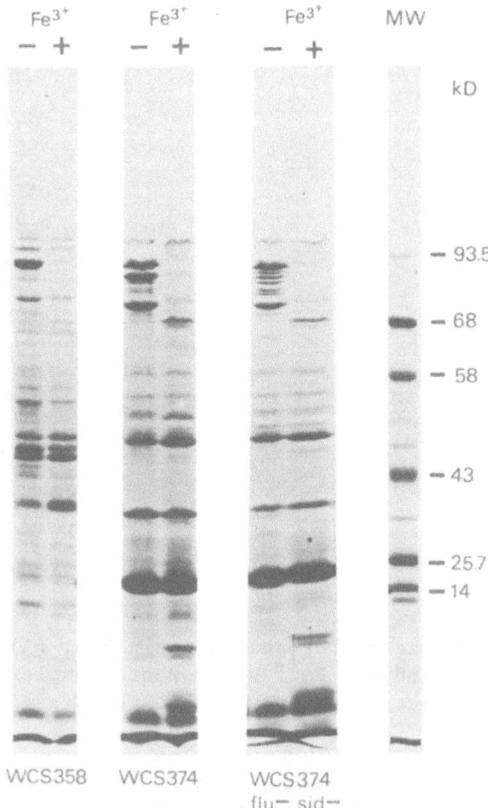

Fig. 1. Cell-envelope protein pattern of WCS358. WCS374 and a flu-sid- mutant of WCS374, grown in KB-medium at 30° with 20 µM $FeCl_3$(+) or without added iron (-). Molecular weight markers are shown in the right lane.

and nalidixic acid is not very suitable for WCS358.

The two strains differ considerably in the protein-composition of their cell-envelopes (See Fig. 1). This was found to be true of most Pseudomonas isolates obtained from roots (Weger et al., 1985) and therefore such a protein pattern is a reliable and fast way to check the identity of the strains under investigation.

The result of iron-limitation on the outer-membrane proteins of WCS358 is the appearance of a strong band of MW 80.000 and for WCS374 the induction of three proteins of MW 80.000, 76.000 and 70.000. One of these bands probably represented the receptor-protein for the iron-siderophore complex.

SIDEROPHORES

At iron (III) concentrations below 20 µM and at temperatures below 35°, a yellow-green fluorescing siderophire was synthesized and excreted by both strains in large quantities. WCS358 excreted up to 200mg per liter cell culture when growing at 25°C. The absorption and fluorescence spectra for the siderophores of both strains were identical, that from

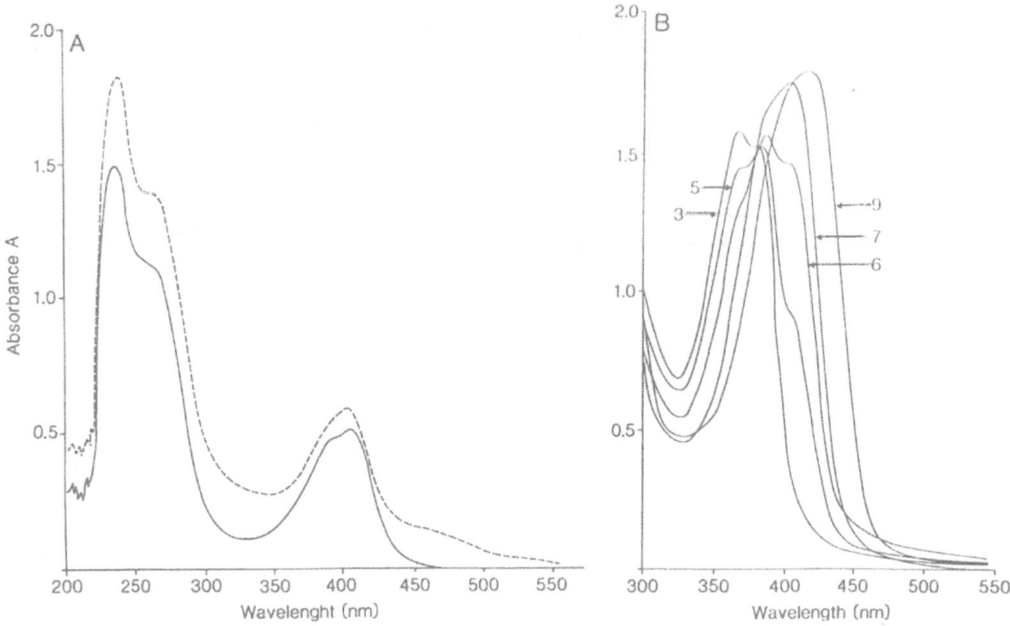

Fig. 2. Absorption spectrum of speudobactin 358. (A). The iron-free complex (_) and the Fe^{3+} siderophore complex ( ); pH 7.0. (B). The effect of pH on the absorption

WCS358 (pseudobactin 358) is shown in Fig.2. The absorption spectrum of the iron-free molecule exhibits two main peaks, at 230 nm and 405 nm respectively (Fig. 2a). At pH values ≤ 5 the peak in the visible region had two maxima, at 366 nm and 382 nm. At pH values ≥ 6 this peak consists of a single maximum which shifts from 388 nm (pH 6), via 405 nm (pH 7) to 414 nm (pH 9) (Fig. 2b). The peak in the ultraviolet region was not influenced by pH changes. The absorption spectrum of the Fe^{3+}-pigment complex showed two maxima; 230 and 405 nm with a shoulder at 450 nm (Fig. 2a). The spectrum did not vary with pH. The fluorescence spectrum of the free pigment showed a maximum of emission at 470 nm, with maximal excitation at 405 nm at pH 7. The physico-chemical characteristics of the pigment were essentially the same as those found for the siderophores pyoverdin (Philson and Llinas, 1982; Meyer and Abdallah, 1978) and pseudobactin (Teintze et al., 1981).

Under iron-limitation siderophore synthesis and excretion closely follow the growth and division of the cells at the stationary phase, siderophore synthesis also level off (Fig. 3). The conclusion is that biosynthesis of the siderophore is completely in line with other cellular metabolic processes. This appears to be in contrast to results obtained with *P. aeruginosa* where the production of pyoverdin Pa apparently starts when the cell culture reaches the stationary phase (Cox and Adams, 1985).

Although the absorption and fluorescence spectra of the WCS358 and WCS374 siderophores are identical, their structure must be different. This is concluded from the observation that WCS374 did not use the siderophore produced by WCS358. Strain WCS358 on the other hand did use the WCS374 compound (and also those from several other strains) for its own iron-acquisition. This was tested both with wild-type and mutant strains in competition-experiments on solid media. The results indicate that the structural and functional relation between the iron-siderophore

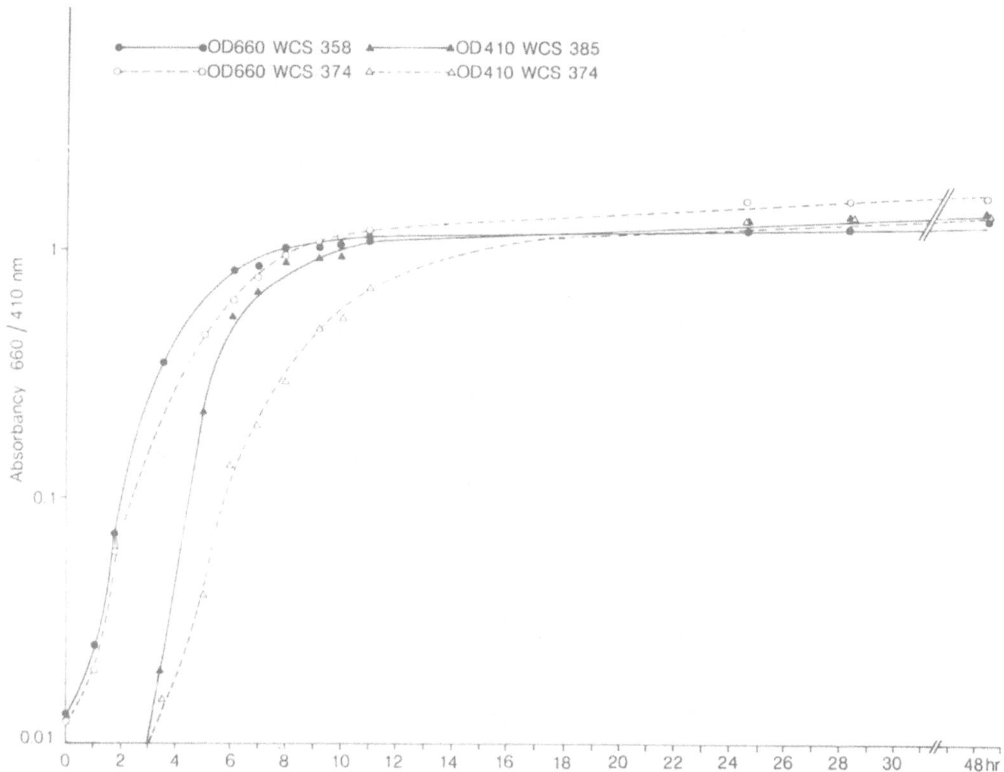

Fig. 3. Cell-growth and siderophore production. Cells were grown in KB-medium at 30°. Cell density was measured at 660 nm and siderophore-concentration by measuring the cell-free supernatant at 410 nm.

complex and its receptor protein is such that certain receptors are able to bind multiple siderophores whereas other receptors are much more specific in their binding and/or transport activities.

In order to understand the biosynthesis of the siderophore one needs to have an insight to its structure. For that reason research was started on the structure of pseudobactin 358, the siderophore of *P. putida* WCS358. Ferric pseudobactin 358 was purified, from culture medium supplemented with $FeCl_3$, via evaporation concentration, 100% $(NH_4)SO_4$ precipitation, phenol-chloroform extraction, precipitation from the organic phase with ether and subsequent extraction with water. A preparation was obtained that on HPLC gel filtration chromatography consisted of a single component with a purity of approximately 95%. This material was used for amino acid analysis. A 6N HCl hydrolysate revealed the presence of the following amino acids and their molecular ratio: Lys(2), Ser(1), Thr(2), Ala(1), Asp(1), HO-Asp(1) and both Orn and Glu in a molar ratio of less than ½. We assume that Orn and Glu are formed from HO-Orn during acid hydrolysis. Using manual Edman degradation the amino acid sequence of the oligopeptide part of pseudobactin 358 was examined. The results taken together strongly indicate the following sequence:

NH2-Lys.Ser.Asp.Thr.Ala.HO-Asp.Thr.Lys.HO-Orn-COOH.

With the chromophore-group attached to the ε NH2-group of the N-terminal lysine.

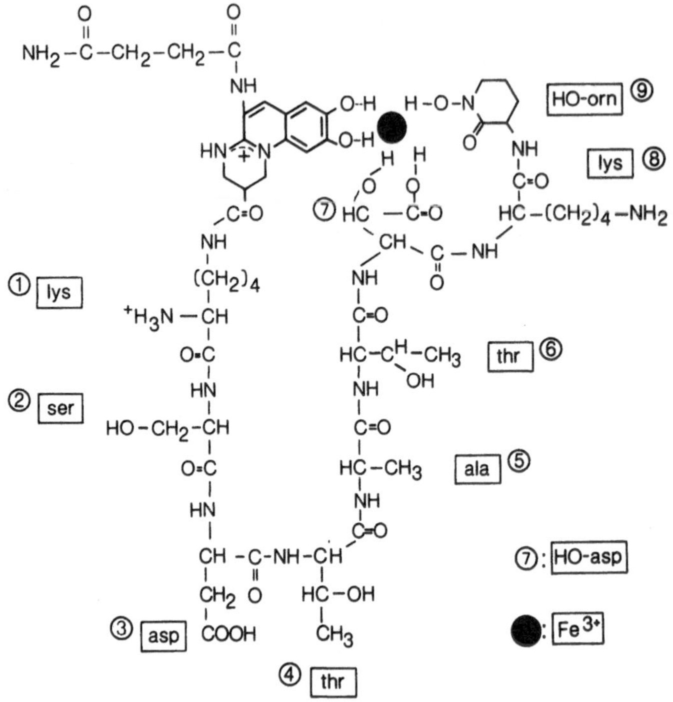

Fig. 4. Tentative structure of pseudobactin 358

Fast atom bombardment mass spectrometry (FAB) of partial acid hydrolysates and nuclear magnetic resonance (NMR) were performed to resolve the entire structure of pseudobactin 358. The results obtained so far combined with the observation that all pseudobactin-like siderophores analyzed have an identical fluorescing group, suggest the tentative scructure shown in Fig. 4.

Comparison with pseudobactin (Teintze et al., 1981) pseudobactin 7SR (Yang and Leding, 1984), and pyoverdin (Wendenbaum et al., 1983) learns that: (1) as in pseudobactin the link between the quinoline-group and the peptide is formed by the ε-NH2-group of lysine whereas in pyoverdin the HO-group of serine is used, (2) the functional build up is very similar; the iron-binding structure in each is constituted by a hydroxamate (HO-ornithine), a catechol (the quinoline group) and a hydroxy-acid group (or modified hydroxamate). Also that apart from lys, HO-asp and HO-orn, the type of amino acid and their number can be very different. Within the conditions of hexadentate binding of Fe^{3+} the number can vary between 6 and 10 and many different amino acids can be used.

Pseudobactin 358 could be used effectively on solid medium to inhibit the growth of several plant pathogenic micro-organisms in the same way as *P. putida* WCS358 itself. Mutants of WCS358 disturbed in siderophore-biosynthesis could be rescued by the addition of the pure compound. This shows that indeed this siderophore is the causal agent of the biological effects observed.

MUTANTS

The strains WCS 358 and WCS 374 were found to be subject to RP4-

dependent conjugation. This opened the way to transfer efficiently DNA from *E. coli* to these strains or to exchange DNA between the two *Pseudomonas* strains. In combination with the well-documented ability of transposon Tn5 (Haas, 1983) to integrate more or less randomly in the Pseudomonas chromosome, this makes it possible to do transposon-mutagenesis in the two growth-stimulating strains. The genetic system that was used to bring Tn5 into WCS358 and WCS374, consists of a pBR322-derived replicon containing transposon Tn5 and the mobilization-site (mob-site) of the broad host-range conjugative plasmid RP4 (pSUP2021). This plasmid can be mobilized by an *E. coli* host that contains the RP4-specific transfer-genes in its chromosomal DNA (Simon et al., 1983). The transferred replicon is unable to function in the acceptor Pseudomonas cell, therefore the only way for the transposon to survive is to insert itself in the acceptor-chromosome. The system was developed for transfor from *E. coli* to *Rhizobium* but we found that it functioned equally well with these *Pseudomonas* strains. With standard conjugation and selection-techniques frequencies of transposon-containing Ps. cells pf 10^{-5} to 10^{-6} were obtained. This compares well with frequencies found in *Rhizobium* (Simon et al., 1983). The kanamycin gene is expressed in *Pseudomonas* (as is normal in *E. coli)* but the streptomycin gene on the transposon was also active. The latter gene is not expressed in *E. coli* but it is in *Rhizobium* (Selvaraj and Iyer, 1984). Integration of Tn5 in the *Pseudomonas* chromosome was tested by hubridization of the DNA of Tn5-containing cells and of DNA from cells without Tn5 with Tn5 DNA (see Fig. 5A). This showed that only a single integration event had occurred and that the position of Tn5 was at random within the chromosome. The size of the Tn5 DNA inserted and its position for each mutant could be tested by using selected chromosomal DNA fragments (obtained via DNA cloning and complementation of mutants) as radioactive hybridization-probes (Fig. 5B, Fig. 6). With this labelled DNA the chromosomal DNA fragments that contain or flank the Tn5 insertion could be made visible. This proved again that for the two mutants tested the transposon

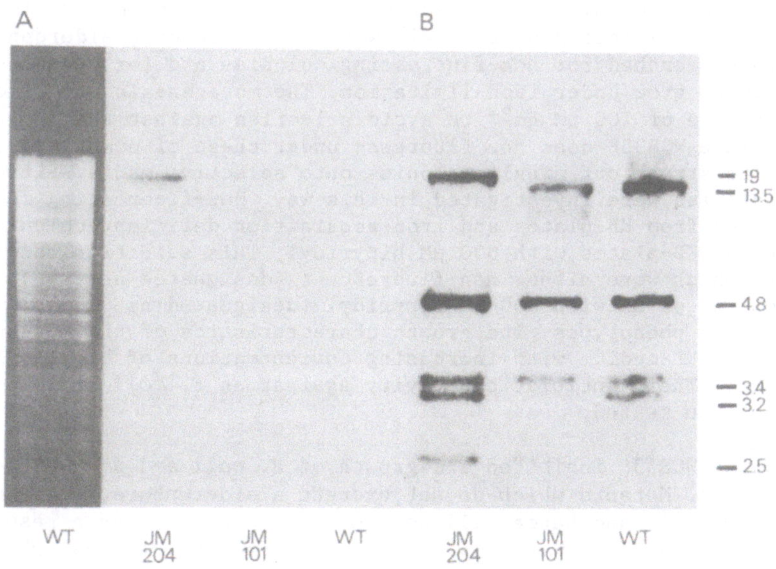

Fig. 5. Restriction enzyme and hydridizationalanalysis of chromosomal DNA (A). Left lane, DNA of WCS 358 digested with *Eco*RI. Middle and right lane: blots of WCS358 mutants JM204 and JM101 and wildtype hybridized with labelled Tn5 DNA. (B) Blots of the same DNA as in (A) but hybridized with labelled DNA of cosmid pMA3.

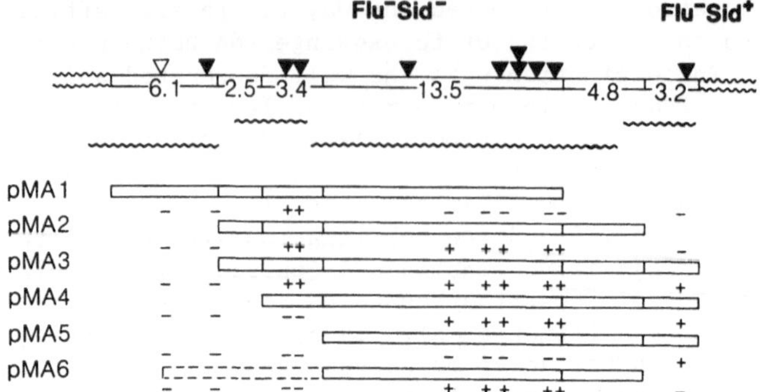

Fig. 6. Physical map of the cosmid inserts of group A, visualizing their regions of overlap and the position of the Tn5 insertions ( ) in the mutant chromosomes. Complementation results of each cosmid with the mutants is given as + or -. Mutants used are 10 flu-sid- strains and one flu-sid+. Potential transcription units are indicated ( ).

had integrated in *Eco*RI fragments of different length and also that the size of the insert was approximately 5500 base-pairs, the length of the complete transposon.

Autotrophic mutants were found among the kanamycin and streptomycin-resistant transconjugants with a frequency of ca 1%. This was another indication that integration into the chromosome had occurred randomly. The Tn5 mutagenesis resulted in a mutant colony bank of strain WCS358 with approximately 8000 kanamycin- (and Streptomycin-) resistant independently obtained transconjugants.

This mutant colony bank was screened for mutants defective in the biosynthesis or excretion of the yellow-green fluorescent siderophore. We therefore searched for non-fluorescing colonies and for colonies that were unable to grow under iron-limitation. The mutagenesis had taken place in the presence of 100 µM Fe^{3+} to avoid selection against the desired mutants. Since WCS358 does not fluoresce under these circumstances, it was necessary to streak out single colonies onto selective media. Fifteen-hundred colonies were investigated in this way. Non-fluorescing colonies were isolated from KB plates and iron-acquisition deficient colonies were identified on KB-plates with 800 µM bipyridyl. This selection resulted in 28 mutants which were either non-fluorescent (designated as Flu-), or unable to grow on KB with 800 µM bipyridyl (designated as Sid-), or which exhibited both phenotypes. The growth characteristics of the mutants were determined on KB-medium with increasing concentrations of bipyridyl (200-800 µM). Their antibiotic activity against an *E. coli* indicator strain was also tested.

Wildtype WCS358 inhibited and growth of *E. coli* and a number of other microorganisms. Mutants which do not excrete a siderophore, will not be able to bind Fe^{3+}, and hence will not act as inhibitors. As a result of these tests the mutants were divided in six different classes, which are listed in Table 2.

Class I, with the Flu^-Sid^+ phenotype, consisted of three mutants that did not fluoresce. They were still able to grow under iron-limitation (KB + 800 µM bipyridyl). Two mutants of this class, JM101 and JM103, started to fluoresce after 48-64 hours of incubation. All three mutants exhibited

Table 2. Siderophore defective mutants of WCS358

Class	phenotype[a]	mutant	fluorescent	max. growth concn. bipyridyl	*in vitro* antibiosis
I	Flu^-Sid^+	JM101	-	800 M	+
		JM102	-	"	+
		JM103	-	"	+
II	Flu^-Sid^-	JM201	-	200	-
		JM202	-	"	-
		JM203	-	"	-
		JM204	-	"	-
		JM205	-	"	-
		JM206	-	"	-
		JM207	-	"	-
		JM208	-	"	-
		JM209	-	"	-
		JM210	-	"	-
		JM211	-	"	-
		JM212	-	"	-
		JM213	-	"	-
		JM214	-	"	-
		JM215	-	"	-
		JM216	-	"	-
		JM217	-	"	-
		JM218	-	"	-
III	$Flu^-Sid^{\pm}$	JM301	-	400	+
IV	$Flu^{\pm}Sid^{\pm}$	JM401	±	600	+
V	$Flu^{\pm}Sid^+$	JM501	±	800	+
		JM502	±	"	+
		JM503	±	"	+
		JM504	±	"	+
VI	$Flu^+Sid^{\pm}$	JM601	+	400	+

[a]: Flu^-: no fluorescence; $Flu^{\pm}$: moderate fluorescence; Flu^+: normal (wt) fluorescence; Sid^-: growth in presence of max 200 µM bipyridyl; $Sid^{\pm}$; growth in presence of 400-600 µM bipyidyl; Sid^+: growth in presence of max. 800 µM bipyridyl. Fluorescence was determined with irradiation of UV-light (360 nm). In all cases mutants were grown on KB-medium.

in vitro antibiosis against the indicator strain *E. coli* S17-1. The eighteen Flu^-Sid^- mutants (class II) neither fluoresced nor grew on KB at concentrations of bipyridyl >200 µM. They exhibited no *in vitro* antibiosis. Class III, comprising one mutant differed only from class I in growth on bipyridyl (Table 2), i.e. growth up to 400 µM bipyidyl instead of 800 µM. The class IV mutant has an intermediate phenotype: slightly affected in both fluorescence and growth under iron-limitation. Class V consists of mutants that fluoresced less clearly than the wild type, but exhibited the same growth qualities. In contrast with this, the mutant of class VI showed fluorescence like the wild type, but was affected in its ability to reverse iron-limitation induced by bipyridyl. The mutants of all classes III - VI were effective inhibitors in the antibiosis test. None of the mutants seemed to have a defect in the uptake of the Fe^{3+} - siderophore complex, and all were able to use the siderophore secreted by the wild type WCS358 under conditions of iron-limitation (not shown).

In a similar way, a mutant bank was constructed for WCS374. The analysis of the mutants obtained and the isolation of the wild type genes is under way.

GENETIC MAP

Gene bank

The broad host range cosmid pLAFR1 (Friedman et al., 1982) was used to construct a genomic library of WCS358. The 21.6 kb vector has a unique *Eco*RI restriction site, encodes resistance against tetracycline, and contains the bacteriophage λ *cos*-site. In the presence of helper cells with the plasmid pRK2013 (which carries the conjugal transfer functions) (Figurski and Helinski, 1979) it can be mobilized to a wide range of gram negative bacteria, including *Pseudomonas* (Ditta et al., 1980).

Total DNA of strain WCS358 was partially digested with *Eco*RI, and fractionated on a 10-30% neutral sucrose gradient. Fragments of 15-33 kb were then ligated in pLAFR1 which was digested to completion with *Eco*RI. The ligation mixture was packaged *in vitro* and subsequently transduced to *E. coli* HB101. The resultant gene colony bank consisted of more than one thousand independent clones. From fifteen clones plasmid DNA was isolated to determine their insert size. *Eco*RI digests of the plasmids were separated by electrophoresis on 0.8% agarose gels. This revealed that the inserts ranged between 17 and 29 kb with an average of 26 kb. Assuming that the Pseudomonas genome has a molecular size of about 5000 kb, the colony bank with slightly over 1000 members is large enough to represent any DNA sequence of the genome with a probability of 99%. This gene bank was used in further experimenting.

Complementation of mutants

The complementation analysis was performed by conjugation (plate matings) of the cosmid colony bank to each of the twenty-eight WCS358 siderophore deficient mutants. All non- and slightly fluorescing- mutants (class I-V) were screened for complementation by restoration of the wild type fluorescence. The fluorescing mutant of Class VI was complemented by restoring the property to grow under iron-limiting conditions.

Fifteen cosmid clones were isolated that were able to produce nalidixic acid- and tetracycline-resistant transconjugants and to complement the defect in fluorescence of the host cell. One clone was able to convert a nalidixic acid- and tetracycline-resistant transconjugant of JM601 into the Sid^+-phenotype. *Eco*RI restriction enzyme analysis of these cosmids revealed thirteen different types. These respective cosmids are listed in Table 3. The complementation data suggest that the cosmids can be divided into five groups, designated as A, B, C, D, and E. Each group complemented a specific mutant, or a set of mutants (also shown in Table 3).

Group A consisted of six clones, which together complemented nine mutants belonging to class I (JM101) and II (JM201, JM203, JM204, JM205, JM209, JM211, JM213, JM214). The Flu^-Sid^- (class II) mutants were not only restored in their fluorescence, but also regained the ability to grow under iron-limitation. The *Eco*RI restriction enzyme analysis showed that the cosmids had several fragments in common, representing regions of

Table 3: Complementing *E. Coli* cosmid and their complemented mutants.

Group	Cosmid	Complemented Mutants	*Eco*RI fragment sizes from insert (kb)[a]	total size insert (kb)
A	pMA1	JM209;JM214	13.5;6.1;3.4;2.5	25.5
	pMA2	JM209:JM214:JM201:JM203:JM204; JM205;JM211:JM213	13.5;4.8;3.4:2.5	24.2
	pMA3	JM209:JM214:JM201:JM203:JM204: JM205:JM211:JM213;JM101	13.5;4.8;3.4;3.2;2.5	27.4
	pMA4	JM201:JM203:JM204:JM205:JM211: JM213:JM101	13.5;4.8;3.4;3.2	24.9
	pMA5	JM101	13.5;4.8;3.2	21.5
	pMA6	JM201:JM203:JM204:JM205:JM211 JM213	13.5;4.8;9.2	27.5
B	pMB1	JM601	13.0;7.5;4.8;1.8	27.1
C	pMC1	JM401	9.6;6.9;5.8;4.2	26.5
	pMC2	JM401	7.0;6.9;5.8;5.2;3.7	28.6
	pMC3	JM401	15.0;5.8;5.2	26.0
D	pMD1	JM103	9.6;9.5;8.0	27.1
E	pME1	JM503	16.0;6.0	22.0
	pME2	JM503	16.0;3.2	19.2

[a]: as determined on agarose gels

overlap between them. This overlap is visualized in the physical map of the six cosmids of group A (Fig. 6). The analysis also revealed that cosmid pMA6 originated from a recombination event in the ligation mixture, since it contains a fragment that does not match with the others. JM209 and JM214 were complemented by pMA1, pMA2 and pMA3. These three cosmids shared a 2.5 kb *Eco*RI fragment, whereas the other cosmids of group A did not. This is an indication that the genetic information on this fragment is needed for complementation of these mutants. JM101 was complemented by pMA3, pMA4 and pMA5. In this case the three cosmids share a 3.2 kb *Eco*RI fragment exclusively. This fragment will thus be necessary for complementation of JM101. Together with pMA6, the cosmids pMA2, pMA3 and pMA4 also complemented the mutants JM201, JM203, JM204, JM205, JM211 and JM213. Therefore it is likely that the *Eco*RI fragment of 4.8 kb is indispensable for complementation of these mutants. Although pMA5 contains the 4.8 kb *Eco*RI fragment, it did not complement the six mutants. The complementation results are summarized in Fig. 6. In the mutants the above-mentioned fragments may contain Tn5 insertions. However, it may well be that information needed for complementation is carried in adjacent *Eco*RI fragments.

The groups B (one cosmid), C (three), D (one) and E (two) each complemented one particular mutant which belonged to separate classes (resp. VI, IV, I, and V, Table 2,3). *Eco*RI restriction enzyme analysis revealed that the cosmids isolated from clones within group C contained overlapping DNA stretches. In particular, they share a 5.8 kb *Eco*RI fragment, which probably is needed for complementation of JM401. The cosmids of group E have an overlap, corresponding with a 16.0 kb *Eco*RI fragment. As pMB1 and pMD1 were the only cosmids of group B and D, respectively, a tentative localization of mutations was not possible here.

The inserts of the cosmids belonging to different groups (A-E) showed no resemblance with each other, and therefore contained DNA coming from different parts of the genome. From a total of twenty-eight mutants, only thirteen could be complemented. The other fifteen were mated with the thirteen clones of groups A - E, and again with the complete cosmid bank, but both with negative results. This may be a reflection of polarity induced by the Tn5 insertions, but one cannot exclude the possibility that our gene bank is incomplete.

The results show that at least five gene clusters, with a minimum of eight genes, are involved in the siderophore biosynthesis of WCS358. Cluster A contains at least four genes, whereas the other four (or more) are separated over the groups B, C, D and E.

Eight Flu$^-$Sid$^-$ mutants and one Flu$^-$Sid$^+$ mutant are complemented by single cosmid of group A (Table 2, 3 pMA3). The complementation data indicate that the Flu$^-$Sid$^-$ mutations are scattered on the fragment represented by the insert, with the Flu$^-$Sid$^+$ mutation being located terminally. This means that the genes of at least a number of the enzymes involved in the biosynthesis are clustered on the genome. In Group A only cosmids were found, that, starting with the 3.2 kb *Eco*RI fragment (Fig. 6), extend in just one direction. Considering the length of the fragments in combination with the complementation results it is suggested that these genes are arranged in 4 separate transcription units.

BIOSYNTHESIS

Little information is available on the biosynthesis of pseudobactin-like siderophores. Two observations published so far might give an indication of the direction of synthesis: (1) the structural homology between pseudobactin and pseudobactin A (Teintze et al, 1981) indicates that one of the last steps in the biosynthesis of pseudobactin occurs in the chromophore-part of the molecule and (2) the similarity in amino acid-composition of pyoverdin and ferribactin (Philson and Llinas, 1983) suggests that first the peptide part is synthesized, followed by the step-wise construction of the chromophore group at the NH_2-terminus of the peptide.

Analysis of the structure of the intact siderophore and of precursor molecules produced by mutant strains can help to identify the different steps in its synthesis. We have observed that mutants of WCS358 excreted material with an absorption spectrum that relates it to pseudobactin 358 and was able sometimes to bind and transport iron.

From the comparison of the order of the mutants on the map in Fig. 6 and the absorption spectrum of what each mutant excreted, one can conclude that from the left, JM212, via JM203 to JM101 on the right, the complexity of the excreted compound increased. Assuming that no specific degradation of precursor-molecules occurred this suggests that the genes within cluster A are involved in the biosynthesis route of the siderophore in such a way that their order of action is from left to right.

The strain WCS374 excreted a yellow-brownish component late in the stationary phase that is probably related to ferribactin (Philson and Llinas, 1983), a possible precursor of the complete siderophore. Several of the WCS374-mutants also synthesized this yellow-brownish precursor but a number of the mutants were not only deficient in the fluorescing molecule but also in this assumed precursor-peptide. These mutants also contain a single transposon-insertion, therefore synthesis of both compounds appears to be controlled by a single gene and consequently there must be a pre-cursor- end product relation between the two.

CONCLUSIONS

Mutants and genes of the plant growth stimulating *Pseudomonas* WCS358 and WCS374 have been obtained and analysed. WCS358 produces a single siderophore but WCS374 has a more complex way of acquiring its iron. In addition to its fluorescing siderophore, it produces a possible precursor, but all SCS374 mutants were still able to grow in the presence of up to 400 µm bipyridyl.

The mutants are being used for the elucidation of the biosynthetic pathway of the siderophores; the genes isolated will be used for the investigation of the iron-regulated induction of the complete enzymatic pathway. Initial experiments show that the mutants defective in sidero-phore-synthesis no longer stimulate the plant-growth and that mutants can be rescued by the wild type strain when growing at the surface of the plant root.

Detailed information about the process of iron-uptake will allow us to manipulate these bacteria in such a way that their interaction with and their stimulation of the plant can be analysed and optimized.

ACKNOWLEDGEMENT

The research described was performed in part with financial support from the Netherlands Technology Foundation (STW) and by the Biomolecular Engineering Programme of the EEC (Contract No. GBI-4-108-NL).

REFERENCES

Cox, C.D., and Adams, P., 1985. Siderophore activity of Pyoverdin for *Pseudomonas aeruginosa*. Infect. Immun. 48:130-138.

Cox, C.D., Rinehart, K.L., Moore, M.L., and Cook, J.C., 1981. Pyochelin: Novel structure of an iron-chelating growth promoter for *Pseudomonas aeruginosa*. Proc. Nat. Acad. Sci., USA. 78:4256-4260.

Ditta, G.S., Stanfield, D.D., Corbin, D., and Helinski, D.R., 1980. Broad host range DNA cloning system for gram-negative bacteria: Construction of a gene-bank of *Rhizobium meliloti*. Proc. Nat. Acad. Sci., USA, 77:7737-7751.

Figurski, D.H., and Helinski, D.R., 1979. Replication of an origin containing derivative of plasmid RK2 dependent on a plasmid function provided in trans. Proc. Nat. Acad. Sci., USA, 76:1648-1652.

Friedman, A.M., Long, S.R., Brown, S.E., Buikema, W.J., and Ausubel, F.M., 1982. Construction of a broad host range cosmid cloning vector and its use in the genetic analysis of *Rhizobium* mutants. Gene, 18: 289-296.

Geels, F.P., and Schippers, B., 1983a. Selection of antagonistic fluorescent *Pseudomonas* spp. and their root colonization and persistence following treatment of seed potatoes. Phytopathol. Zeitung, 108: 193-207.

Geels, F.P., and Schippers, B., 1983b. Reduction of yield depressions in high frequency potato cropping soil after seed tuber treatments with antagonistic fluorescent *Pseudomonas* spp. Phytopathol. Zeitung, 108:207-214.

Haas, D., 1983. Genetic Aspects of biodegradation by pseudomonads. Experimentia, 39:1199-1213.

Hider, R.C., 1984. Siderophore mediated absorption of iron. In: Struct. Bond., 58:25-87. Springer, Verlag, Berlin.

Kloepper, J.W., Leong, J., Teintze, M., and Schroth, M.N., 1980. Enhanced plant growth by siderophores produced by plant growth promoting rhizobacteria. Nature, 286:885-886.

Marugg, J.D., Van Spanje, M., Hoekstra, W.P.M., Schippers, B., and Weisbeek, P.J., 1985. Isolation and analysis of genes involved in siderophore biosynthesis in the plant growth stimulating *Pseudomonas putida* strain WCS358. J. Bacteriol. (In press).

Meyer, J.M., and Abdallah, M.A., 1978. The fluorescent pigment of *Pseudomonas fluorescens:* Biosynthesis, purification and physicochemical properties. J. Gen. Microbiol., 107:319-328.

Philson, S.B., and Llinas, M., 1982. Siderochromes from *Pseudomonas fluorescens*. Isolation and characterization. J. Biol. Chem., 257: 8081-8085.

Philson, S.B., and Llinas, M., 1983. Siderochromes from *Pseudomonas fluorescens*. Structural homology as revealed by NMR spectroscopy. J. Biol. Chem., 257:8086-8090.

Selvaraj, G., and Iyer, V.N., 1984. Transposon Tn5 specifies streptomycin resistance in *Rhizobium* spp. J. Bacteriol., 158:580-589.

Simon, R., Priefer, U., and Puhler, A., 1983. A broad host range mobilization system for *in vivo* genetic engineering: transposon mutagenesis in gram-negative bacteria. Biotechnology, 1:784-791.

Teintze, M., Hossain, M.B., Barnes, C.L., Leong, J., and Van Der Helm, D., 1981: Structure of ferric pseudobactin, a siderophore from a plant growth promoting *Pseudomonas*. Biochemistry, 20:6446-6457.

Wendenbaum, S., Demange, P., Dell, A., Meyer, J.M., and Abdallah, M.N., 1983. The structure of pyoverdine Pa, the siderophore of *Pseudomonas aeruginosa*. Tetrahedron Lett., 24:4877-4880.

Weger, L.De., Boxtel, R., Van Burg, B., Van Der Gruters, P., Geels, P., Schippers, B., and Lugtenberg, B., 1985. Outer membrane proteins of plant growth stimulating root colonizing *Pseudomonas* spp. (submitted).

Yang, C.C., and Leding, J., 1984. Structure of Pseudobactin 7SR1, a siderophore from a plant-deleterious *Pseudomonas*. Biochemistry, 23:3534-3540.

GENOTYPIC DIVERSITY OF FLUORESCENT PSEUDOMONADS AS REVEALED BY SOUTHERN HYBRIDIZATION ANALYSIS WITH SIDEROPHORE-RELATED GENE PROBES

E.C. Lawson, C.B. Jonsson and B.C. Hemming

Rhizobiology Group, Biological Sciences Department
Monsanto Life Sciences Research Center
Chesterfield, Missouri, U.S.A.

INTRODUCTION

Interest in the siderophores of fluorescent pseudomonads has been highlighted because of their antagonism toward the agent of the plant diseases, fusarium wilts and take-all diseases of wheat; namely, *Fusarium oxysporum* f. sp. *lini* and *Gaeumannomyces graminis* var. *tricici* as well as other aspects of plant disease (Burr et al., 1978; Howell and Stipanovic, 1979; Howell and Stipanovic, 1980; Kloepper and Schroth, 1981; Kloepper et al., 1980; Scher and Baker, 1982). The mechanism invoked includes the denial of Fe(III) for the growth of these pathogens by specific siderophores from fluorescent pseudomonad strains. It has been shown also that certain pseudomonads aggressively colonize the roots of plants and may be considered as beneficial, deleterious or neutral to the plant (Schroth and Hancock, 1981; Schroth and Hancock, 1982; Suslow and Schroth, 1982).

Production of antibiotics, numbering over 50 distinct compounds has been documented for the genus, *Pseudomonas* (Leisinger and Margraff, 1979). As inferred in the frequent appellation given a particular subgeneric group "the fluorescent pseudomonads", the singular characteristic common to this group is the production of diffusible, extracellular, water-soluble, UV-fluorescent pigments. Regulation of pigment biosynthesis by iron concentrations and chelation and transport of Fe(III) of the yellow-green fluorescent pigment from specific strains of *Pseudomonas fluorescens* has demonstrated this pigment (pyoverdine $_{p}$f) to be the siderophore of this species (Lenhoff, 1963; Schroth and Hancock, 1981; Schroth and Hancock, 1982). The determination of the molecular structure by x-ray crystallography of pseudobactin, the trivial name for the Fe(III) specific transport chelatant of the high-affinity iron assimilation system of *Pseudomonas* sp. strain B10, was significant in its unveiling of the unique features of this type of siderophore (Teintze et al., 1981) as well as its commonality to the classes of microbially-produced siderophores (Neilands, 1981). Pseudobactin was shown to consist of a linear hexapeptide, L-Lys-D-three-β-OH-Asp-L-Ala-D-allo-Thr-L-Ala-D-ζ-N-OH-Orn, with the lysyl ε-NH2 group participating in an amide linkage to a fluorescent dihydroxyquinoline moiety. Strain B10 has been described as of the *P. fluorescens-putida* affiliation.

An NMR spectroscopic study addressing the degree of structural diversity among siderophores produced by fluorescent pseudomonads revealed that *P. fluorescens* ATCC 13525 produced two iron binding compounds, ferribactin and a pyoverdine-type molecule (Philson and Llinas, 1982). The proton and ^{13}C NMR data confirmed that ferribactin, previously described as a decapeptide (Maurer et al., 1968) is a nonapeptide containing one residue each of D-serine, L-serine, L-glutamine (or glutamic acid), glycine, D-tyrosine, ζ-N-OH-D-ornithine, ζ-N-OH-L-ornithine and 2 moles of L-lysine. The structure of pyoverdine $_{p}$a, a siderophore from *Pseudomonas aeruginosa* (ATCC 15692, PAO 1) was elucidated from data obtained by FAB mass spectrometry and by NMR, which revealed a linear octapeptide bound to a dihydroxyquinoline derived chromophore, but differing in its amino acid composition from pseudobactin or ferribactin (Wendenbaum et al., 1983). The similarity of the chromophoric group of the siderophores produced by pseudomonads and a fluorescent peptide isolated from iron-limited cultures of *Azotobacter vinelandii* (Knosp et al., 1984; Page and Huyer, 1984) was also noted in these studies (Philson and Llinas, 1982; Wendenbaum et al., 1983). The structure of the yellow green fluorescent peptide produced by *A. vinlandii* strain O includes the sequence:

Chromophore-aspartyl-homoseryl-seryl-homoseryl-citrullinyl-seryl-glycyl-B-hydroxyaspartate (Page and Huyer, 1984) and has been shown to be a siderophore of this strain (Knosp et al., 1984).

By rRNA-DNA hybridization studies, five groups of genetic homologies have been identified in the genus *Pseudomonas* (Palleroni, 1975). All of the fluorescent strains including the fluorescent phytopathogenic species are in a single subgroup (Palleroni, 1975). Some phytopathogenic strains, such as a pathovar of the relatively nutritionally fastidious *Pseudomonas syringae* group produce N-hydroxylated amino acid-containing peptide phytotoxins, syringomycin and syringotoxin (Hemming and Strobel, 1983). Unlike siderophore-generated inhibition, the antimycotic activity of most toxin producing strains increases dramatically when grown in media containing iron (Hemming et al., 1982). Syringomycin has a broad spectrum antifungal activity in addition to its phytotoxic nature. A preliminary report suggested that syringomycin functions in iron assimilation for its producer strain (Gross, 1982). The completion of further study on the regulation of syringomycin synthesis in *Pseudomonas syringae* pv. *syringae* (hereafter referred to as *P. s. syringae*) has invalidated the preliminary report (Gross, 1985) and confirmed the earlier finding of an iron stimulated effect on syringomycin production, previously shown for a number of strains. The determination of the amino acid composition of the siderophore from a syringomycin producing strain would be of interest in determining the possible structural relationship between syringomycin and the hydroxamate siderophore of the same strain in addition to providing a foundation for research on iron regulatory interactions of these compounds.

The application of genetic approaches to the elucidation of the biosynthetic genes of fluorescent pseudomonads suggests an alternative but complementary route to examination of the diversity of fluorescent siderophores by structural means. Studies directed toward establishment of the genetic organization of siderophore biosynthesis have been established for at least two species, *P. fluorescens-putida* strain B10 and *P. s. syringae* strain JL2000 (Loper et al., 1984; Moores et al., 1984; Orser et al., 1983). These species are of some significance because of possible applications involving their use as biocontrol agents (Lindow, 1982; Moores et al., 1984; Weller, 1983; Weller, 1984; Weller and Cook, 1983). Genomic libraries of both strains were constructed in the broad host range cloning vector, pLAFR1, developed by Friedman et al.,(1982) from pRK290 (Ditta et al., 1980). The libraries were used in genetic complementation analyses to identify siderophore related genes. In the case of the pseudobactin producing strain, B10, it was concluded from the pattern of complementation of nonfluorescent mutants that a minimum of 12 genes arranged in four gene clusters is required for its biosynthesis (Moores et al., 1984). In a similar fashion, the *P. s. syringae* experiments generated four recombinant cosmids containing DNA fragments involved in some unidentified facet of fluorescent pigment production. These cosmids were found to be structurally distinct by *Eco*RI restriction analysis and differed also in the mutants for which fluorescence restoration could be demonstrated. The molecular nature of the siderophore from the parental strain has not been determined nor has the existence of further gene clusters been eliminated (Loper et al., 1984).

The number of different N-hydroxylated amino acid-containing peptide products of the fluorescent pseudomonads suggests that the siderophore/ fluorescent pigments or similar compounds from various species or ecotypes may exhibit some functional specificity. Such specificity might be exploited as a tool for identification. However, production and utilization of siderophores in other bacterial genera are not necessarily tightly correlated to taxonomic groups (Neilands, 1981); whereas the fluorescent phenotype of *Pseudomonas* spp. is a useful taxonomic character. Selective pressures of an ecological niche might be reflected in the type and quantity of pigment produced. Therefore, the structural identity of an isolate's siderophore, as governed by its genetic elements and their expression, might be exploited by correlation to functional tests for inhibitory activities. The availability of *P. s. syringae* cosmids harboring DNA fragments related to fluorescent pigment production has now permitted us to examine a large number of phenotypically diverse fluorescent pseudomonads by the colony hybridization technique followed by Southern hybridization analysis of positive isolates. Such probes permit the search for genetic homologies and divergence for correlation with taxonomic data from phenotypic tests and bioasay measurements. These cosmids and their subclones have proven useful in identification of strains and in deciphering the degree of genetic diversity in siderophore production in these bacteria.

Table 1. Designation of complemented mutant strains and size of *Eco*RI fragments of siderophore-related genes from *P. s. syringae* JL2000 harbored as pLAFR1 cosmid clones in *E. coli* HB101 (Loper et al., 1984; Orser et al., 1983).

Clone[a] No.	(No. of *Eco*RI fragments)	sizes/total kb					Strain No.[b] of Complemented Mutant
pLAFR1	(1)	21.6				/21.6	-
pSFL10	(3)	14.4	11.3			/25.7	JL2140
pSFL11	(3)	9.5	9.4	5.4		/23.3	JL2003
pSFL12	(3)	17.7	8.2	6.4		/32.3	JL2023
pSFL14	(6)	9.7	6.0	3.75	1.55 1.25	/25.75	JL2087
pMON1501	(1)	4.2 in pLAFR1				/25.8	JL2003

[a] Clone pMON1501 is a *Bgl*II/*Eco*RI (double digestion product of pSF11) fragment subcloned into pLAFR1.

[b] These strains are rifampicin resistant, nonfluorescent mutants of strain JL2000 representing four complementation groups (Loper et al., 1984).

MATERIALS AND METHODS

Bacterial strains and plasmids

The bacterial strains received as a gift from Dr. M.N. Schroth have been previously described as *E. coli* HB101 clones harboring siderophore-related DNA fragments of *P. s. syringae* JL2000 inserted in the broad-host range cosmid, pLAFR1, which complement nonfluorescent mutants to restore fluorescent pigment production (Loper et al., 1984). Relevant fragment sizes of the clones and a description of an additional subclone further generated in this study are found in Table 1.

Representative pseudomonads from which total genomic DNA was isolated and used in hybridization studies are described in Table 2. These strains are listed by the designations given in this study along with known synonyms and identifications as received from the indicated donors. Such designation of strains were required for ease of operation of computer programs for cluster analyses and for other computer-generated graphic material. Table 3

Table 2. Representative pseudomonad strains used in this study.

Strain no. designation	Identification (as supplied by donor)	Strain no.	Donor[a]
B10	*P. fluorescens-putida*	(same)	M.N. Schroth
Ps6	*P. syringae*	(same)	A.K. Vidaver
7SR1	*P. (deleterious)* sp.	(same)	C. Orser
PsCol	*P. syringae*	(same)	A.K. Vidaver
PsC1B	*P. syringae*	(same)	A.K. Vidaver
146A1	(soybean isolate)	-	this study
3732	*P. aeruginosa*	(same)	this study
392B1	(soybean isolate)	-	this study
520A3	(soybean isolate)	-	this study
536A3	(soybean isolate)	-	this study
736C2	(soybean isolate)	-	this study
1295G2	(soybean isolate)	-	this study
1489F2	(soybean isolate)	-	this study
1767E2	(soybean isolate)	-	this study
JL2000	*P. syringae* pv. *syringae*	(same)	M.N. Schroth
5013Z1	*P. ovalis (P. putida)*	IF03738	C. Walsh
5015Z1	*P. pseudoalcaligenes*	JM326	J.F. Kane
5018Z1	*P. stutzeri*	JM706 Phe-2	J.F. Kane
5063Z1	*P. syringae* pv. *tagetes*	-	M. Sasser
5075Z1	*P. syringae* pv. *tabaci*	BR2(pBRW1)	P.D. Shaw
5076Z1	*P. syringae* pv. *tabaci*	PT28014	P.D. Shaw
5077Z1	*P. syringae* pv. *angulata*	Pa52001	P.D. Shaw
5078Z1	*P. syringae* pv. *striafaciens*	PS1	P.D. Shaw
5079Z1	*P. syringae* pv. *glycinea*	P659000	P.D. Shaw
5080Z1	*P. syringae* pv. *coronafaciens*	PC27001	P.D. Shaw
5081Z1	*P. putida*	ATCC 15070	ATCC
5082Z1	*P.* sp.	B-15132	NRRL
5083Z1	*P.* sp.	B-15133	NRRL
5084Z1	*P.* sp.	B-15134	NRRL
5085Z1	*P.* sp.	B-15135	NRRL
5086Z1	*P. aeruginosa*	ATCC 7700	ATCC

[a]ATCC = American Type Culture Collection, Rockville, Maryland
NRRL = Northern Regional Research Laboratories, Peoria, Illinois.

Table 3. API test profiles and identification of soybean rhizosphere isolates of this study

Strain No.	API Profile	API Data base identification	Frequency (clinical environs)
0119A1	2206046F	Fl. Ps. group excell. id.	1:1765
0121A1	2206046F	Fl. Ps. group excell. id.	1:1765
0295D1	2206040F		
0437D1	2204446F		
0523A3	2200000F	Fl. Ps. group excell. id.	1:8
0701E1	2206046F	Fl. Ps. group excell. id.	1:1765
0785C2	2206046F	Fl. Ps. group excell. id.	1:1765
0820C1	2206046F	Fl. Ps. group excell. id.	1:1765
0821C1	2202040F		
0938C1	2206044F	Fl. Ps. group very good id.	1:215
1072D2	2202046F	Fl. Ps. group excell. id.	1:706
1097D2	0202002F	CDC group V E-2	
1141F1	2202044F	Fl. Ps. group excell. id.	1:260
1204G1	2204004F	Fl. Ps. group, *P. fluorescens, P. aerug.* or *putida*	
1606F2	2204044F	Fl. Ps. group excell. id.	1:83
1753G1	2206044F	Fl. Ps. group very good id.	1:215
1781E2	2202004F	Fl. Ps. group acceptable id.	1:8
1894D3	2200004F	Fl. Ps. group excell. id.	1:8
1905D3	2200000F		

abbreviations: F= fluorescent on King's medium B
Fl. Ps. = fluorescent pseudomonad

contains an additional list of rhizosphere isolates collected from soybean plants (*Glycine max* [L] Merill, in the majority of cases from cultivar 'Williams') and phenotypically characterized by the rapid API 20E test method (API Analytab Products, Inc., Lenexa, KS). The soybean bacteria isolates reported in this paper are a diverse subset of a collection of more than 327 fluorescent pseudomonads isolated from soybean plants obtained from eight different Midwestern U.S. locations which have been analyzed by numerical analysis employing at least 58 distinctive phenotypic tests on each isolate (Hemming et al., 1984). Fluorescent pigment production was determined for these strains under 360 nm U.V. light on King's medium B (King et al., 1954) or M9 media (Maniatis et al., 1982).

Hybridization and DNA Manipulation

Bacterial isolates were point inoculated onto sterile nitrocellulose BA85 membranes (82 mm circles, 0.45 µm porosity, Schleicher and Schuell, Keene, New Hampshire) on Luria broth agar medium and incubated at 30°C overnight for *in situ* hybridization experiments. Preparation of filters for DNA-DNA colony hybridization and preparation of *in vitro* labeled clones DNA was completed as described by Maniatis, et al., (1982) and the accompanying procedures supplied by the nitrocellulose manufacturer. Radiolabeled cosmid probes were made by nick translation using $[^{32}P]$ αdCTP with specific activity of 3000 Ci/mM (Amersham Corp., Arlington Heights, IL). Labeled probe material was separated from unincorporated nucleotides by Sephadex G]50 gel filtration. The large collection of fluorescent pseudomonads was screened by colony hybridization using a mixture of all four cosmid clones, pSFL10, pSFL11, pSFL12, and pSFL14. Positive isolates were then examined using single probes of each cosmid or subclone and agarose gel electrophoresis of *Eco*RI digested genomic DNA after the manner of Southern (1975). Satisfactory probes contained between 10^7 and 10^8 cpm/µg DNA. Genomic DNA suitable for digestion by restriction endonucleases was prepared from small volumes of stationary phase cells via a rapid SDS-lysis, phenol extraction and ethanol DNA precipitation procedure (Drahos, Brackin and Barry, manuscript in preparation).

T_4 DNA ligase and restriction enzymes *Bgl*II and *Eco*RI were obtained from Bethesda Research Laboratories, Inc. (Gaithersburg, MD) or from New England Biolabs (Beverley, MA). DNA polymerase I (Cat # 104485) was obtained from Boehringer-Mannheim (Indianapolis, IN). Electrophoresis was performed with 0.7% agarose slab gels in Tris-acetate buffer (40 mM Tris, lmM EDTA, pH 8.2 with glacial acetic acid) at 40 mA for 3-5 hrs. or typically until the bromophenol blue tracking dye had migrated approximately 8 cm. Gels were stained in 2 ug/ml ethidium bromide for 10 min and the DNA was visualized by ultraviolet transillumination. *Hind*III digested bacteriophage lambda DNA was end labeled for use as molecular size markers by the procedure of Maniatis, et al., 1982. Southern hybridizations were conducted at 65°C for 30 hrs. The filters were washed for 2 hrs. at 65°C in 0.2 x SSC and 0.1% SDS before air drying. Autoradiographic exposures were made with Kodak XAR-5 film using a (Dupont) lightning-plus intensifying screen. Exposure times were in the range of 12-16 hrs.

Plasmid pSFL11 was subcloned by restriction digestion with *Bgl*II, *Eco*RI and *Bgl*II/*Eco*RI, isolation of the resulting fragments and religation into pLAFR1 using standard procedures (Maniatis et al., 1982). Cosmid clones having single and multiple inserts were mated into the nonfluorescent mutant *Pseudomonas* strain JL2003 by methods previously described by Loper et al.,(1984).

Application of Cluster Analysis

Numerical analysis was completed on an IBM 3081 VM/CMS computer system with the cluster analysis package, Clustran 2.1 (released January, 1982), by application of the simple matching coefficient (Ssm) and unweighted average linkage (Wishart, 1978). Banding patterns from autoradiograms were converted to a binary format in the following fashion. Graph paper possessing 1 mm calibration lines was placed under autoradiograms atop a light box. Commencing from the top of the gel, each 2 mm segment down the gel represented a binary state test and coded 1 or 0 for the presence or absence of a band, respectively. In this fashion typically 32 such tests were regenerated from a gel and the banding pattern for each isolate was recorded. Preliminary editing and summarizing of data to remove non-discriminating tests (i.e., no band present for any strain in a discrete 2 mm section of the gel) was performed using EDITMAT, a computer program written to edit data matrices (Walczak and Krichevsky, 1980). By these means, a large number of strains and banding patterns can be systematically analyzed to rapidly identify and group isolates having similar patterns.

RESULTS AND DISCUSSION

The cosmid clones of this study were confirmed in their capacity to restore fluorescent pigment production in their respective mutants (see Table 1) as reported by Loper et al., (1984). The cosmids, pSFL10, pSFL11, pSFL12 and pSFL14 are distinct by *Eco*RI endonuclease restriction patterns; moreover, there appears to be very little, if any DNA homology between them as shown by the Southern hybridization procedure. An autoradiogram

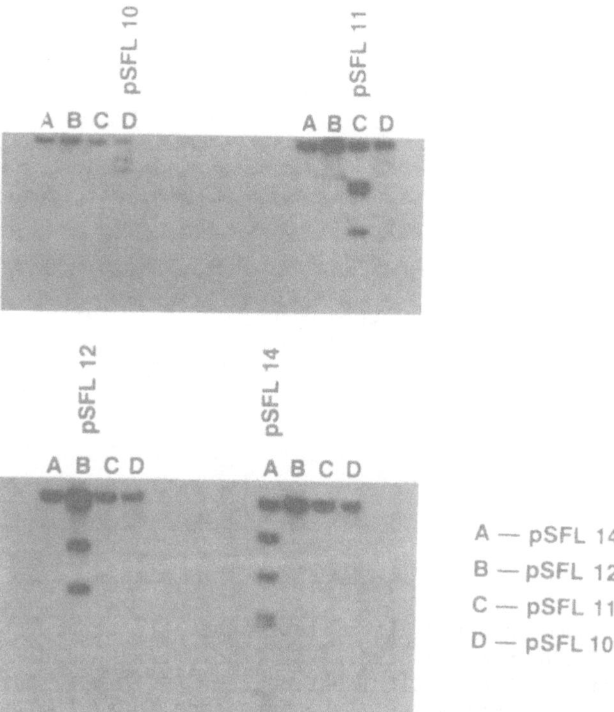

Fig. 1. Autoradiograms prepared after electrophoretic separation of *Eco*RI digests of DNA from the 4 siderophore complementation cosmids (pSFL10, pSFL11, pSFL12 and pSFL14) and subsequently with individually nick-translated hybridized cosmids DNAs. Lane assignments for digestion are given by letter of the alphabet as indicated.

produced by probing each cosmid digest after *Eco*Rl restriction with each of the four ^{32}P labeled cosmids is shown in Fig. 1. The sizes of the respective cosmid clones are presented in Table 1. A 4.0 kb size subclone of pSFL11 was generated in a *Bgl*II/*Eco*RI digestion which could still complement mutant JL2003. When the four cosmids were used as mixed probes in colony hybridization experiments to screen over 327 rhizosphere fluorescent pseudomonad isolates, only 97 showed significant hybridization. This method, however, may be influenced by many factors related to lysis of the cells and the amount of interfering debris of particular strains.

Positive isolates of the colony hybridization screen were examined by the Southern hybridization procedure. Under the stringent hybridization conditions employed, only fragments containing base sequences similar to those in the probe DNA will be shown on the autoradiogram. The intensity of the labeling is a function of the exposure duration, probe specific activity and base sequence divergence. Sequence divergence from the probe DNA and genomic DNA among the different isolates may cause an increase in the number of labeled fragments observed or an increased number of unpaired bases resulting in the decrease of hybrid DNA, such that the fragment is less intensely labeled. A typical autoradiogram of end labeled *Hind*III restricted bacteriophage DNA and genomic preparations of 19 strains of fluorescent pseudomonads, cut with *Eco*RI and hybridized with pSFL14 is presented in Fig. 2. The pseudomonads are designated by strain number and

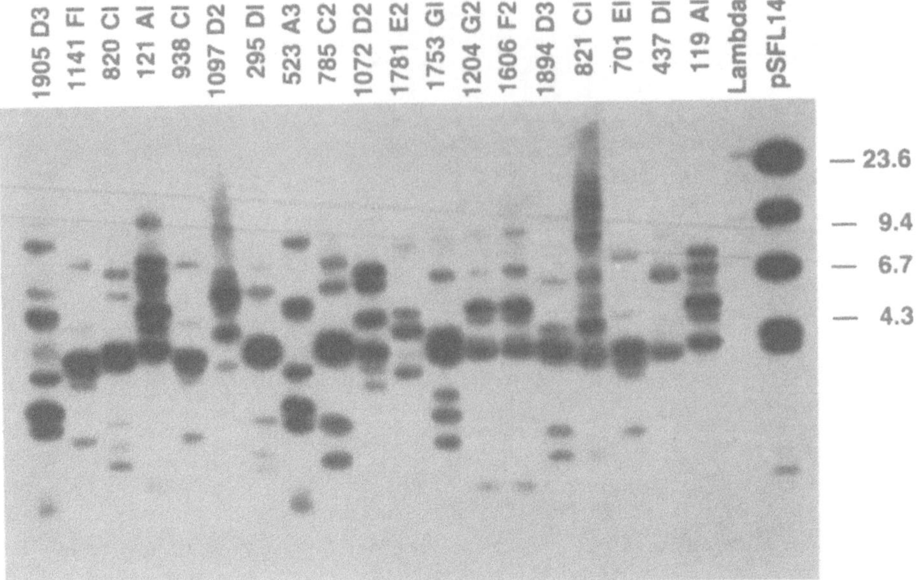

Fig. 2. Autoradiogram prepared after electrophoretic separation of *Eco*RI digests of DNA from 19 strains of fluorescent pseudomonads and hybridization with pSFL14 labeled with ^{32}P. *Hin*dIII digested lambda fragment sizes are given in megadaltons. The pseudomonad strains are listed in Table 3.

are described in Tables 2 and 3. Three similar gels were run concomitantly and prepared under the same conditions yet probed with the remaining cosmids, pSFL12, pSFL11, and pSFL10. The cosmid, pSFL10, was found not to hybridize under the conditions employed to any of these strains (data not shown); whereas, the autoradiograms produced from pSFL12 and pSFL11 are presented in Figs. 3 and 4, respectively. These results indicate the DNA of pSFL14 shows a great amount of sequence homology to all strains examined. The strains examined represent a wide range of fluorescent pseudomonads as discussed in the material and methods section. The API test results (Table 3) indicate 13 distinct test profiles were represented in the 19 strains presented. The cosmid pSFL14 was considered a candidate for possibly harboring gene sequences of siderophore regulatory functions (Loper et al., 1984), which may explain the broad nature of its hybridization to all strains examined. However, the nature of this DNA has not yet been established to function in this manner.

An additional 18 strains probed individually with pSFL14 and pSFL10 provide a contrast between the specificity exhibited by these siderophore-related cosmid clones (Figs. 5 and 6). The cosmid, pSFL10, strongly hybridizes to the genomic DNA of strain no. 5063Z1, a *Pseudomonas syringae* pv. *tagetes* isolate pathogenic to marigold and zinnia plants (Fig. 6). The original cosmid bank was made from a *P. s.* pv. *syringae* isolate (Loper et al., 1984); therefore it is expected that some sequence homology might exist. It should be noted that other *P. syringae* isolates did not show as great a homologous nature (isolates Ps6, GN2, PsC1B, etc.; Fig. 6). The broad specificity of pSFL14 is again demonstrated by those additional fluorescent pseudomonads (Fig. 5).

Six strains of an additional 5 pathovars of *P. syringae* including an

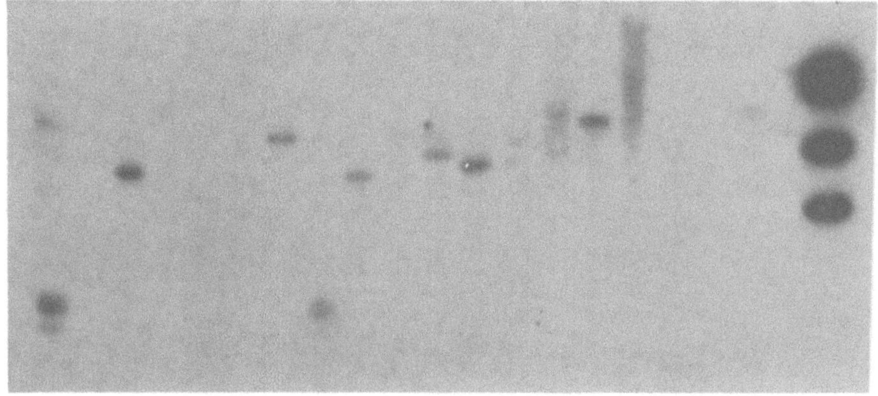

Fig. 3. Autoradiogram prepared after electrophoretic separation of *Eco*RI digests of DNA from 19 strains of fluorescent pseudomonads and hybridization with pSFL12 labeled with ^{32}P. The strains are ordered as in Fig. 2 and are fou,d listed in Table 3.

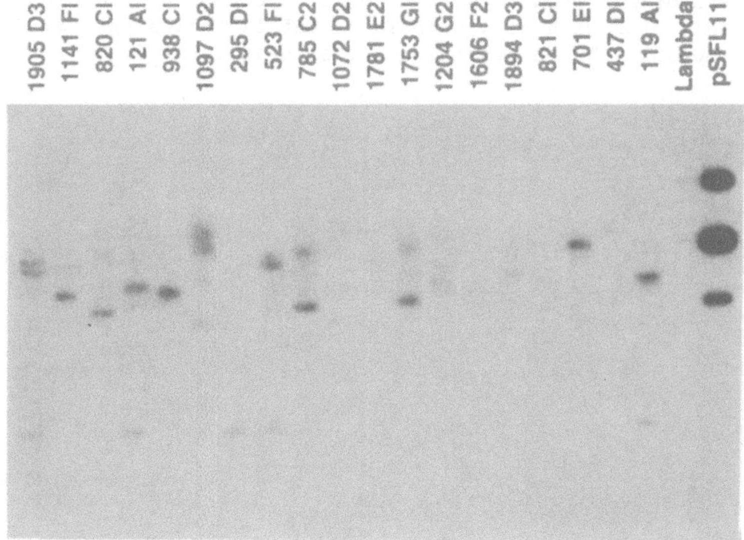

Fig. 4. Autoradiogram prepared after electrophoretic separation of *Eco*RI digests of DNA from 19 strains of fluorescent pseudomonads and hybridization with pSFL11 labeled with ^{32}P as in Figs. 2 and 3.

additional eight known fluorescent pseudomonad isolates, a few of which produce siderophores of known or partially known structure, were examined with pSFL10 and pSFL14 (Figs. 7 and 8). Under the hybridization conditions employed in this study, the vector pLAFR1 did not hybridize to any pseudomonad genomic preparations; however, a band visible in Figs. 7 and 8 of the *E. coli* strain JM101 has been demonstrated to be complementary to some sequences of the vector.

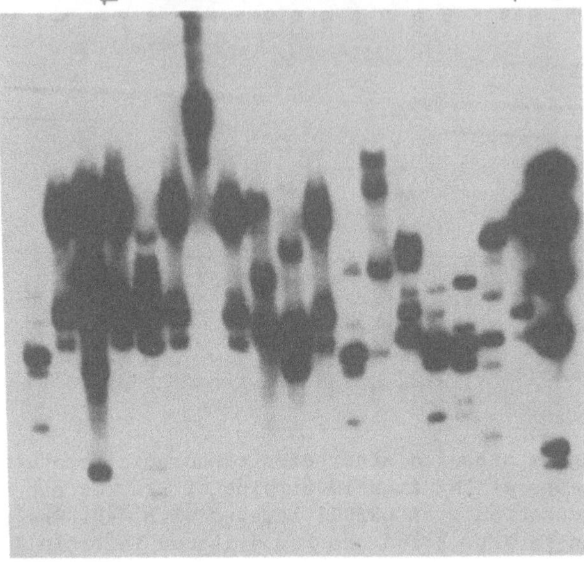

Fig. 5. Autoradiogram prepared after electrophoretic separation of *Eco*RI digests of DNA from 18 strains of fluorescent pseudomonads and hybridization with nick translated pSFL14 including *P.s. targetes* (No. 5063Z1). These strains are listed in Table 2.

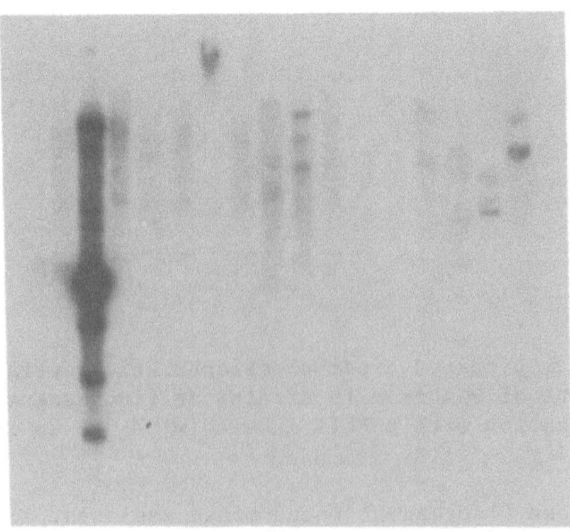

Fig. 6. Autoradiogram prepared after electrophoretic separation of *Eco*RI digests of DNA from 18 strains of fluorescent pseudomonads including *P.s. targetes* (5063Z1) and hybridization with pSFL10 in contrast to pSFL14 hybridization in Fig. 5. These strains are listed in Table 2.

Illustrative of the application of cluster analysis to grouping similar banding patterns is the presentation of the banding patterns for isolates of the two gels (shown in Figs. 7 and 8) in a dendogram format (Figs. 9 and 10). The *P. syringae* isolates, numbers 5075Z1 through 5078Z1

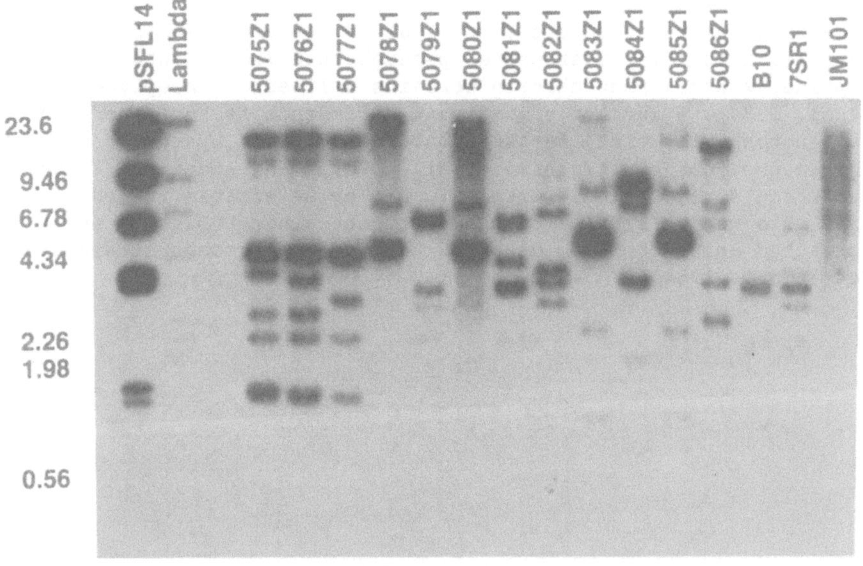

Fig. 7. Autoradiogram prepared after electrophoresis separation of *Eco*RI digests of DNA from an additional 14 strains of fluorescent strains including 5 pathovars of *P. syringae* and hybridization with pSFL14. These strains are listed in Table 2.

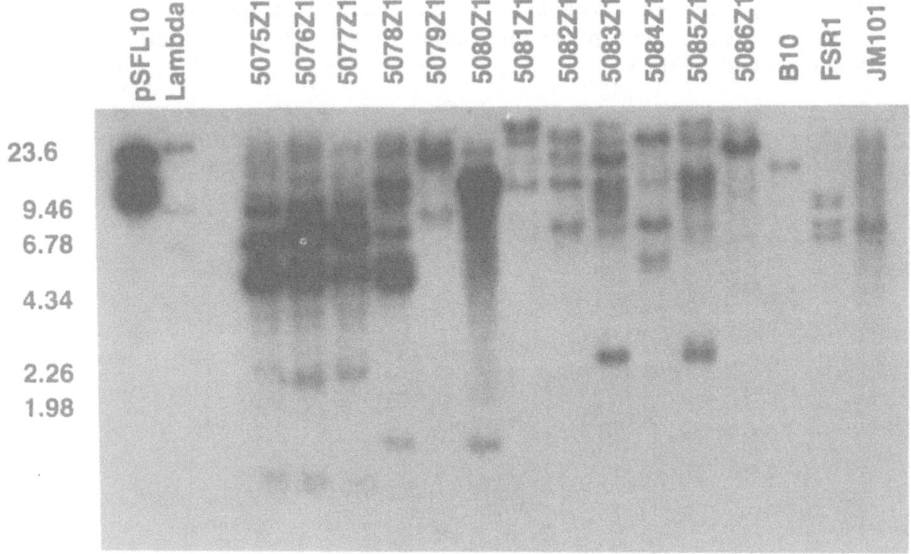

Fig.8. Autoradiogram prepared after electrophoresis separation of *EcoRI* digests of DNA from the strains indicated in Fig. 7 but with hybridization with pSFL10.

inclusive and 5079Z1, show strong homology to both probes. The banding patterns for the two 'tabaci' pathovars and *P. s.* pv. *angulata* although distinct are remarkably similar; therefore, these isolates cluster closely in the respective dendograms. This methodology is only applicable when the probes utilized show no homology to another.

The siderophore structure determination for *Pseudomonas* strain 7SR1 has revealed that this siderophore, pseudobactin 7SR1, contains a cyclic octapeptide, whereas pseudobactin contains a linear hexapeptide, (Yang and Leong, 1984). A further significant difference is the ester linkage of the fluorophoric moiety in pseudobactin 7SR1 as compared to the amide linkage in pseudobactin (Teintze et al., 1981; Yang and Leong, 1984). Whether these structural differences would be detectable in genotype patterns of a nature similar to those presented here (Figs. 7 and 8) could be approached using the cosmids produced from *Pseudomonas* strain B10 (Moores et al., 1984).

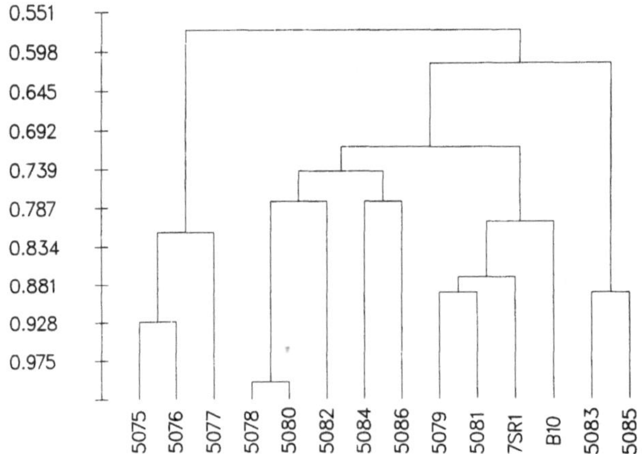

Fig. 9. The dendogram resulting from cluster analysis of the *Eco*R1 genomic restriction patterns shown in Fig. 7 for 14 pseudomonads when probed pSFL14.

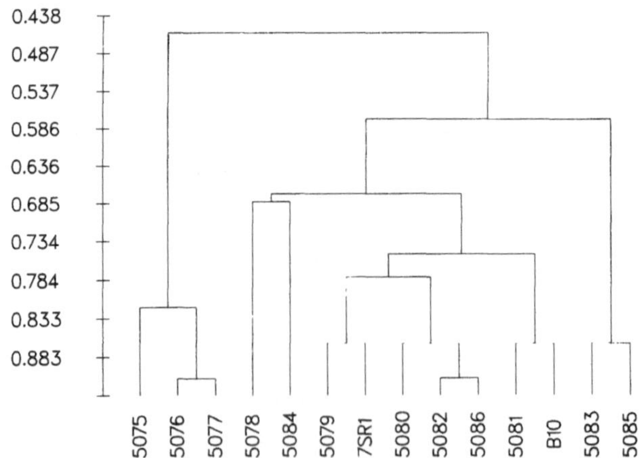

Fig. 10. The dendogram resulting from cluster analysis of the *Eco*RI genomic restriction patterns shown in Fig. 8 for 14 pseudomonads when probed with pSFL10.

SUMMARY

Southern hybridization analysis has identified the cosmid, pSFL10, as a possible species or subspecies specific probe. Further examination of various *P. syringae* pathovars will be required to validate this present claim. In addition to pSFL10, other siderophore complementation clones (particularly pSFL14) have demonstrated the capacity to distinguish pathovars of *P. syringae* by unique hybridization banding patterns (Fig. 7). Expanding the data base on this aspect is currently in progress to examine the possible application of these clones as a method for rapid identification of *P. syringae* pathovars. The specificity exhibited by the various cosmids of this study and the banding patterns observed support the conclusion that genotypic organization of siderophore-related DNA is highly diversified. The application of numerical analysis to restriction fragment patterns appears justified. When probes are distinct from one another, a combination of more than one probe may possibly be combined to produce a functional dendogram. Whether structural differences are reflected in the genotype patterns observed here will require further characterization of the specific genes involved.

ACKNOWLEDGEMENTS

We acknowledge the help of Ms. Vicki Grant and Ms. Barbara Schiermeyer for the typing of this manuscript. Gratitude is extended to Dr. David J. Drahos for his support and reading of the manuscript drafts. We are indebted to Dr. M.A. Schroth and associates for providing the necessary cosmids without which this study could not have been completed.

REFERENCES

Burr, T.J., Schroth, M.N. and Suslow, T., 1978. Increased potato yields by treatment of seed pieces with specific strains of *Pseudomonas fluorescens* and *P. putida*. Phytopathology, 68:1377-1383.

Ditta, G., Stanfield, S., Corbin, D., and Helinski, D.R., 1980. Broad host range DNA cloning system for Gram-negative bacteria: Construction of a gene bank of *Rhizobium meliloti*. Proc. Nat. Acad. Sci., USA, 77:7347-7351.

Friedman, A.M., Long, S.R., Brown, S.E., Buikema, W.J., and Ausubel, F.M., 1982. Construction of a broad host range cosmid cloning vector and its use in the genetic analysis of *Rhizobium* mutants. Gene, 18:289-296.

Fukasawa, K., Got, M., Sasaki, K., and Hirata, Y., 1972. Structure of the yellow-green fluorescent peptide produced by iron-deficient *Azotobacter vinelandii* strain O. Tetrahedron, 28:5359-5365.

Gross, D.C., 1982. Evidence that syringomycin, produced by *Pseudomonas syringae* pv. *syringae,* is a ferric siderophore. Phytopathology, 72:941 (Abstr.).

Gross, D.C., 1985. Regulation of syringomycin synthesis in *Pseudomonas syringae* pv. *syringae* and defined conditions for its production. J. Appl. Bacteriol., 58:167-174.

Hemming, G., Drahos, D., Brackin, J., and Jonsson, C., 1984. Characterization of HCN-producing soybean rhizosphere bacterial isolates. Phytopathology, 74 (8): A45 (Abstract).

Hemming, B.C., Orser, C., Jacobs, D.L., Sands, D.C. and Strobel, G.A., 1982. The effects of iron on microbial antagonism by fluorescent pseudomonads. J. Plant Nutr., 5 (4-7):683-702.

Hemming, B.C., and Strobel, G.A., 1983. Bacterial disease and the iron status of plants in: "Genetic Aspects of Plant Nutrition," M.R.(Saric and B.C. Loughman, Eds.). Martinus Nijhoff/Dr. W. Junk, Pubs, The Hague.

Howell, C.R., and Stipanovic, R.D., 1979. Control of *Rhizoctonia solani* on cotton seedlings with *Pseudomonas fluorescens* and with an antibiotic produced by the bacterium. Phytopathology, 69:480-482.

Howell, C.R., and Stipanovic, R.D., 1980. Suppression of *Pythium ultimum*-induced damping-off of cotton seedlings by *Pseudomonas fluorescens* and its antibiotic, pyoluteorin. Phytopathology, 70:712-715.

King, E.O., Wards, M.K., and Raney, D.E., 1954. Two simple media for the demonstration of Pyocyanin and Fluorescin. J. Lab. Clin. Med. 44:301-307.

Kloepper, J.W., and Schroth, M.N., 1981. Relationship of *in vitro* antibiosis of plant growth-promoting rhizobacteria to plant growth and the displacement of root microflora. Phytopathology, 71:1020-1024.

Kloepper, J.W., Schroth, M.N., and Miller, T.D., 1980. Effects of rhizosphere colonization by plant growth-promoting rhizobacteria on potato plant development and yield. Phytopathology, 70:1078-1082.

Knosp, O., von Tigerstrom, M., and Page, W.J., 1984. Siderophore-mediated uptake of iron in *Azotobacter vinelandii*. J. Bacteriol., 159 (1):341-347.

Leisinger, T., and Margraff, R., 1979. Secondary metabolites of the fluorescent pseudomonads. Microbiol. Revs., 43 (3):422-442.

Lenhoff, H.M., 1963. An inverse relationship of the effects of oxygen and iron on the production of fluorescin and cytochrome c by *Pseudomonas fluorescens*. Nature, London, 199:601-602.

Lindow, S.E., 1982. Population dynamics of epiphytic ice nucleation positive bacteria on frost-sensitive plants and frost control by means of antagonistic bacteria in: "Plant Cold Hardiness," (A. Sakai and P.H. Li, Eds.), Academic Press, New York.

Loper, J.E., Orser, C., Panopoulos, N.J., and Schroth, M.N., 1984. Genetic analysis of fluorescent pigment production in *Pseudomonas syringae* pv. *syringae*. J. Gen. Microbiol., 130:1507-1515.

Maniatis, T., Fritsch, E.F., and Sambrook, J., 1982. "Molecular Cloning, A Laboratory Manual," Cold Spring Harbor, New York.

Maurer, B., Müller, A., Keller-Schierlein, W., and Zahner, H., 1968. Ferribactin, a siderochrome from *Pseudomonas fluorescens* Migula. Arch. Mikrobiol., 60:326-339.

Meyer, J.M., and Abdallah, M.A., 1978. The fluorescent pigment of *Pseudomonas fluorescens:* biosynthesis, purification and physicochemical properties. J. Gen. Microbiol., 107:319-328.

Meyer, J.M., and Hornsperger, J.M., 1978. Role of pyoverdine, the iron-binding fluorescent pigment of *Pseudomonas fluorescens,* in iron transport. J. Gen. Microbiol., 107:329-331.

Moores, J.C., Magazin, M., Ditta, G.S., and Leong, J., 1984. Cloning of genes involved in the biosynthesis of pseudobactin, a high-affinity iron transport agent of a plant growth-promoting *Pseudomonas* strain. J. Bacteriol., 157 (1): 53-58.

Myers, D.F., and Strobel, G.A., 1983. *Pseudomonas syringae* as a microbial antagonist of *Ceratocystis ulmi* in the apoplast of American elm *(Ulmus americana)*. Trans. Br. Mycol. Soc. 80 (3):389-394.

Neilands, J.B., 1981. Microbial iron compounds. Ann. Rev. Biochem., 50:715-731.

Orser, C., Staskawicz, B.J., Loper, J.E., Panopoulos, N.J., Lindow, S.E., Dahlbeck, D., and Schroth, M.N., 1983. Cloning of genes involved in bacterial ice nucleation and fluorescent pigment/siderophore production. In: "Molecular Genetics of the Bacteria-Plant Interaction," (A. Pühler, Ed.). Springer-Verlag, Berlin.

Page, W.J., and Huyer, M., 1984. Derepression of the *Azotobacter vinelandii* siderophore system, using iron-containing minerals to limit iron repletion. J. Bacteriol. 158 (2): 496-502.

Palleroni, N.J., 1975. General properties and taxonomy of the genus *Pseudomonas*. In: "Genetics and Biochemistry of *Pseudomonas*," (P.H. Clarke and M.H. Richmond, Eds.). John Wiley and Sons, London.

Philson, S.B., and Llinas, M., 1982. Siderochromes from *Pseudomonas fluorescens*. II. Structural homology as revealed by NMR spectroscopy. J. Biol. Chem., 257:8086-8090.

Scher, F.M., and Baker, R., 1982. Effect of *Pseudomonas putida* and a synthetic iron chelator on induction of soil suppressiveness to Fusarium wilt pathogens. Phytopathology, 72:1567-1573.

Schroth, M.N., and Hancock, J.G., 1981. Selected topics in biological control. Ann. Rev. Microbiol., 35:453-476.

Schroth, M.N., and Hancock. J.G., 1982. Disease-suppressive soil and root-colonizing bacteria. Science, 216:1376-1381.

Southern, E., 1975. Detection of specific sequences among DNA fragments separated by gel electrophoresis. J. Mol. Biol, 98:503.

Suslow, T.V., and Schroth, M.N., 1982. Rhizobacteria of sugar beets: Effects of seed application and root colonization on yield. Phytopathology, 72:199-206.

Suslow, T.V., and Schroth, M.N., 1982. Role of deleterious rhizobacteria as minor pathogens in reducing crop growth. Phytopathology, 72: 111-115.

Teintze, M., Hossain, M.B., Barnes, C.L., Leong, J. and Van der Helm, D., 1981. Structure of ferric pseudobactin, a siderophore from a plant growth-promoting *Pseudomonas*. Biochemistry, 20:6446-6457.

Walczak, C.A., and Krichevsky, M.I., 1980. Computer methods for describing of Pyocyanin and Fluorescin. J. Lab. Clin. Med., 44:301-307.

Wendenbaum, S., Denmange, P., Dell, A., Meyer, J.M. and Abdallah, M.A., 1983. The structure of pyoverdine Pa, the siderophore of *Pseudomonas aerugonisa*. Tetrahedron Lett. 24 (44):4877-4880.

Weller, D.M., 1983. Colonization of wheat roots by a fluorescent pseudomonad suppressive to take-all. Phytopathology, 73:1548-1553.

Weller, D.M., 1984. Distribution of a Take-All Suppressive Strain of *Pseudomonas fluorescens* on Seminal Roots of Winter Wheat. Appl. Environ. Microbiol., 48(4):897-899.

Weller, D.M., and Cook, R.J., 1983. Suppression of take-all of wheat by seed treatments with fluorescent pseudomonads. Phytopathology, 73:463-469.

Wishart, D., 1978. "CLUSTAN User Manual," 3rd edition, Program Library Unit, Edinburgh University, Edinburgh.

Yang, C.-C., and Leong, J., 1984. Structure of pseudobactin 7SR1, a siderophore from a plant deleterious *Pseudomonas*. Biochemistry, 23(15):3534-3540.

BIOLOGICAL CONTROL OF PHYTOPATHOGENS BY *PSEUDOMONAS* SPP.: GENETIC ASPECTS OF SIDEROPHORE PRODUCTION AND ROOT COLONIZATION

F. O'Gara, P. Treacy, D. O'Sullivan, M. O'Sullivan and P. Higgins

Microbiology Department
University College Cork
Ireland

INTRODUCTION

Interest in plant-microbial interactions has received new impetus in recent years. This has resulted mainly from an increased awareness of the benefits to be gained for crop production from the exploitation of beneficial bacteria. The role of microorganisms in influencing plant productivity is a widely researched and broad topic. However, when focussing on the involvement of microorganisms in crop production attention generally tends to concentrate on a number of well recognised plant-microbial interactions. In the context of beneficial interactions, the role of symbiotic nitrogen fixing *Rhizobium* spp. is well recognised as an important area because of the clearly proven contribution these bacteria make to crop production. In a converse situation, the necessity to investigate the nature of interactions between pathogenic microorganisms and plants is also well recognised due to the reduction in yields that result as a consequence of plant pathogens. More recently however, with the increased awareness of the potential for exploiting microorganisms for biotechnological purposes, attention has been focussed on the use of microorganisms for biological control purposes. The concept of biological control is based on using a beneficial microorganism to control the activity of undesirable populations. It is not a new concept and is considered to occur naturally, being an important factor in explaining how plant diseases are usually not epidemic.

Seed inoculation with plant growth promoting *Rhizobacteria* have been exploited to increase plant yields (Kloepper et al., 1980; Schroth and Hancock, 1981; Weller and Cook, 1983), but collective results of experiments on biological control of plant disease in field trials through inoculation have not been clear cut in producing consistently significant effects. A major challenge in the exploitation of microorganism for biological control purposes lies in the ability to transfer laboratory scale success to field conditions. Success in this task will depend largely on having a clear understanding of the basic biology of microorganisms involved in biological control and how they interact with plants. The availability of genetic engineering technology provides the necessary tools to achieve this understanding and to programme microorganisms for improved use as biological control agents.

Table 1. Growth of *Pseudomonas* spp. in the presence of EDDA

Strains	EDDA Concentration (mg/ml)					
	0	0.31	0.62	1.25	2.5	5.0
B2/6	++	++	++	++	++	++
E2/6	++	++	++	++	-	-
M4/10	++	++	++	-	-	-
E1/1	++	++	+	-	-	-
M6/3	++	++	++	++	+	-
M11/4	++	++	++	+	-	-
R11/2	++	++	++	+	-	-
R15/2	++	++	++	++	+	-
R5/1	++	++	++	+	+	+
R21/1	++	-	-	-	-	-
C2/1	++	++	++	+	+	-
F1/4	++	-	-	-	-	-
C2/2	++	+	+	-	-	-
A24/4	++	++	++	+	+	-
R8/3	++	++	+	+	+	-
R8/4	++	+	+	+	-	-
R9/3	++	+	+	+	-	-
R8/1	++	++	++	+	-	-
D2/7	++	++	+	+	-	-
R11/3	++	++	+	+	-	-
D1/7	++	++	+	+	-	-
E1/4	++	++	+	+	-	-
F2/5	++	++	+	+	-	-
C1/1	++	++	+	+	-	-
R9/4	++	++	+	+	-	-
Erwinia	++	+	+	+	-	-
P. fluorescens	++	++	++	+	+	

Isolates were streaked on sucrose-asparagine plates containing the different concentrations of EDDA and incubated at 25°C for 2-3 days.
++ : strong growth. + : moderate growth. - : no growth

<u>Siderophore producing</u> *Pseudomonas* <u>spp</u>.

Bacterial strains from the *Pseudomonas fluorescens-putida* group are currently receiving considerable attention as biological control agents (Burr et al., 1978; Schroth and Hancock, 1983). These strains produce iron chelating siderophores which are considered to play a key role in biological control.

The siderophores are capable of complexing Fe^{+++} and essentially making it unavailable for competing pathogens. The most suitable bacteria for use as biological control agents should therefore be capable of producing siderophore(s) with a high affinity for iron. Bacterial isolates with this particular characteristic were sought from the roots of sugar beet plants. As the microbial population on plant root and in the rhizosphere is complex and diverse, it was decided to selectively enrich for the recovery of these bacteria. By incorporating an iron chelating agent in the growth media, only the growth of microorganisms capable of competing for the available iron is

Table 2. Growth inhibition of microbial pathogens by *Pseudomonas* spp. B2-6

	Inhibition
Geotrichium candidum	+
Aspergillus niger	+
Rhizoctonia solani	++
Fusarium oxysporum	+
Botrytis cinerea	++
Phoma beta	++
Penicillium claviforme	++
Erwinia caratovora	+++

Inhibition tests were performed on sucrose asparagine minimal medium (Garibaldi and Neilands, 1955). Siderophore diffused from *Pseudomonas* B2/6 inoculated on to the centre of the plates. Plates were then sprayed with suspensions of the fungal or bacterial pathogens and incubated at 25°C.

+ = inhibition zone less than 10 mm in diameter
++ = inhibition zone greater than 10 mm in diameter
+++ = inhibition zone greater than 25 mm in diameter

permitted. Bacterial isolates prepared from the root system of sugar beet were plated on to sucrose-asparagine media (Scher and Baker, 1982) containing the synthetic iron chelators ethylene-diamine-Di-[O-hydroxy-phenyl acetic acid] (EDDA, 0.3 mg/ml) and 2,2 dipyridyl (500 µm-6 mM). The majority of isolates capable of growing on the higher concentration of these chelators produced a yellowish fluorescent pigment. Standard classification tests indicated that they were *Pseudomonads*.

Detailed analysis based on the physiological and biochemical characteristics devised for the genus *Pseudomonas* (Stanier et al., 1966) grouped them in the *P. fluorescens* and *P. putida* species. However, a number of the isolates were atypical in that they did not conform to either of these groups. The growth of the isolates in the presence of the iron chelators showed considerable variation. Twenty five isolates were screened for their growth response to a range of different concentrations of EDDA (0-5 mg/ml; Table 1). While the majority of the isolates showed some growth at a concentration of 1.25 mg/ml EDDA, only 28% could grow at an EDDA concentration of 2.5 mg/ml and 4% at a concentration of 5 mg/ml. Only those isolates capable of growing at the higher concentrations were selected for further study.

A siderophore from *Pseudomonas* B10 termed Pseudobactin has been purified and characterized (Teintze et al.,1981). The adsorption spectrum for this

Table 3. Transfer frequency of RP_4 and Tn5 into siderophore producing *Pseudomonas* spp.

Strain	RP_4 transfer frequency	Tn5 transfer frequency
F2/2	1.0×10^{-1}	2.5×10^{-3}
D2/7	1.5×10^{-3}	$< 1 \times 10^{-8}$
M11/4	1.2×10^{-2}	6.6×10^{-5}
A24/4	1.0×10^{-4}	1.0×10^{-6}
R8/3	3.6×10^{-7}	$< 1 \times 10^{-8}$
R8/1	6.2×10^{-6}	$< 1 \times 10^{-8}$

Exconjugants were selected on sucrose asparagine medium containing kanamycin at 25 μg/ml. Transfer frequency is expressed per recipient.

siderophore shows a maximum absorption peak in the region of 400 nm. The adsorption spectra of culture filtrates of the sugar beet isolates also showed maximum adsorption profiles in the 410 nm range. When these *Pseudomonas* isolates were grown in iron enriched media the characteristic adsorption peak in the 410 nm range was not visible.

The ability of the siderophore producing *Pseudomonas* spp. isolated from sugar beet roots to inhibit the growth of common bacterial and fungal pathogens was investigated (Table 2). For these experiments isolate B2/6 which grows in the presence of the highest concentration of EDDA was chosen. As shown in Table 2 this strain was capable of inhibiting a range of common bacterial and fungal pathogens in '*in vitro*' plate tests. No significant inhibition was observed in these tests when the growth media was enriched with exogenous iron.

Plasmid Transfer in *Pseudomonas* spp.

The application of genetic engineering technology to siderophore producing *Pseudomonas* spp. to improve or modify specific traits desirable for biological control purposes is dependent on an available gene transfer system. Broad host range plasmids have been exploited to promote gene transfer (Holloway et al., 1979) and construct cloning vectors (Bagdasarian et al., 1981; Ditta et al., 1980) suitable for use in a number of *Pseudomonas* spp. The siderophore producing *Pseudomonas* spp. isolated from sugar beet roots were screened for their ability to accept incompatibility P (RP_4), and Q (pMM34) group plasmids and transposon Tn5 from the Tn5-Mob system (Simon, 1984). A selection of strains chosen showed considerable variation in the transfer frequency of RP, and Tn5 (Table 3). Exconjugates could not be detected for some of the isolates and in some instances deletions were observed in the P-group plasmids that established. The transfer frequency for the Inc-Q plasmid varied between 10^{-2} and 10^{-4} per recipient.

These isolates were also screened for the presence of indigenous plasmids. A number of *Pseudomonas* spp. have been shown to harbour plasmids which may vary in size up to 300 M daltons (Curiale and Mills, 1977; Hansen and Olsen, 1978; Rosenberg et al., 1982). Isolates were screened for plasmid DNA using the methods of Eckhardt (1978) and Kado and Liu (1981). These procedures are particularly suited for detecting very large plasmids.

Table 4. Characterisation of Siderophore Mutants

Class	Subclass	Fluorescence		Growth in	Inhibition of *Erwinia*	
		2 days	7 days	Presence of EDDA	No Fe^{3+}	S/A + Fe^{3+}
I	(i) A24-4 M6-3	-	-	-	-	-
	(ii) B2-6	-	-	+++	-	-
II	(i) A24-4 M6-3	-	+++	-	+	-
	(ii) B2-6	-	+++	+++	+	-
w/t		+++	+++	+++	+++	-

- = Negative result; +++ = Maximum levels; + = Reduced levels; S/A = Minimal sucrose asparagine medium (no added Fe^{3+}).

From 50 isolates screened, plasmid DNA could only be detected in two isolates. The techniques used were capable of resolving large plasmids (300-400 M daltons) from *Rhizobium* strains and could detect RP_4 when transferred into plasmid free isolates. Therefore it appears that only a minority of these strains harbour indigenous plasmids.

Mutants defective in siderophore production

The antagonistic response of fluorescent *Pseudomonas* spp. against microbial pathogens is largely attributed to the effect of the siderophore(s) complexing available iron. The main evidence for this is based on the fact that growth inhibition can be reversed by the addition of exogenous iron to culture media (Kloepper et al., 1980). However, in a number of instances evidence suggests that antibiotic like substances may also contribute to the inhibitory response (Leisinger and Margraff, 1979). The isolation of mutants specifically defective in the biosynthesis of siderophore(s) provides a direct and unambiguous approach to understanding the nature of this inhibitory response. Furthermore isolation of mutants is a prerequisite step to study the genetic organization and regulation of the genes involved in siderophore biosynthesis.

Mutants defective in the synthesis of the fluorescent siderophores were isolated in a number of *Pseudomonas* spp. from sugar beet roots. Both chemical and transposon mutagenesis with Tn5 was exploited. Potential mutants were isolated as fluorescent negative clones following N-Methyl-N nitro-nitrosoguanidine (N.T.G.) and Tn5 mutagenesis. The properties of mutants isolated following N.T.G. treatment (30 μg/ml for 15 min) are presented in Table 4. For all three strains (A24-4, M6-3, B2-6), the frequency of fluorescent negative (Flu^-) mutants was approximately 5×10^{-3}. A class of mutant which was fluorescent negative and failed to grow in the presence of EDDA was unable to inhibit *Erwinia* in contrast to the wild type strain. A similar mutant class has also been isolated in *Pseudomonas* B10 (Moores et al., 1984) and *Pseudomonas putida* WCS358 (Marugg et al., 1985). A second class of mutant in B2-6 which was fluorescent negative and unable to inhibit *Erwinia* but capable of growing in the presence of EDDA was also isolated. This type of mutant has not been described previously and an understanding of the nature of this mutation will require further investigation. Collectively the results on the siderophore mutants confirm that the growth inhibitory response of these isolates is dependent on the synthesis of siderophore(s).

Root colonization

In practical terms the ability of siderophore producing *Pseudomonas* spp. to colonize roots in high numbers is of fundamental importance for the effective use of these bacteria as biological control agents. Root colonization can be broadly divided into two phases. Firstly, there are the early events leading to the attraction and interaction of bacteria with the roots. This is followed by the "persistence phase" where the particular bacteria utilize root exudates to multiply and survive. The ideal biological control system can be envisaged as a complex 'bioreactor-type' system. The siderophore producing bacteria supported on the root surface utilize available nutrients to sustain microbial metabolism and excrete siderophores into the surrounding rhizosphere.

The method employed to isolate fluorescent *Pseudomonas* from the roots of plants may have an influence on the type of population recovered. For example, in this study a number of the siderophore producing *Pseudomonas* spp. used were isolated from roots following selection on EDDA. While this would provide a specific selection for bacteria producing siderophore(s) with a high affinity for iron it would not provide a selection for aggress-

Table 5. Colonization of Sugar Beet Roots by *Pseudomonas* spp.

Strain	Colonization on roots in autoclaved soil	Colonization on roots in untreated soil
M11/4	1.2×10^6 (60%)	1×10^4 (4.6%)
B2/6	-	3.3×10^6 (54%)
M6/3	1.9×10^4 (2%)	$< 1 \times 10^{-1}$ (< 0.01%)
A24/4	1.0×10^5 (78%)	$< 1 \times 10^{-1}$ (< 0.01%)

Figures in parentheses represent the population of the inoculum *Pseudomonas* spp. as a percentage of the total bacterial count.

-ive colonization. Therefore we investigated the colonization ability of some isolates on sugar beet plants. Four strains that were capable of growing on relatively high concentrations of EDDA (Table 1) were chosen. Root colonization was investigated in autoclaved and untreated soils (Table 5). Surface sterilized beet seeds were dipped in a suspension of the *Pseudomonas* strains (approximately 10^9 cfu/ml) genetically marked with antibiotic resistance markers (A24-4 Rif Sp; M6-3 Nal; M11-4 Chl; B2-6Nal, Chl.). Seeds were then allowed to germinate and develop for 4-6 weeks in the soil preparations in large test-tubes. Bacterial populations on excised and washed roots were determined by the standard plate count procedure on complex and selective media. The results obtained (Table 5) indicate that the colonization ability of the strains were quite different. In autoclaved soil strains M11-4 and A24-4 could be recovered in high numbers but not from plants grown in untreated soil. Strain B2-6 could be recovered in high numbers from plants grown in untreated soil indicating that it could compete and colonize in the presence of indigenous soil microorganisms. Strain M6-3 showed poor colonization ability even in autoclaved soil and may be a useful strain to study factors influencing colonization. For example, preliminary studies on the metabolic flexibility of this strain demonstrate that it is lacking protease activity. This raises the possibility of an involvement of protease activity during the colonization process with these strains. Studies on the colonization of roots by *Pseudomonas* strains which had different abilities for gelatin hydrolysis failed to establish a correlation between this activity and colonization ability (Scher et al., 1984). However, as *Pseudomonas* strains may produce more than one type of protease (London et al., 1984; Stapleton et al., 1984) further investigations based on biochemical genetics will be necessary to unambiguously determine if protease(s) are involved in aggressive root colonization by fluorescent *Pseudomonas*.

ACKNOWLEDGEMENTS

This work was supported by grants HEIC 76/83 and SGI/84 from the National Board for Science and Technology.

REFERENCES

Bagdasarian, M., Lurz, R., Franklin, F.C.H., Bagdasarian, M.M., Frey, J., and Timmis, K.N., 1981. Specific-purpose plasmid cloning vectors II. Broad host range, high copy number, RSF1010-derived vectors, and a host-vector system for gene cloning in *Pseudomonas*. Gene 16:237-247.

Burr, T.J., Schroth, N., and Suslow, T.V., 1978. Increased potato yields by treatment of seed pieces with specific strains of *Pseudomonas fluorescens* and *P. putida*. Phytopathology, 68:1377-1383.

Curiale, M.S., and Mills, D., 1977. Detection and characterization of plasmids in *Pseudomonas glycinea*. J. Bacteriol., 131:224-228.

Ditta, G., Stanfield, S., Corbin, D., and Helsinki, D.R., 1980. Broad host range DNA cloning system for Gram-negative bacteria: construction of a gene bank of *Rhizobium meliloti*. Proc. Natl. Acad. Sci. U S A., 77:7347-7351.

Eckhardt, T., 1978. A rapid method for the identification of plasmid deoxyribonucleic acid in bacteria. Plasmid, 1:584-588.

Garibaldi, J.A., and Neilands, J.B., 1955. Isolation and properties of ferrichrome A. J. Am. Chem. Soc., 77:2429-2436.

Hansen, J.B., and Olsen, R.H., 1978. Inc P2 group of *Pseudomonas*, a class of uniquely large plasmids. Nature, 274:715-717.

Holloway, B.W., Krishnapillai, V., and Morgan, A.F., 1979. Chromosomal genetics of *Pseudomonas*. Microbiol. Rev., 43(1):73-102.

Kado, C.I., and Liu, S.T., 1981. Rapid procedure for detection and isolation of large and small plasmids. J. Bacteriol., 145:1365-1373.

Kloepper, J.W., Long, J., Teintze, M., and Schroth, M.N., 1980. Enhanced plant growth by siderophores produced by plant growth-promoting rhizobacteria. Nature, 286:885-886.

Leisinger, T., and Margraff, R., 1979. Secondary metabolites of the fluorescent *Pseudomonads*. Microbiol. Rev., 43:422-442.

London, C.J., Griffith, I.P., and Kortt, A.A., 1984. Proteinases produced by *Pseudomonads* isolated from sheep fleece. Applied Environ. Microbiol., 47:75-79.

Marugg, J., Van Spanje, M., Schippers, B., and Weisbeek, P.J., 1985. The iron uptake system of plant growth-stimulating *Pseudomonas*. In: "Genetic Engineering of Plants and Microorganisms Important for Agriculture". (Eds. E. Magnien, and D. de Nettancourt) Martinus Nijhoff Junk Publ.

Moores, J.C., Magazin, M., Ditta, G.S., and Leong, J., 1984. Cloning of genes involved in the biosynthesis of pseudobactin, a high-affinity iron transport agent of a plant growth-promoting *Pseudomonas* strain. J. Bacteriol., 157:53-58.

Rosenberg, C., Casse-Delbart, F., Dusha, I., David, M., and Boucher, C., 1982. Megaplasmids in the plant-associated bacteria *Rhizobium meliloti* and *Pseudomonas solanacearum*. J. Bacteriol., 150:402-406.

Scher, F.M., and Baker, R., 1982. Effect of *Pseudomonas putida* and a synthetic iron chelator on induction of soil suppressiveness to *Fusarium* -wilt pathogens. Phytopathology, 72:1567-1573.

Scher, F.M., Ziegle, J.S., and Kloepper, J.W., 1984. A method for assessing the root-colonizing capacity of bacteria on maize. Can. J. Microbiol., 30:151-157.

Schroth, M.N., and Hancock, J.G., 1981. Selected topics in biological control. Annu. Rev. Microbiol., 35:453-476.

Schroth, M.N., and Hancock, J., 1983. Disease-suppressive soil and root colonizing bacteria. Science, 216:1376-1382.

Simon, R., 1984. High frequency mobilization of gram-negative bacterial replicons by the *in vitro* constructed Tn5-Mob transposon. Mol. Gen. Genet., 196:413-420.

Stanier, R.Y., Palleroni, N.J., and Doudoroff, M., 1966. The aerobic *Pseudomonads*; a taxonomic study. J. Gen. Microbiol., 43:159-271.

Stapleton, M.J., Jagger, K.S., and Warren, R.L., 1984. Transposon mutagenesis of *Pseudomonas aeruginosa* exoprotease genes. J. Bacteriol., 157: 7-12.

Teintze, M., Hossain, M.B., Barnes, C.L., Leong, J., and Van der Helm, D., 1981. Structure of ferric pseudobactin, a siderophore from a plant growth promoting *Pseudomonas* B10. Biochemistry, 20:6446-6457.

Weller, D.M., and Cook, R.J., 1983. Suppression of take-all of wheat by seed treatments with fluorescent *Pseudomonads*. Phytopathology, 73: 463-469.

PARTICIPANTS

ABDALLAH, Dr. M.A., Institut de chimie, Universite Louis Pasteur de Strasbourg, 1, rue Blaise Pascal, 67008 Strasbourg, Cédex, Strasbourg, France.

ALABOUVETTE, Dr. C., Department de Pathologie Vegetale, INRA, Station de Recherches, Sur La Flore Pathogene Dans Le Sol, 17 rue Sully, 21034 Dijon Cédex, France.

ALSTRÖM, Mrs. S., Swedish University of Agricultural Sciences, Department of Plant and Forest Protection, P.O. Box 7044, S-750 07 Uppsala, Sweden.

BAKER, Professor R., Department of Plant Pathology and Weed Sciences, Colorado State University, Fort Collins, Colorado 80523, U.S.A.

BAKKER, A.W., Willie Commelin Scholten Phytopathological Laboratory, Javalaan 20, 3742CP Baarn, The Netherlands.

BAKKER, P.A.H.M., Willie Commelin Scholten Phytopathological Laboratory, Javalaan 20, 3742CP Baarn, The Netherlands.

BARASH, Professor I., Tel-Aviv University, George S. Wise Faculty of Life Sciences, Department of Botany, Ramat-Aviv, Tel-Aviv, Israel.

BIENFAIT, Dr. H.F., University of Amsterdam, Laboratory for Plant Physiology, Kruislaan 318, 1098 SM Amsterdam, The Netherlands.

BROWN, Dr. A.E., Department of Agriculture for Northern Ireland, Faculty of Agriculture and Food Science, The Queen's University of Belfast, Newforge Lane, Belfast, BT9 5 PX, N. Ireland.

BYERS, Professor B. Rowe, Department of Microbiology, The University of Mississippi Medical Center, 2500 North State Street, Jackson, Mississippi, 39216-4505, U.S.A.

CAMPBELL, Dr. R., Department of Botany, University of Bristol, Woodland Road, Bristol, BS8 1UG, U.K.

DECOCK, Dr. P.C., "Freshfield", 430 North Deeside Road, Cults, Aberdeen, AB1 9ER, Scotland, U.K.

DE WEGER, Dr. L., Department of Plant Molecular Biology, Botanical Laboratory, State University, Nonnensteeg 3, 2311 VJ Leiden, The Netherlands.

EXPERT, Dr. D., Ministére de L'Agriculture, Institut National Agronomique P-G, Laboratoire de Pathologie Végétale, 16 rue Claud Bernard, 75231 Paris Cédex 05, France.

FUNCK-JENSEN, Dr. D. The Royal Veterinary and Agricultural University, Department of Plant Pathology, Thorvaldsenvej 40, DK-1871 Copenhagen V, Denmark.

GARRETT, Dr. C.M.E., Plant Pathology Department, East Malling Research Station, East Malling, Maidstone, Kent, ME19 6BJ, U.K.

GREAVES, M.P., Long Ashton Research Station, Weed Research Division, Begbroke Hill, Yarnton, Oxford OX5 1PF, U.K.

HADAR, Dr. Y., Department of Plant Pathology and Microbiology, The Levi Echkol School of Agriculture, The Hebrew University of Jerusalem, Rehovot 76-100, P.O. Box 12, Israel.

HEDGES, Dr. R.W., Plant Genetic Systems, n.v., J. Plateaustraat 22, B-9000 Gent, Belgium.

HEMMING, Dr. B.C., Rhizobiology Group/Biological Sciences, Corporate Research and Development Staff, Mailzone AA2G, Monsanto Life Sciences, Research Center, 700 Chesterfield Village Parkway, Chesterfield, MO 63198, U.S.A.

HENRY, L., De Danske Sukkerfabrikker, Technological Research Laboratories, Biotechnology Section, 1 Langebrogade, P.O. Box 17, DK-1001 Copenhagen K, Denmark.

HIDER, Dr. R.C., University of Essex, Department of Chemistry, Wivenhoe Park, Colchester, CO4 3SQ, U.K.

HIGGINS, Dr. P., Department of Microbiology, University College, Cork, Ireland.

HORNBY, Dr. D., Crop Protection Division, Rothamsted Experimental Station, Harpenden, Herts. AL5 2JQ, U.K.

KLOEPPER, Dr. J.W., Allelix Inc., 6850 Goreway Drive, Mississauga, Ontario, L4V 1P1, Canada.

LEMANCEAU, Dr. P., E.N.I.T.H.A., Rue le Nôtre, 49045 Angers Cedex, France.

LEONG, Dr. S.A., Plant Disease Research Unit, Department of Plant Pathology, University of Wisconsin, 1630, Linden Drive, Madison, Wisconsin, 53706, U.S.A.

LE PAGE, Dr. R., Biotechnica Ltd., 5 Chiltern Close, Llanishen, Cardiff, U.K.

MARTINS, J.M.S., Department of Plant Pathology, Estacao Agronomica Nacional, 2780 Oeras, Portugal.

MARUGG, Dr. J.D., State University of Utrecht, Department of Molecular Cell Biology, Padualaan 8, P.O. Box 80.056, 3508 TB Utrecht, The Netherlands.

MEYER, Dr. J.M., Laboratoire de Biochimie Microbienne, University Louis Pasteur, 4 rue Blaise Pascal (Esplanade), 6700 Strasbourg, France.

MISAGHI, Professor I.J., The University of Arizona, College of Agriculture, Department of Plant Pathology, Building 36, Tucson, Arizona 85721, U.S.A.

NEILANDS, Professor J.B., Department of Biochemistry, University of California, Berkeley, California, 94720, U.S.A.

O'GARA, Dr. F., Department of Microbiology, University College, Cork, Ireland.

PANAGOPOULOS, Professor C.G., Laboratory of Plant Pathology, Athens College of Agricultural Sciences, GR-118 55 Athens, Greece.

PATON, Dr. F.J., Agricultural Genetics Co. Ltd., Unit 154/155, Cambridge Science Park, Milton Road, Cambridge CB4 4BH, U.K.

POWELL, Dr. K.A., Imperial Chemical Industries plc, Agricultural Division, Jealott's Hill Research Station, Bracknell, Berks. RG12 6EY, U.K.

REID, Professor C.P.P., Department of Forest and Wood Sciences, Colorado State University, Fort Collins, Colorado 80523, U.S.A.

RENWICK, Dr. A., Department of Botany, University of Bristol, Woodland Road, Bristol, U.K.

RHODES, Dr. D.J., The Queen's University of Belfast, Department of Mycology and Plant Pathology, Newforge Lane, Belfast, N. Ireland.

SCHER, Dr. F.M., Allelix Inc., 6850 Goreway Drive, Mississauga, Ontario, L4V 1P1, Canada.

SCHIPPERS, Professor Dr. B., Willie Commelin Scholten Phytopathological Laboratory, Javalaan 20, 3742CP Baarn, The Netherlands.

SCHROTH, Professor M.N., College of Natural Resources, Agriculture Experiment Station, Department of Plant Pathology, University of California, 147 Hilgard Hall, Berkeley, California, 94720, U.S.A.

SUSLOW, Dr. T., Plant Pathology/Biocontrol Group, Advanced Genetic Sciences, 6701 San Pablo Avenue, Oakland, CA94608, U.S.A.

SWINBURNE, Dr. T.R., Crop Protection Division, East Malling Research Station, East Malling, Maidstone, Kent ME19 6BJ, U.K.

SZANISZLO, Professor P.J., Department of Microbiology, The University of Texas at Austin, Austin, Texas 78712, U.S.A.

VAN DER HOFSTAD, Dr. G.A.J.M., State University of Utrecht, Department of Molecular Cell Biology, Padualaan 8, P.O. Box 80.056, 3508 TB Utrecht, The Netherlands.

VON ALTEN, Dr. H., Universität Hannover, Institut für Pflanzenkrankheiten und Pflanzenschutz, Herrenhäuser Strasse 2, 3000 Hannover 21, W. Germany.

WEINBERG, Professor, E.D., Medical Sciences Program, Indiana University, School of Medicine, Myers Hall, Bloomington, Indiana 47405, U.S.A.

WEISBEEK, Dr. P.J., State University of Utrecht, Department of Molecular Cell Biology, Padualaan 8, P.O. Box 80.056, 3508 TB Utrecht, The Netherlands.

WELLER, Dr. D.M., USDA-ARS Root Disease and Biological Control Research Unit, 367 Johnson Hall, WSU, Pullman, Washington, 99164-6430, U.S.A.

WINKELMANN, Professor Dr. G., Institut für Biologie I, Mikrobiologie I, Universität Tubingen, Auf der Morgenstelle 1, D-7400 Tubingen, W. Germany.

WOOD, Professor R.K.S., F.R.S., Imperial College of Science and Technology, Department of Pure and Applied Biology, Prince Consort Road, London SW7 2BB, U.K.

WOLFFHECHEL, H., The Royal Veterinary and Agricultural University, Department of Plant Pathology, Thorvaldsenvej 40, DK-1871, Copenhagen V, Denmark.

INDEX